INDUSTRIAL ROBOTS

VOLUME 2 / APPLICATIONS

SECOND EDITION

William R. Tanner
Editor

Published by:

Robotics International of SME
Society of Manufacturing Engineers
Marketing Services Department
One SME Drive
P.O. Box 930
Dearborn, Michigan 48128

INDUSTRIAL ROBOTS
VOLUME 2 / APPLICATIONS

Copyright 1981 by the
Society of Manufacturing Engineers
Dearborn, Michigan 48128

Second Edition

Library of Congress Catalog Card Number: 81-51986

International Standard Book Number: 0-87263-071-4

Manufactured in the United States of America

SME wishes to express its acknowledgement and appreciation to the following publications for supplying the various articles reprinted within the contents of this book.

Assembly Engineering
Hitchcock Publishing Co.
Hitchcock Building
Wheaton, Illinois 60187

Automotive Industries
The Chilton Company
Chilton Way
Radnor, Pennsylvania 19089

Die Casting Engineer
455 State Street
Des Plaines, Illinois 60016

Foundry Management and Technology
Penton/IPC Publications
Penton Plaza
Cleveland, Ohio 44114

Iron Age
The Chilton Company
Chilton Way
Radnor, Pennsylvania 19089

Material Handling Engineer
Penton/IPC Publications
614 Superior Avenue West
Cleveland, Ohio 44113

Metal Progress
American Society for Metals
Metals Park, Ohio 44073

Modern Machine Shop
600 Main Street
Cincinnati, Ohio 45202

Modern Materials Handling
Cahners Publishing Co.
Division of Reed Holdings, Inc.
270 St. Paul Street
Denver, Colorado 80206

Plastics Design and Processing
Lake Publishing Corporation
Box 159
700 Peterson Road
Libertyville, Illinois 60048

Plastics Engineering
Society of Plastics Engineers
656 W. Putnam Avenue
Greenwich, Connecticut 06830

Precision Metal
Penton/IPC Publications
614 Superior Avenue West
Cleveland, Ohio 44113

Production
Bramson Publishing Company
Box 101
Bloomfield Hills, Michigan 48013

Robotics Today
Society of Manufacturing Engineers
One SME Drive
P.O. Box 930
Dearborn, Michigan 48128

Unimation Application Notes
Unimation Inc.
Shelter Rock Lane
Danbury, Connecticut 06810

Grateful appreciation is also expressed to:

International Fluidics Services Ltd.
35-39 High Street
Kempston, Bedford MK42 7BT
England

Society of Automotive Engineers
400 Commonwealth Drive
Warrendale, Pennsylvania 15096

Society of Die Casting Engineers, Inc.
455 State Street
Des Plaines, Illinois 60016

PREFACE

As we enter the 1980's, robots are far from the "mechanical men" in the science fiction laboratory. Today's industrial robots are part of the "here and now" technology and are designed to accomplish many industrial tasks.

It wasn't always so.

Immediately following their development in the 1960's and the early 1970's, industrial robots experienced limited application.

However, within the past years, as labor costs rose, safety and environmental regulations took hold and robotic capabilities became more sophisticated, the robots became valuable pieces of production equipment. Manufacturers put robots on the job to consistently produce quality products, to work in a hazardous environment or to perform heavy duty work which humans cannot do. Since the mid-70's, the number of robots striving "hand-in-hand" with workers to manufacture better products has shown a steady growth both in numbers and in diverse applications.

Today's robots are available in a wide variety of configurations and capabilities. And with increased robotic capabilities, the growth pattern for robots will continue. Thus, knowledge gained in the field of robotics is likely to be knowledge that proves valuable now and in the future.

As the number of robot types increased, so too did the amount of information available about them. Since there is such a wealth of robot information, SME's Manufacturing Update Series is presenting two volumes on industrial robots. This is the second of those volumes. Volume I is concerned with the robotic fundamentals and capabilities.

Volume II details many of the specific uses of robots by some leaders in many industries. It contains an informative collection of technical papers, application notes and the best trade journal articles on the subject.

The book is made up of 13 chapters.

Chapter One, Material Handling, provides some tips on increasing production in palletizing and explores some of the robot applications in loading boxes, conveyor line tracking, glass manufacturing and handling harrow discs. Chapter Two, Machine Loading, explores a growing robot application. Sequence of operation and costs are explored in several case histories.

The book's third chapter, Die Casting, presents a systems approach to robot use in this field, and outlines the comparative costs of automation in die casting. Chapter Four, Investment Casting, discusses how robots help lower costs and improve quality in this field. The chapter presents the operational aspects of robot-controlled mold making and explores the automated investment casting shelling process.

Press Loading, Chapter Five, discusses loading and unloading of presses in the stamping line. Loading and unloading of presses is one of the robot's earliest applications.

In the book's sixth chapter, robots in hot forging operations are discussed. The chapter entitled Forging and Heat Treating, explores the cost cutting function and economic justification of the industrial robots.

Chapter Seven, Foundry, outlines the robot's role in foundry work. The chapter discusses sensory feedback, cutting performance and foundry mold preparation.

Chapter Eight, Plastic Molding, explains how robots can improve quality and reduce scrap in the plastic molding processes. Product selection, costs and equipment are discussed.

For more than a decade, robots have been used to produce spot welds in production lines. Among the operations discussed in Chapter Nine, Welding, are parts welding at both Chrysler and Volvo.

In Chapter 10, Machining, robotic drilling, profiling and deburring are discussed. Chapter 11, Finishing, explains how robots are used in finishing metal office furniture, and automotive parts and how they are used in the appliance industry.

A relatively new application, Inspection, is discussed in Chapter 12. The chapter briefly outlines the design and installation of automatic dimensional inspecting systems.

The book's final chapter, Assembly, explores cost-effective programmable systems and the state-of-the-art in adaptable-programmable assembly systems.

Each of the many installments is authored by an expert in the field. I wish to express my gratitude to each of these authors and to the publications which were very generous in supplying some of the material you will find on the pages of this volume. These publications include: *Assembly Engineering, Automotive Industries, Die Casting Engineer, Foundry Management and Technology, Iron Age, Material Handling Engineer, Metal Progress, Modern Materials Handling, Modern Machine Shop, Precision Metal, Plastics Design and Processing, Plastics Engineering, Production, Robotics Today* and *Unimation Application Notes.* My appreciation is also extended to *International Fluidics Services, Ltd.,* the *Society of Automotive Engineers* and the *Society of Die Casting Engineers.*

Finally, my thanks to Bob King and Judy Stranahan of the SME Marketing Services Department for their help in preparing this book.

William R. Tanner
President
Productivity Systems, Inc.
Editor

SME

The informative volumes of the Manufacturing Update Series are part of the Society of Manufacturing Engineers' effort to keep its Members better informed on the latest trends and developments in engineering.

With 50,000 members, SME provides a common ground for engineers and managers to share ideas, information and accomplishments.

An overwhelming mass of available information requires engineers to be concerned about keeping up-to-date, in other words, continuing education. SME Members can take advantage of numerous opportunities, in addition to the books of the Manufacturing Update Series, to fulfill their continuing educational goals. These opportunities include:

- Chapter programs through the over 200 chapters which provide SME Members with a foundation for involvement and participation.

- Educational programs including seminars, clinics, programmed learning courses and videotapes.

- Conferences and expositions which enable engineers to see, compare, and consider the newest manufacturing equipment and technology.

- Publications including Manufacturing Engineering, the SME Newsletter, Technical Digest and a wide variety of books including the Tool and Manufacturing Engineers Handbook.

- SME's Manufacturing Engineering Certification Institute formally recognizes manufacturing engineers and technologists for their technical expertise and knowledge acquired through years of experience.

In addition, the Society works continuously with the American National Standards Institute, the International Standards Organization and other organizations to establish the highest possible standards in the field.

SME Members have discovered that their membership broadens their knowledge throughout their career.

In a very real sense, it makes SME the leader in disseminating and publishing technical information for the manufacturing engineer.

TABLE OF CONTENTS

CHAPTERS

1 MATERIAL HANDLING

2 MACHINE LOADING

TABLE OF CONTENTS (Cont.)

5 PRESS LOADING

6 FORGING AND HEAT TREATING

TABLE OF CONTENTS (Cont.)

7 FOUNDRY

8 PLASTIC MOLDING

9 WELDING

TABLE OF CONTENTS (Cont.)

12 INSPECTION

13 ASSEMBLY

INDEX

CHAPTER 1
MATERIAL HANDLING

Commentary

Material handling is one of the most common applications of industrial robots and operations range from very simple to very involved. Robots are used to advantage in handling heavy or fragile parts as well as parts that are very hot or very cold. Robots are used to palletize and depalletize parts; often utilizing contact sensors in the hand tooling.

Material handling may involve loading and unloading moving conveyors. In some applications, one or more axes of the robot may be synchronized with the conveyor.

Robots equipped with multiple tooling can handle more than one part at a time. A variety of tooling may be used, including vacuum cups, magnets and mechanical grippers.

Robots perform well in these applications. They offer advantages in handling heavy loads, reaching overhead, handling hot parts and in relieving people from tedious, repititious or difficult tasks.

Presented at the Robots III Conference, November 1978

Universal Transfer Device Application

By Martin D. Smith
Ford Motor Company

INTRODUCTION

The popularity of Industrial Universal Transfer devises (UTD's) over the past few years has been building momentum, not only in the automotive industry, but in most every other industrial corner as well. The reasons for this are notable, but more importantly, the tremendous impact which has been created by even the individual application shall eventually be a primary cause in the re-orientation of our industrial society.

The individual application, rather obviously, has its own character and significance, each demonstrating its own case and point. With that case and point comes a lesson which, to some degree, can be helpful to others interested in similar applications.

This work effort shall concern itself with a specific application which should reveal a number of interesting concepts. The subject application was the first U.T.D. installation in the Transmission & Chassis Division of Ford Motor Company, which consists of some seven (7) plant facilities. The Livonia Plant, the recipient of the first U.T.D., has a work force of over 5,000 and is expanding. In a plant of this size, there are conditions and problems unique to it which all have affected, or have been effected by the installation in some manner. These points shall be discussed along with the application engineering involved with the job.

Extensive us of the U.T.D. within the automotive industry has been restrained primarily to assembly and die casting operations. This can be mainly attributed to the large number of operations which are condusive to U.T.D. applications. Within the Transmission and Chassis Division, however, the typical operation is somewhat different in nature, and while there are a considerable number of potential U.T.D. applications, they are not as "proven out" as the spot welding and die cast operations found in other plants.

The original drive to install a U.T.D. within the T & C Division of Ford Motor Company led to many proposals and eventually a single project which brought success in both application and labor relations. The first robot was installed in March of 1978, after approximately nine (9) months of engineering and development. The objectives in mind were very simple: to prove robot success within the Division and to realize manpower savings.

The installation of the unit included various stages, which are: proposal, analysis, engineering, personnel preparation, tryout, and production. Each of these stages is important to the others and all in some way contributed to the ultimate success of the project.

PROPOSAL, ANALYSIS AND ENGINEERING

As management desires began to build to purchase a U.T.D. within the Livonia T & C Plant since they were already widely used in other Ford Divisions, research was conducted to find the most viable robot applications. Various manufacturers analized the plant's potential and a list of possible U.T.D. applications was developed. Since many of the possibilities required considerable rearrange of existing equipment, the list was narrowed down to include only those operations which could be automated with little or no change to the area.

Other important considerations before selection were the problems of maintenance and machine complexity. Hourly personnel reaction and ability to handle a new piece of equipment, such as a U.T.D., was a concern of our engineering personnel. The extent of the problem depended purely upon the choice of a robot, as there are both medium and high technological units available. Therefore, the interest in choosing the first robot application within the plant was to install a unit with controls that would be somewhat familiar to the plant personnel involved.

After discussing several of the remaining proposals with prospective manufacturers, a decision was made to concentrate on a monorail loading operation in the transmission case machining department. The installation could be done relatively easily and would yield a cost savings of (1) man/shift or approximately $50,000 anually after the payout period. While each manufacturer admitted that this type of operation had never been accomplished before, they were confident that it could be handled with no serious problems.

The operation began with parts being automatically discharged to roller conveyor from an air test machine over which the parts flowed by gravity to an unload station. A man would then pick the transmission case up and load to a moving monorail adjacent to him, moving at 25 FPM. The required capacity is 360 pcs/hr. The specific application was definately not a simple one, and there were many obstacles to be overcome. To begin with, there are (3) models of transmission cases which vary in dimension in certain locations, which creates the question of how and where to grasp the case. The requirement of part orientation for pick-up was a serious concern along with other potential problems such as consistant monorail carrier shape, passing carriers already loaded with parts, potential hang-ups or accidents, and repeatability.

With these potential problems in mind, the various robot manufacturers went to work to come up with a proposed plan of attack. Each vendor had a different approach to the solution and each represented, in some way, the use of a robot with different capabilities. As mentioned before, it was our concern to pick the unit that would cause the least amount of maintenance problems to the hourly trade personnel if possible. This would be a large consideration in making a final decision.

TRANSMISSION CASE

WT.: 22 LBS.

MAT L: DIE CAST ALUMINUM

The most difficult period during the course of events was the analysis of the various proposals. Knowing full well that each vendor would not admit to any reservations after the proposal submission, it was our Plant Engineering Department's task to determine who could handle the job best. An added problem to the issue was the fact that the range of robot costs was between $23,000 for a medium technology unit and $60 to $70,000 for a more complex unit. The question then arose as to what unit was required to do the job adequately. The first unit to be installed in the plant was very important to the future of robots within the T & C Division, therefore, success was a requirement and the final choice between proposals would be crucial.

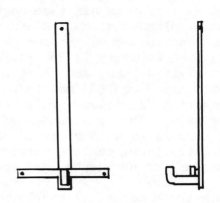

MONORAIL CARRIER

It was decided to go with the low bidder, but in order that the company be protected from an in-house disaster, the purchase order was designed to include the requirement of a simulation set-up at the vendor's plant to duplicate the Livonia Plant conditions. The success of the vendor simulation would determine whether Ford Motor Company would purchase the unit, making the company liable only for the design/build costs of the gripper tooling required. The simulation tryout was to be conducted for a period of three (3) hours without a failure, which may be somewhat less than desirable, however, it was felt that a continuous run of this length would reveal any weakness in the system.

Another important factor in the purchase of the unit was the fact that it did not meet with J.I.C. specifications or plant specifications, which comparably would be considered more rigid than most. However, rather than requiring the vendor to comply with these restrictions, it was our concern to acquire an "off the shelf item" so that other units purchased in the future would be interchangable, except in the case of gripper tooling.

While the vendor developed his design, it was our task to insure that all the variables that could cause problems later were approached and solved now.

The possibility of deformed monorail carriers was of particular concern since it was not known exactly what tolerance would have to be maintained to insure repeatability. Research was conducted to see what type of irregularities existed in the carriers and what range of accuracy could be preserved. It was found that the carriers were fairly consistant in form and that there should be no serious problem.

Also, the transmission case not only came in (3) models, but between two suppliers there were a total of (21) die molds, each with minor differences to parts of the case not important to the transmission assy. The only way it could be determined if a problem was present, was to use each of the castings during the research and development stages.

Various other potential problems were analized in much the same way and systematically the list was narrowed down and added to, as new potential problems were discovered.

TRYOUT

After several months of exchanging information with the manufacturer, the first tryout session was scheduled and conducted. It was our objective, during the approval tryout, to produce as many irregular conditions as possible in order to see what problems could arise in a production atmosphere.

A number of system flaws were found at this time, which would have caused serious problems if installed in the plant. Some may feel that a robot should be brought into the plant and designed or fitted to the application in-house. While this may be acceptable in certain applications, it would not have been the case with this installation. First of all, the production interruptions could not be tolerated in order that the system be perfected, but more importantly, the labor reaction to a new piece of automation such as the U.T.D. that did not work would have been negative to the extent that the future employment of the robot within the plant would have been seriously jeopardized.

Because of these considerations, further development took place. While there were problems with the system at this point, it was clearly evident that the application was attainable and that success was within reach.

At the second tryout session, significant improvement was made, however, there was still the possibility of accidents. Both the vendor's engineering staff and our own staff looked into minimizing any production accident and several more changes were incorporated into the system.

The system which was finally approved was functional in every way. It consists of a very simple network of limit switch sensors which control the logic system through the use of a step programmer. Transmission cases, after automatically loaded to a roller conveyor from an air test transfer machine, flow by gravity to an orientor. When a limit switch senses part pressure, the orientor cycle orientes and pushes the part into the load position. At this location, another switch tells the robot that a part is ready for pick-up. At this time, the robot arm will extend and clamp onto the case. If an empty monorail carrier is sensed, the robot will pick up the case and position at a pre-load location where still another limit switch will sense the moving monorail carrier. At this time, the robot will proceed to load the part to the carrier. If a "successful load" sensor is triggered, the unit will then repeat these motions. If the "successful load" switch is not made, the unit will stop cycling and a warning light is triggered to alert personnel.

Clamping is done on two symetrical part surfaces facing the robot unit
in the orientor. As stated before, there are (3) models of cases presently. Two are high volume parts and the third is a service part of which
only 300 are required per year. The service case configuration is much
different than that of the other two and is not handled by the robot,
but is manually loaded. The distance between the clamping surfaces on
the other two cases are different by 1 1/2 inches, but is not a problem
in gripping as the hydraulic cylinder stroke is adequate for both
dimensions.

In order to make ready for the robot installation, a number of items had
to be completed. First, a launch plan was laid out with schedules,
labor requirements and installation drawings. One of the more important
areas of plant preparation, however, was the approach upon labor relations.

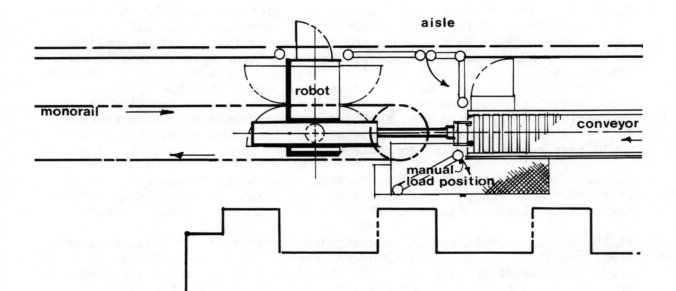

ROBOT INSTALLATION PLAN

It is important that management creates a positive attitude among the
hourly people who will operate and maintain the new equipment. This
can be done by involving them in the project by providing proper introduction to the new equipment, job intent and proper training in the
operation and maintenance of the unit. This concept became very important with our installation. The robot was installed in an off line
position to conduct training sessions for all maintenance personnel.

Upon installation at the job location and during debug operations, production personnel were also trained in robot operation and trouble shooting. During this initial start-up period the labor atmosphere was somewhat tense, probably because no one knew at the time what reactions might occur.

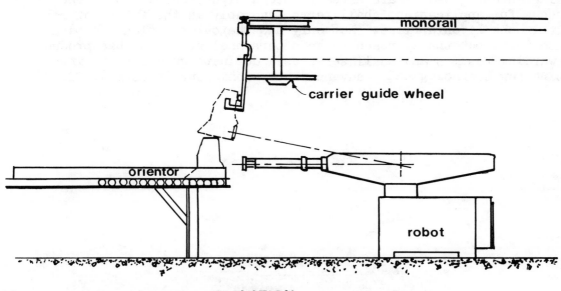

ELEVATION

From the first day of start-up, the unit loaded cases successfully. There were many adjustments which had to be made over a two (2) week period, but the unit did perform. During this period, hourly employes made several signs to show that they were not totally pleased with the idea, but after the system continued to perform without any serious problems, they became quite impressed and actually interested in the unit. The maintenance personnel were very pleased with the simplicity of the electrical and hydraulic system. This also added to the ease with which the unit went in.

If the system was incorporated at any of the earlier stages, many additional days would have been required to debug the unit, which would have had tremendous detrimental effects upon labor relations. The point here is simply that all necessary engineering to make the system go in smoothly from the beginning must be accomplished before the production and skilled trades personnel ever come in contact with it. This is particularly important for the first unit installed within a plant.

IN PRODUCTION

After seven months of production use, 16 hours per day, the U.T.D. system has had a 98% up time average overall. With the exception of some minor electrical and mechanical problems, the unit has proved to be a complete success in both application and labor relations.

Because of this advancement, numerous proposals through out the plant are being investigated or are already into design stages. This, the first successful application, shall serve favorably in the future of robots within the Livonia Plant. Not only has the careful planning been important, but the approach towards personnel relations has produced rewarding results and shall serve as a definate guideline in incorporating new technological advancements within our industrial field.

**THE UNIT PURCHASED OPERATES THROUGH
A SIMPLE ELECTRICAL AND HYDRAULIC SYSTEM**

Reprinted from Robotics Today, Fall 1980

Glass tubes—31 at a time—being picked up by a robot in a palletizing operation. The current palletizing rate is some 100 tubes per minute.

Increasing Production in Palletizing

Palletizing glass tubes that are used for fluorescent lamps is a delicate job that can get quite tedious. But at two plants in Ohio, robots are doing it 24 hours a day

THE ELIMINATION of repetitious jobs that few workers found rewarding, improved efficiency, and greater adaptability to surges in production—these are but a few of the benefits realized by an electric tube manufacturer since it automated glass-tube palletizing in two Ohio plants. Key to the automation: general-purpose Unimate ® robots from Unimation Inc., Danbury, CT.

The open-end uncoated glass tubes, which nominally measure 92" (2337 mm) in length, are produced for fluorescent lamps. Palletizing is performed 24 hours a day, seven days a week. A typical pallet load is 25 layers of tubes, each layer having 31 tubes. Initially, the tube-handling rate for palletizing was approximately 70 tubes per minute; currently, rates average from 100 to 110 tubes per minute for each robot. The tubes are subsequently shipped by truck to another plant for coating and assembly. Here's a closer look at the robot applications.

After the last production operation—automatic forming of a shoulder and neck on both ends of each tube—the flourescent tubes are conveyed to the work station of the robot. The robot's task is to build up a pallet load of tubes and, once a pallet is full, to trigger another conveyor that moves the filled pallet away while bringing an empty pallet into position for loading.

At one of the plant installations, hot glass tubes with necked ends are carried away on a conveyor with the tubes located in parallel positions for air cooling and delivery to the robot's pickup point. The parallel tubes are automatically moved closer together into a queue at the end of the conveyor. The head of the queue advances into the pickup area as another tube joins the rear of the queue.

When the queue contains 31 tubes—the capacity of the pickup area—the head tube has moved to a point where it trips a limit switch. The switch performs two functions. It signals the robot that it can pick up a load, and it operates a gate that halts tube number 32. The following tubes form a buffer storage until the robot has palletized the other tubes (they will subsequently form the next group of 31).

The robot picks up all 31 tubes in the pickup area, turns 90° to its left, and deposits the layer of tubes on a stake-sided pallet that can accommodate 26 layers. Conical plugs on the robot's gripper fit into each end of every tube to retain them during transfer.

The tubes are released by the gripper a fraction of an inch above the last pallet layer. This short drop aids the nesting of tubes in each layer and assures that accumulated stacking height deviations do not cause the robot to press a new layer against those below it. There has been a notable decrease in the breakage rate since the robots have been installed.

A complete transfer of 31 tubes takes approximately 10 seconds. The present rate of 100 tubes per minute also reflects the time needed for a 31-tube queue to form and a filled pallet to be replaced by an empty one.

At the other glass tube plant, the setup is slightly different. The pallet is located between the robot and the pickup area. So instead of making a turn, the robot simply reaches across the pallet, grasps a layer of 23 tubes, and moves the tubes back over the pallet. Finally, the robot lowers the tubes onto a 24-layer pallet. The current production rate is 110 tubes per minute.

A counter keeps track of the number of layers placed on a pallet. After the last layer is set in place, an electrical signal energizes the conveyor. It moves the filled pallet away and brings an empty pallet into position. Once the empty pallet is in position, it operates a limit switch that stops the conveyor and signals the robot to proceed. The robot will function only when enough tubes for a full layer have accumulated at the pickup point.

The release points over the pallet are programmed into the robot in height increments; each point is slightly higher for successive layers. Also programmed is a lateral staggering of the release points so that each tube on an upper layer rests between two tubes on the layer below, except for alternate end tubes.

The palletizing operations at the plants are quite successful. In 1978, the tube manufacturer determined that its original 10-year-old robots had to be retired since maintenance costs justified such a move. At that point, alternatives for the robots were evaluated. For example, modern conveyors were considered, and one gravity-operated conveyor was located that would be more economical than robots.

The conveyor setup would make the operation a combined manual and gravity-conveyor arrangement. However, one critical point eliminated the conveyor-based system: an anticipated production-rate increase to 150 tubes per hour could not be handled by it.

So the company replaced the older robots with current models of the Unimate robots. Uptime for the new units has been in excess of 98%, overall. And the robots have the capability to handle increases in the palletizing rate. ∎

Presented at the Robots IV Conference, October 1979

Experiences in Applying Robots in Light Industry and Future Needs

By Richard Becker
Chesebrough-Pond's, Inc.

The process of making clinical thermometers involves several steps which entail shrinking, expanding, and moving the mercury about in the thermometer so as to drive out entrained gases. Formerly a manual operation, these steps were automated by the installation of a Unimate 2100 robot.

A more common manual operation, that of removing empty jars from a case and placing them on the infeed end of a filling line, was automated by the installation of a PACER II robot.

A discussion of each installation is presented, as well as desires addressed to the manufacturers for the type of robot and peripheral equipment needed to expedite the introduction of robots to light industry, and more quickly impact productivity gains.

INTRODUCTION

One of the products produced by Chesebrough-Pond's is mercury-in-glass, clinical (fever) thermometers. These are produced at our Watertown, New York, facility, in the largest plant of its type in the world.

The process of making highly accurate and repeatable thermometers involves some 160 manufacturing steps. Certain of these steps entail shrinking, expanding, and moving the mercury about in the thermometer in a process similar to that of kneading dough, so as to drive entrained gases out of the mercury. This process utilizes hot and cold water baths, vibration, and centrifuging.

Formerly, our employees would manually operate these various pieces of equipment and manually transfer the thermometers from one step to the next in handfuls of about 50 each such as is shown in Figure 1. These employees stood on their feet most of the day while exposing their hands to hot and cold water. These unfavorable working conditions prompted us to look for a better way to do the job. In early 1976, we automated this portion of the thermometer making process by the use of a Unimate Model 2100 robot with 5 degrees of freedom, and redesigned process equipment.

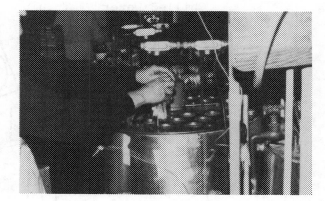

FIGURE 1
Manual Thermometer Processing

EQUIPMENT

The general layout of the automated system is shown in Figure 2. The Unimate is located in the center, with its entire working envelope occupied by various auxiliary equipment of our own design. This equipment includes warm and cold water baths, a vibrating table, holding and turning tables, several centrifuges, and both incoming and outgoing conveyors. The centrifuges, vibrator, and both conveyors are interlocked with the robot so that, for example, the centrifuges will signal that they have stopped and their lids have opened before the robot will attempt to reach into the machine. In like fashion, the incoming boxes of thermometers must trip a switch which indicates to the robot that the box is ready for pick up. On the outgoing conveyor, a photo eye is blocked when a finished box is placed on the conveyor. This photo eye must then be cleared before the Unimate will place another box on the conveyor. The boxes referred to are made of perforated stainless steel, and hold about 1100 thermometers each.

All loading stations are equipped with beveled openings so as to accommodate any inaccuracies in either box position in the robot gripper, or box placement.

The robot gripper is designed to handle either one or two boxes at a time, and it does both. Centrifuge loading and unloading must be done one box at a time, while transferring boxes from one water bath to another is done two at a time. Figure 3 shows the robot loading a centrifuge and Figure 4 shows loading into a hot water bath.

The turning table is used for setting the boxes down while the hand is reconfigured as needed before the next equipment loading step. In the water baths, for example, the thermometers must be placed bulb end down, while in the centrifuges the bulb end must be facing either radially out or in, depending on the particular step in the process.

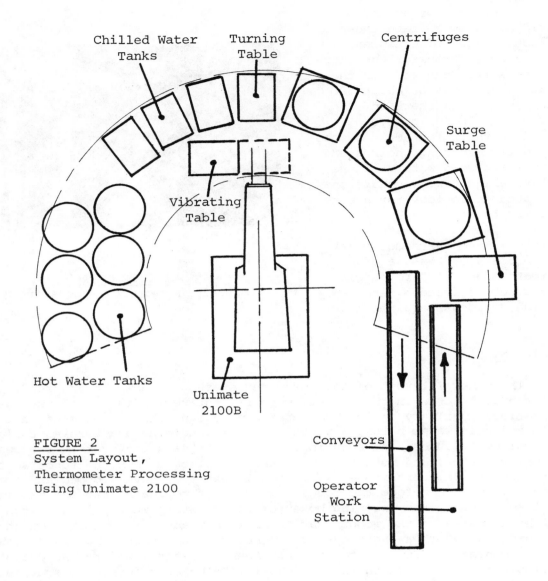

Chilled Water Tanks

Turning Table

Centrifuges

Surge Table

Vibrating Table

Hot Water Tanks

Unimate 2100B

Conveyors

Operator Work Station

FIGURE 2
System Layout,
Thermometer Processing
Using Unimate 2100

FIGURE 3 - Unimate gripper
loading thermometer boxes
into centrifuge

<u>FIGURE 4</u> - Unimate gripper
loading thermometer boxes
into hot water bath

At certain steps in the cycle, time delays are programmed into the robot to allow warm water to drain from the boxes before they are placed in the cold water baths. This lessens the heat load on a recirculating chiller which supplies the cold water.

The centrifuges are equipped with speed switches and hardened steel cams and plungers. When the speed switches sense that the centrifuges have slowed to about 2 rpm, the plunger is actuated and engages itself in a detent in the cam. The cam is locked to the centrifuge drive shaft, hence, the centrifuges are stopped in the same position every time so the boxes can be found by the robot.

One operator runs the entire system. This employee loads thermometers into the ingoing boxes, unloads completed thermometers, and keeps an eye on the overall system. The operator has been trained in and understands the equipment, so can correct most problems that may occur. Throughput of this system is one box every 4.5 minutes, or 245 thermometers per minute.

In preparing for the installation of this robot, we first sent four of our maintenance personnel to the training course at Unimation's headquarters in Danbury, Connecticut. Because of this in-house capability, our non-severe environment, and the fact that we are unfailingly performing the recommended maintenance checks, we have experienced 98% plus uptime with our unit.

REASONS FOR

RESULTS

The system has been operating since early 1976 and has logged over 16,000 operating hours since that time, running on two shifts. Total cost of the installation was $90,000, with $55,000 of that going for the robot. Direct labor savings, including fringe benefits, for this operation have amounted to over $900,000 to date and the employees who formerly performed this work have been transferred to more desirable and higher paying jobs.

Question: If same employees are paid more what salary savings??

The thermometers produced by this system are of a better quality than before because the motions and times spent in the various process steps are now more consistent. Because of this, and the fact that the robot very rarely drops thermometers, scrap and rework rates have been reduced, thus adding additional tens of thousands of dollars to the payback ledger.

Employee acceptance of this first robot presented us with no problem because we kept our people fully informed on our efforts through the course of the project, and, in fact, made many of them active participants in the project. We also built a full scale painted wood model of the robot months prior to its delivery. This not only helped with our layout design, but it also aided in getting people in the plant comfortable with the idea of a large (for this plant) moving piece of equipment like this on our factory floor. After completion of the project, we held an open house so that families of our employees could come and see what the robot was all about.

II

A SECOND INSTALLATION

Since then, we have installed a second robot in another of our plants. The job this robot does is more typical of our operations than the robot installation previously described. This unit is a PACER II robot, as manufactured by Production Automation Corporation in Livonia, Michigan. This robot removes plastic Vaseline^R jars from cases and places them on the infeed end of a jar filling line.

EQUIPMENT

The general layout of this system is shown in Figure 5. Cases of empty jars are manually transferred from a pallet to the system infeed conveyor. This job is done by the same employee who palletizes the filled and sealed cases. The cases of unfilled jars are conveyed through a right angle transfer and then into a standard four flap opener where all four unsealed top flaps are opened. The case is then discharged onto an indexing conveyor which is equipped with flap hold down rails. Various stops and photo eyes on this conveyor signal the robot when the case is in position so the jar unloading cycle can be begun.

The PACER robot is a two axis, hydraulically operated, computer controlled machine; equipped with a multiple fingered vacuum gripper. This gripper removes an entire layer of jars at a time by reaching down into the jars and applying vacuum to the jar bottoms. The jars are then placed on a standard single filing machine, and conveyed to a filler. Each finger of the gripper is equipped with a spring loaded detent so that it will push upward if it encounters an obstacle. In this way, the occasional upside down jar will not damage the gripper. In addition, the vacuum system is equipped with a pressure sensor which stops the robot if an entire layer of jars is not acquired for any reason. Figure 6 shows the vacuum gripper, the conveyor flap hold down rails, and the case stop and position sensing photo eye

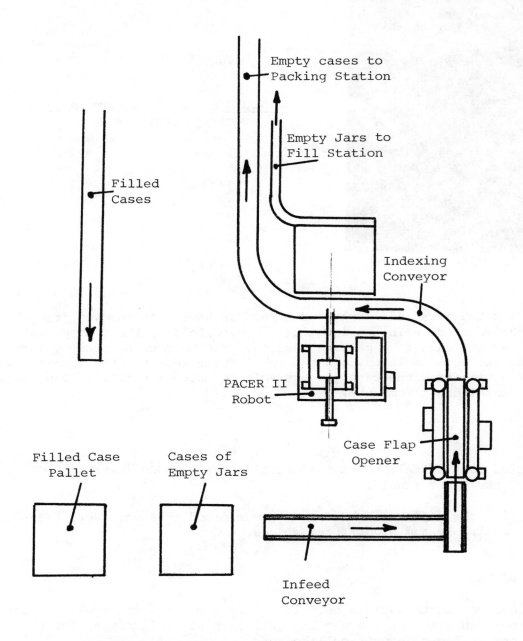

FIGURE 5 - PACER robot system used in unloading jars from cases

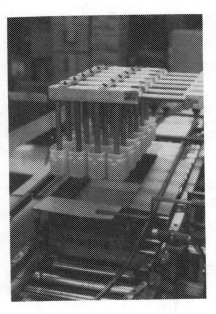

FIGURE 6 - PACER vacuum gripper
acquiring empty jars

After the jars and corrugated separator have been removed from the case, the robot signals the indexing conveyor to drop the conveyor stops and convey the case to a packing station. The next case of empty jars is then indexed into position, and the robot cycle repeats. By removing 24 jars at a time from the case, we are able to match the filling line speed of 150 containers per minute.

RESULTS

Total cost of this robot installation was $50,000, with $31,000 of that for the robot itself.

Again, the employees who formerly performed this work were transferred to more desirable and higher paying jobs.

FUTURE NEEDS

Potential not realized

The robot installations described above, while successful, are not truly representative of the numbers of potential applications for this technology in light industries like ours.

The small, human sized, PUMA-type robot will have a much larger impact on productivity gains and job enrichment than anything we have yet done. One of the main attractions of this robot is that, since it is human sized, it can be used in factories which were originally designed for humans with a minimum amount of disruption to existing work stations. We have had a PUMA since last April, and have identified and bench tested numerous possible applications throughout our many factories.

However, since the implementation of robotic technology is a relatively recent development, our company, like others, has thus far committed only a few staff members to these efforts. Installation of robots,
therefore, is proceeding at the slow pace made necessary by the need to
search out or design peripheral equipment, including the feeding and
orienting devices needed to put blind robots to work. Although the day of
real-time vision equipped robots is not far off, we would still like to
see not only more manufacturers of PUMA type robots, but we'd like to see
these same manufacturers offer a companion line of feeding and orienting
equipment.

Single source supply of this type would expedite the introduction of
robots into our factories and, hence, more quickly impact the productivity
gains that our country so desperately needs to remain competitive in this
changing world.

A Vision—Controlled Robot for Part Transfer

Steven W. Holland, Lothar Rossol,
Mitchel R. Ward, and Robert Dewar
General Motors Corp.
Warren, MI

ABSTRACT

A vision-based robot system capable of picking up parts randomly placed on a moving conveyor belt is described. The vision subsystem, operating in a visually noisy environment typical of manufacturing plants, determines the position and orientation of parts on the belt. The robot tracks the parts and transfers them to a predetermined location. This system can be easily retrained for a wide class of complex curved parts and demonstrates that future systems have a high potential for production plant use.

IN MANY MANUFACTURING ACTIVITIES, parts arrive at work stations by means of systems that do not control part position. Since present robots require parts to be in precisely fixed positions, their use is precluded at these work stations. To automate these part handling operations, intricate feeding devices that precisely position the parts are required. Such devices, however, are often uneconomical and unreliable.

Robot systems equipped with vision represent an alternative solution. This paper describes CONSIGHT, a vision-based robot system for transferring unoriented parts from a belt conveyor to a predetermined location. CONSIGHT:

- determines the position and orientation of a wide class of manufactured parts including complex curved objects,
- provides easy reprogrammability by insertion of new part data, and
- works on visually noisy picture data typical of many plant environments.

As a result of these characteristics -- and because the vision subsystem does not require light tables, colored parts or other impractical means for enhancing contrast -- CONSIGHT systems have a high potential for production plant use.

From the outset CONSIGHT was targeted for plant installation. The experimental version, described here in detail, has been succeeded by a production prototype. And, although the prototype was implemented on different hardware and had to be partially reprogrammed, the overall structure of the system has remained intact. At each phase careful consideration was given to succeeding phases and close cooperation was maintained among the groups responsible for the different versions of CONSIGHT.

Much of the motivation for this work was provided by early vision and robot tracking research done at SRI International.

A FUNCTIONAL OVERVIEW

CONSIGHT functions in two modes: a setup mode and an operational mode. During setup, various hardware components are calibrated and the system is programmed to handle new parts. Once calibrated and programmed for a specific part, the system can be switched to operational mode to perform part transfer functions.

Part transfer operates as follows: The conveyor (Fig. 1) carries the randomly positioned parts past a vision station which determines each part's position and orientation on the moving belt. This information is sent to a robot system which tracks the moving parts on the belt, picks them up, and transfers them to a predetermined location. The above sequence operates continuously with no manual intervention except for placing parts on the conveyor.

Fig. 1. CONSIGHT conveyor, camera, and robot

CONSIGHT is capable of handling a continuous stream of parts on the belt, so long as these parts are not touching. The maximum speed limitation is imposed by neither the vision system nor the computer control, but by the cycle time of the robot arm.

Calibration for CONSIGHT is a simple procedure taking about 15 minutes and may be performed by an operator who need not understand or be aware of the mathematical procedures involved. Calibration need be done only during initial setup or after physically moving the robot, conveyor or camera.

Programming CONSIGHT for new parts is also a simple procedure which retains much of the "teaching by showing" concept of current robot systems.

A SYSTEM OVERVIEW

Organization - CONSIGHT is logically partitioned into independent vision, robot, and monitor subsystems. Intersystem communication was designed to allow easy substitution of new subsystems with minimal impact. For example, the vision subsystem reports only a unique point (i.e., x and y coordinate) for each part, and an orientation. Neither the location of the point on the part, nor the reference from which to measure orientation, is specified. Other vision modules [2,3] can be substituted with little effect on the control program, the robot system, the part programming methods, or the calibration methods. Equally important, another robot, a different monitor subsystem, or different control computers could be substituted without a great effect on other CONSIGHT subsystems. This modular organization has been very beneficial during the implementation of the production prototype (described later), however, this is not to say that replacing one robot with another is by any means a trivial task.

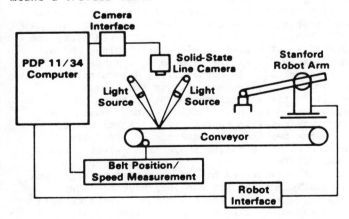

Fig. 2. CONSIGHT hardware schematic

Hardware - The hardware of the experimental system consists of a a Digital Equipment Corporation PDP 11/34 computer operating under the RSX-11S real-time executive, a Reticon RL256C 256x1 line camera, a Stanford Arm made by Vicarm [4], and a belt conveyor which is instrumented with an encoder for position and speed measurement (Fig. 2).

Software - The software organization for CONSIGHT reflects the three major modules of the system (Fig. 3).

The monitor coordinates and controls the operation of CONSIGHT and also assists in calibration and reprogramming for new parts. The monitor queues part data and the system is thus capable of handling a continuous stream of parts on the belt.

The vision subsystem uses structured light [5,6,7] in which two projected light lines, focused as one line on the belt, are displaced by objects on the belt. The line camera, focused on the line, detects the silhouette of passing objects. When it has seen the entire object, the vision subsystem sends to the monitor the object's position and a belt position reference value.

The robot subsystem executes a previously "taught" robot program to transfer the part from the conveyor to a fixed position. It accepts information concerning the part's location on the moving belt and uses this data to update the "taught" program. It then monitors belt position and speed to track the part along the moving belt, pick up the part and transport it to a predetermined location.

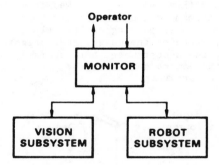

Fig. 3. CONSIGHT software organization

THE VISION SUBSYSTEM

The vision subsystem detects objects passing through its field of view and reports their position and orientation to the monitor program. Parts may follow in an unending stream. It is also permissible for several parts to be within the field of view simultaneously. Parts which are overlapping or touching each other are ignored and allowed to pass by the robot for subsequent recycling.

The vision subsystem employs a linear array camera. The linear array images a narrow strip across the belt perpendicular to the belt's direction of motion. Since the belt is moving, it is possible to build a conventional two-dimensional image of passing parts by collecting a sequence of these image strips. The linear array consists of 256 discrete diodes, of which 128 are used in the system described here. Uniform spacing is achieved between sample points (both across and down the belt) by use of the belt position detector which signals the computer at the appropriate intervals to record the camera scans of the belt.

The two main functions of the vision subsystem are object detection and position determination.

Object Detection - A fundamental problem which must be addressed by computer vision systems is the isolation of objects from their background. If the image exhibits high contrast, such as would be the case for black objects on a white background, the problem is handled by simple thresholding. Unfortunately, natural and industrial environments seldom exhibit these characteristics. For example, foundry castings blend extremely well with their background when placed on a conveyor belt. Previous approaches for introducing the needed contrast, such as the use of flourescent painted belts or light tables [8] would severely restrict the number of potentially useful applications of vision-based robot systems. We developed a unique lighting arrangement which accomplishes the same results without imposing unreasonable constraints on the working environment.

A slender tungsten bulb and cylindrical lens are used to project a narrow and intense line of light across the belt surface. The line camera is positioned so as to image the target line across the belt. When an object passes into the beam, the light is intercepted before it reaches the belt surface (Fig. 4). When viewed from above, the line appears deflected from its target wherever a part is passing on the belt. Therefore, wherever the camera sees brightness, it is viewing the unobstructed belt surface; wherever the camera sees darkness, it is viewing the passing part (Fig. 5).

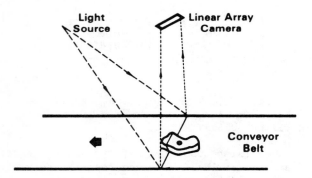

Fig. 4. Basic lighting principle

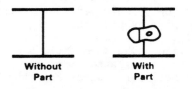
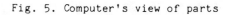

Fig. 5. Computer's view of parts

Unfortunately, a shadowing effect causes the object to block the light before it actually reaches the imaged line, thus distorting the part image. The solution is to use two (or more) light sources all directed at the same strip across the belt (Fig. 6). When the first light source is prematurely interrupted, the second will normally not be. By using multiple light sources and by adjusting the angle of incidence appropriately, the problem is essentially eliminated.

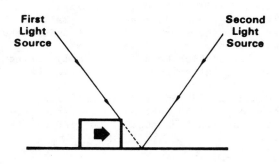

Fig. 6. Improved lighting arrangement

The described lighting arrangement produces a height detection system. Any part with significant thickness will be "seen" as a dark object on a bright background. The computer's view is a silhouette of the object. Note that while the external boundary will appear sharp, some internal features, such as holes, are still subject to distortion or occlusion due to the shadowing effect. These internal features may optionally be ignored or recorded under software control.

It is the vision computer's responsibility to keep track of objects passing thru the vision system. Since several pieces of an object or even different objects may be beneath the camera at any one time, the continuity from line to line of input must be monitored. The conventional binary image segmentation schemes are the 4-connected and the 8-connected algorithms. Both of these algorithms result in ambiguous situations [9] when deciding on the inside/outside relationship between some binary regions. A clever solution to that problem is the use of 6-connected regions [10], that is, connectivity is permitted along the four sides of a picture element and along one of the diagonals. At the expense of a minor directional bias in connectivity determination, the inside/outside ambiguity is resolved. In addition, the algorithms which implement the segmentation for 6-connected regions remain simple and symmetric with respect to black and white, since the 6-connected algorithm artificially introduces the pleasing properties gained through hexagonal tessalation.

The 6-connected binary segmentation algorithm is readily adapted for run-length coded input, that is, where only the transition points between black and white segments are recorded. This is a significant advantage. The straight forward binary segmentation algorithm requires that the intensity of the neighbors for each pixel be examined. The execution time is therefore "order n squared" where "n" is the linear camera resolution. Since the number of black/white transitions across a line is essentially independent of the resolution for these types of images, the execution time is reduced to "order n" for the algorithm using the run-length coding scheme.

Once the passing objects have been isolated from the background, they may be analyzed to determine their position and orientation relative to a reference coordinate system.

<u>Position Determination</u> - For each object detected, a small number of numerical descriptors is extracted. Some of these descriptors are used for part classification -- that is, deciding which part it is. Others are used for position determination.

For position specification, we describe the part's position by the triple (x, y, theta). The x and y values are always selected as the center of area of the part silhouette. For most parts, this represents a well-defined point on the part. There is no convenient method for uniquely assigning a theta value to all parts. However, one useful descriptor for many parts is the axis of the least moment of inertia (of the part silhouette). For long, thin parts, this can be calculated accurately. The axis must still be given a sense (i.e., a direction) to make it unique. This is accomplished in a variety of ways and is part specific. The internal computer model for the part specifies the manner in which the theta value should be computed. For example, one method available for giving a sense to the axis value is to select the moment axis direction which points nearest to the maximum radius point measured from the centroid to the boundary. Another technique uses the center of the largest internal feature (e.g., a hole) to give direction to the axis. Several other techniques are also available.

Parts which have multiple stable positions require multiple models. Parts whose silhouettes do not uniquely determine their position cannot be handled.

Reprogramming the vision system for a new part requires entering a description of a new model. Each model description includes information to determine if a detected object belongs to the class defined by the model and also prescribes how the orientation is to be determined.

The vision system sees the world through a narrow slit. As objects pass by the slit, statistics concerning that object are continuously updated. Once these statistics have been updated, the image line is no longer required. Consequently, storage need only be allocated for a single line of the two-dimensional image, offering a major reduction in memory requirements.

The block of statistics describing an object in the field of view is referred to as a component descriptor. The component descriptor records information for every picture element which belongs to that component. This includes the following.

1. external position reference
2. color (black or white)
3. count of pixels
4. sum of x-coordinates
5. sum of y-coordinates
6. sum of product of x- and y-coordinates
7. sum of x-coordinates squared
8. sum of y-coordinates squared
9. min x-coordinate and associated y-coordinate
10. max x-coordinate and associated y-coordinate
11. min y-coordinate and associated x-coordinate
12. max y-coordinate and associated x-coordinate
13. area of largest hole
14. x-coordinate for centroid of largest hole
15. y-coordinate for centroid of largest hole
16. an error flag

Considerable bookkeeping is required to gather the appropriate statistics for passing objects and to keep multiple objects segregated. The primary data structure used for this purpose is the "active line." The active line records the location of each black and white segment beneath the camera and also the objects to which they are connected. For every new segment which extends a previously detected object, the statistics are simply updated. If the segment is the start of a new object which is not in the active line, it will be added to the active line and have a new component descriptor initialized for it. When an object in the active line is not continued by at least one segment in the new input line, it must have passed completely through the field of view. The block of statistics is then complete and may be used to identify the object and compute its position and orientation.

Fig. 7 illustrates the coordinate reference frames used by the programs. It is convenient to consider the x-origin of the vision system to be permanently attached to the belt surface moving to the left. Normally, this would cause the current x-position beneath the camera to climb toward infinity. To avoid this, all x distances are measured relative to the first point of object detection. This in turn induces a complication. The complication occurs when two appendages of one object begin as two separate objects in the developing image. It then becomes necessary to combine the component descriptors for the two appendages into a single component descriptor which reflects the combination of the two. Moments, however, have been referenced to two different coordinate systems (i.e., the x-origin for each was taken as the point of first detection). The required shifting of moments is accomplished by applying the Parallel Axis Theorem.

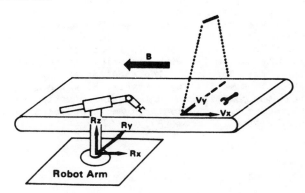

Fig. 7. CONSIGHT coordinate systems

Once parts have passed completely through the field of view, final position determination is made. The required computations proceed asynchronously with respect to the processing of the new lines of picture data. The results must be provided to the monitor subsystem before the part travels past the robot's pickup window. To coordinate the scheduling of these final computations, a queue of completed component descriptors is maintained. Component descriptors are removed from the head of the queue and processed as time permits.

Since the belt on which the parts rest is moving, the vision system records the current belt position whenever a new object appears in the developing image. This belt position reference value is obtained from the belt position/speed decoder. Since the leading edge of each part defines the origin of the coordinate system to be used for that part, the position of that part at some future time can then be readily determined by checking the current belt position and adjusting for any belt travel since the initial reference was recorded.

THE ROBOT SUBSYSTEM

The robot programming subsystem is implemented as two independent tasks. One task is required for robot program development and is necessary only during the programming and teaching phase. The second task, the run-time control system, is required both during the teaching phase and during robot program execution. It interprets and executes a robot program and controls the robot hardware. The execution of this task is controlled by special requests sent from other tasks.

A robot program consists of statements specifying: a position to which the robot should move (setpoint), an operation the robot should perform, or the environment for subsequent execution. Positions to which the robot moves are either taught by moving the robot manually and recording the position or are programmed by entering the specific Cartesian coordinates of a point in space from the keyboard.

In addition to this basic programming support, tracking and real-time program modification were developed for CONSIGHT. Tracking provides the ability to execute a robot program relative to some moving frame of reference. Program modification provides the ability to modify, in real-time, the robot program from another program and thereby dynamically modify the robot's path.

Tracking is implemented by defining new reference coordinate systems called FRAMEs [11]. Normally the robot operates in a Cartesian coordinate system [R] with its origin at the base of the robot (Frame 0). The robot's Cartesian position is described by a matrix [P] which defines the position and orientation of the hand in [R]. The arm solution program then determines a joint vector [J] from [P].

$$[P] \longrightarrow [J]$$

If however, we want to define [P] relative to a different coordinate system (frame) whose position in [R] is defined by a transformation [F], then the solution program must perform the following:

$$[F] \ [P] \longrightarrow [J]$$

Frames provide a means of redefining the frame of reference in which the robot operates. The robot may be programmed relative to one frame of reference and the resulting program executed relative to a different frame of reference. For example, a robot may be programmed to load and unload a testing machine. If the testing machine is moved, the entire program can be updated by simply redefining the frame specifying the position of the testing machine without re-programming each individual position point in the program. The overall effect is to translate/rotate every position to which the robot moves.

In addition to having a position, a frame is defined with a velocity and a time reference. This position, velocity, and time reference are used to predict the frame's position. Each time the run-time system performs an arm solution, (i.e., transforming the position matrix into the corresponding joint angles), it first computes a predicted position for the current frame.

Program modifications are special asynchronous requests sent to the run-time system from other tasks. Via these requests, an external program can modify a robot's programmed path, read the robot's position, start/stop robot program execution, and interrogate status -- all while the robot is operating. These requests greatly expand the capability of the robot system without the use of a powerful robot programming language. Much of the logical, computational, and input/output capabilities of an algorithmic language (Fortran) are retained for programming and controlling the robot external to the normal robot programming and control system.

In CONSIGHT, the part position determined by the vision subsystem defines the position and orientation of a frame and the belt direction and speed define the frame velocity. The approach, pickup, and departure points are all programmed relative to this frame. The robot subsystem does not directly interface to the belt encoder for belt position and velocity data, but receives the data via a request in the same way that the vision data is furnished. Thus, the rate at which the belt position and velocity data are updated is controlled by the monitor program and is a function of the expected variability of the belt speed. The approach, pickup, and departure points are dependent on the type of part being picked up as well as its position and must be modified for each cycle of the robot.

In the production prototype system the Stanford robot was replaced by a production robot with a tracking capability but which does not include the Frame concept. Thus, some of the monitor programs had to be modified to include some of the functions provided throught the use of the frames.

THE MONITOR

The monitor coordinates the operation of the vision and robot subsystems during calibration and part programming as well as during the operation phase. Calibration is required during the initial setup or whenever the camera, the robot or the conveyor have been moved. Part programming is required only when modifying the system to handle a new part.

Calibration - Calibration is the process whereby the relationship between the vision coordinate system [V] and the robot coordinate system [R] is determined (Fig. 7). In particular we want to compute the position of a part [r] in [R] given the part position [v] in [V]. Taking into account belt travel, this computation is represented by the following equation:

$$[r] = [T] [v] + s b [B]$$

In the equation above, [T] is the transformation between [V] and [R], s is a scale factor relating belt distance to robot distance, b is the distance of belt travel, and [B] is the belt direction vector relative to [R]. Thus, [v] and b are the independent variables, and [T], s, and [B] are the unknowns to be determined by the calibration procedure.

To determine [B] and s, a calibration object is placed on the belt within reach of the robot hand. The hand is manually centered over the calibration part, the hand position is read, and a belt encoder reading is taken. The conveyor belt is started and the part is allowed to move down the belt. The robot hand is again centered over the calibration part. A second hand position and belt encoder reading are taken. The monitor system can now compute the belt direction vector [B] and a scale factor s converting belt encoder units to centimeters.

Determining the coordinate transformation between [V] and [R] completes the calibration. The procedure assumes that the plane [Vx, Vy] is parallel to the plane [Rx, Ry] and that scaling is the same in both the x and the y directions.

A calibration object is again placed on the belt and allowed to pass by the vision station. The object position [v1] is determined by vision and a belt reading is taken. The part moves within reach of the robot and the conveyor is stopped. The robot hand is centered over the calibration object, the robot position [r1] is read and the distance of belt travel b1 is computed. This procedure is repeated with the calibration object placed at a different position on the belt resulting in a second set of data, [v2], b2, and [r2]. Combining these two sets of data points into the form above yields:

$$[r1, r2] = [T] [v1, v2] + s [B] [b1, b2]$$

This gives us 4 equations for determining the 4 unknowns in the transformation matrix [T].

Part Programming - Part programming is the process whereby the system is taught to recognize and pick up a new part. To do this, the vision subsystem must have been separately programmed to recognize the new part as described earlier in the

vision section. To program the robot for a new part is to define the gripper position for part pickup. Normally the pickup position is offset from the part reference position determined by vision. Thus, once the vision subsystem locates a part and computes its position, the actual robot pickup position must still be computed.

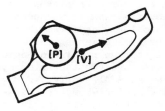

Fig. 8. The pickup problem

Fig. 8 illustrates this problem in a general way. The part position and rotation [v] have been computed by the vision subsystem. The robot needs to know the gripper position and orientation [p] as a function of [v]. The problem is further complicated since [v] is computed with respect to the vision coordinate system [V] and [p] must be computed with respect to the robot's coordinate system [R].

The procedure consists of passing a part by the vision station where [v] is computed and a belt reading is taken. The part moves down the belt to within reach of the robot and the belt is stopped. The monitor subsystem computes the current position and orientation of the part in the robot's coordinate system [R] and sends this position and the part direction to the robot subsystem as the definition of a frame. The robot hand is then placed on the part at the desired position for the grasp and pick up. This hand position is computed by the robot subsystem relative to the newly defined frame and is sent to the monitor subsystem. Thus, the pickup offset is determined relative to the part position and orientation as determined by the vision subsystem.

Later, during the operation phase the procedure is reversed. The part position [v] is again computed relative to [R] and used to define a frame in which the robot operates during the pickup phase of the robot program. The pickup position is then simply the offset determined by the above procedure. The robot subsystem automatically determines the desired robot position at any instant from the frame position, the pickup offset, and the belt position and velocity.

The Operation Phase - The final function of the monitor subsystem is to control the operation of the CONSIGHT system. Fig. 9 illustrates the overall logic of this portion of the monitor. Although the logic is straightforward, the monitor system deals with two rather subtle problems.

The first involves gripper orientation. Since the robot has a limited range of motion in all of its joints, it has a limited workspace. In particular the outermost joint, which controls hand orientation when the hand is pointing down, has only 330 degrees of rotation. However, in order to pick up randomly oriented parts, the gripper must be capable of handling all orientations. Since the gripper is symmetric about 180 degrees, we can effectively achieve all rotations by rotating the outer joint by ±180 degrees for some orientations. The monitor system determines when the orientation needs to be changed for a particular pickup position. If the pickup is modified, the subsequent setpoints may have to be similarly modified so that all parts arrive at the same orientation.

The orientation problem described above arises whenever the robot path is dynamically defined. The general problem, for grippers that are not symmetric or when the gripper is not pointing straight down, is not solvable with commercially available robots. Robots that provide wrists with greater flexibility (greater joint range) are required.

Error recovery is the second problem. The possibilities for parts out of reach, impossible robot positions, collisions, etc. increase significantly when the robot's path is being changed for each cycle. Errors detected by either the vision or the robot subsystems are reported to the monitor. The monitor then takes action, usually restarting the robot. Like the orientation problem above, however, the dynamic nature of the system makes a good general solution to error handling difficult. Further research and experience with installed systems is needed to determine how to best approach the error handling problem.

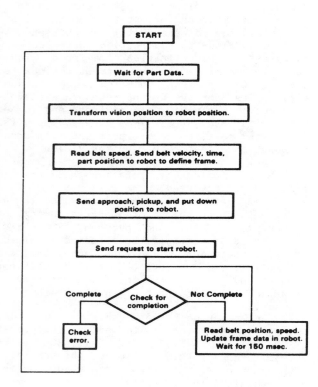

Fig. 9. Overview of the operational phase

PROTOTYPE PRODUCTION SYSTEM

The previous sections described the overall structure and concepts of CONSIGHT and presented details of the implementation of the experimental system. The next phase was to develop a prototype production system. The objectives of this phase were to:

1. Production harden the system. Replace laboratory components with industrial components.

2. Reduce ultimate system cost.

3. Improve reliabilty of system.

4. Reduce the robot cycle time to 5 seconds for a simple pickup, transfer, putdown operation.

To achieve the above objectives required major hardware changes and recoding of software to reflect the hardware changes and to improve the speed. Because CONSIGHT was developed for plant installation from the outset, few functional changes in software were required. Fig. 10 is a schematic of the system. The overall hardware structure resembles the software structure of the experimental system.

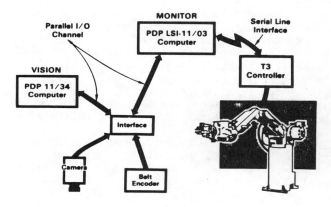

Fig. 10. Production prototype hardware.

The T3 robot, manufactured by Cincinnati Milacron, is computer-controlled and has a line tracking capability similar to that developed for the experimental system. However, in order to meet the functional and operational specifications for CONSIGHT, two enhancements were made to the standard robot. The normal maximum operating speed of 1270 mm/s was increased to 2500 mm/s. Secondly, an adaptive branch function was provided for external modification of the robot program [12]. Both of these features were added to the system by Cincinnati Milacron at GM's request and are now commercially available.

The system interface is a hardware device which accepts data from the camera, automatically thresholds it and sends only the transition points between black and white segments to the vision program. Additionally the interface passes position data from the vision computer to the monitor computer and interfaces to the belt encoding device.

The single control computer of the experimental system has been replaced by a PDP 11/34 vision computer, an LSI-11 monitor computer, and a robot controller which is an integral part of the T3. These changes have greatly improved the cycle time and the reliability of the system.

The monitor subsystem is implemented on an LSI-11 microcomputer. The software is more complex for the production system than for the experimental system. The part programming (described earlier) is implemented in the monitor whereas in the experimental system the frame concept in the robot system automatically handled much of the mathematics. The queueing of multiple parts is also handled by the monitor whereas in the experimental system standard operating system requests were used for data queueing. Operationally, the calibration and operational portions of the monitor are identical to those of the experimental system.

The production prototype system was first demonstrated picking up symmetric parts with only a single part on the conveyor. Further developments of the prototype have included part queueing, part sorting, pick up of non-symmetric parts, refinement of the vision hardware, robot checkout and reduced cycle time.

The gripper orientation problem described earlier is more severe on the T3 since the roll axis has only 240 degrees rotation. Manufacturing Development has added an additional actuator which will flip the gripper by 180 degrees. This additional roll range solves the gripper orientation problem when the gripper is pointing down.

The error handling of the prototype is similar to that of the experimental system. Normally when an error occurs or an unrecognized part is detected, the part is skipped and allowed to fall into a bin at the end of the conveyor. Although not elegant, this error handling technique is not unusual in production situations where humans also, on occasion, are unable to perform their assigned tasks.

Work is continuing on further refining and packaging of the vision hardware and software, developing special robot hands, and extensive system checkout. The current prototype system is operating at a cycle time of 5 seconds and at belt speeds of up to 20 cm/s.

SUMMARY

CONSIGHT differs from previous vision-based robot systems in its potential for practical production use. The vision subsystem, based on structured light, does not require high scene contrast. Both the vision and the robot subsystems are easily reprogrammed for new parts, in fact, the simplicity of "teaching-by-doing" is retained in spite of the complexities of the vision-controlled robot motions.

The advent of low-cost computer-controlled robots, such as the PUMA arm [13], will initiate a new era of sensor-controlled robot systems such as CONSIGHT.

REFERENCES

1. S. Holland, L. Rossol, M. Ward, "CONSIGHT-I: A Vision-Controlled Robot System for Transferring Parts from Belt Conveyors," in "Computer Vision and Sensor-Based Robots", New York, N.Y.: Plenum Press, 1979.

2. W. Perkins, "Model-based vision system for scenes containing multiple parts," in "5th International Joint Conference on Artificial Intelligence", pp. 678-684, Cambridge, August, 1977.

3. M. Baird, "Sequential image enhancement technique for locating automotive parts on conveyor belts," in "5th International Joint Conference on Artificial Intelligence", pp. 694-695, Cambridge, August, 1977.

4. M. Ward, "Specifications for a computer controlled manipulator," GM Research Publication GMR-2066, February, 1976.

5. G. J. Agin, "Representation and discrimination of curved objects," Stanford University Artificial Intelligence Project Memo, AIM-173, October 1972.

6. M. Oshima, Y. Shirai, "A scene description method using three-dimensional information, Progress Report of 3-D Object Recognition," Electrotechnical Laboratory, Japan, March, 1977.

7. R. J. Popplestone, et al., "Formation of body models and their use in robotics," University of Edinburgh, Scotland.

8. G. J. Agin, "An experimental vision system for industrial application," in "Proceedings, 5th International Symposium on Industrial Robots", p. 135, Chicago, September, 1975.

9. R. Duda, P. Hart, "Pattern classification and scene analysis," Wiley-Interscience Publication, p. 284, 1973.

10. G. J. Agin, "Image processing algorithms for industrial vision," Draft Report, SRI International, March, 1976.

11. R. Paul, "Manipulator path control," in "1975 International Conference on Cybernetics and Society", pp. 147-152, September 1975.

12. H. R. Holt, "Robot decision making", in "2nd North American Industrial Robot Conference", November, 1977.

13. R. Beecher, "PUMA: Programmable universal machine for assembly," in "Computer Vision and Sensor-Based Robots", New York, N.Y.: Plenum Press, 1979.

Unimation Application Notes*

Loading Boxes

Through the use of a standard 2000B, 5 axes *Unimate* industrial robot, the hazardous task of handling bottles of photographic chemicals has been eliminated.

APPLICATION DESCRIPTION

The highly toxic photographic chemicals are contained in 4 or 5 quart capacity plastic bottles. The bottles are filled, capped and marshalled to a transfer station via conveyor at a rate of 20 per min.

The *Unimate* robot picks up two bottles at a time and transfers them to boxes of various sizes accepting one, two, four, or six bottles. When the transfer of product is to a box accepting only one bottle, two boxes are presented back to back on the load conveyor and the robot fills both at the same time. If no boxes are present, the *Unimate* robot loads the bottles onto an auxiliary pallet until boxes are provided. The pallet pattern is 6 x 7 bottles requiring 24 transfers per pallet.

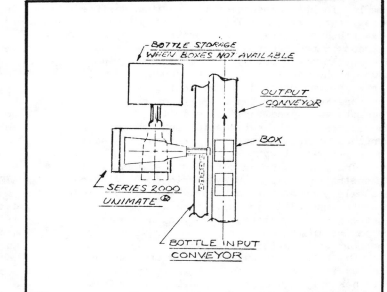

SPECIAL HAND AND FEATURES

The hand and fingers are constructed of a special acid resistant stainless steel. Neoprene pads, which are also acid resistant, are used to prevent the bottles from slipping out of position. Due to the large number of loading arrangements and the programming time required for this application, a 501B1 Cassette Program Storage and Verification Unit was provided. This unit allows storage of *Unimate* robot programs for replay as required, and provides flexibility for future product design.

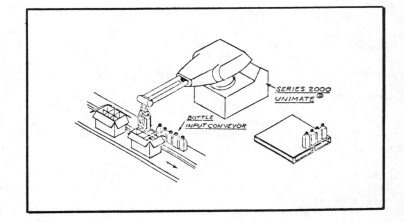

JUSTIFICATION

Serious injury can result to personnel when compression of the bottles causes the caps to come off and the chemical to spill. When this occurs production is stopped while the operator proceeds to shower immediately. The *Unimate* robot handles the task with ease, eliminating this health hazard, as well as the production normally lost during such a crisis. Single bottle weight is 10-15 lbs. The transfer time of two bottles from conveyor to box or pallet is approximately 6 seconds. The *Unimate* performs this operation continuously on a two shift per day basis.

Unimate and Unimation are registered trademarks of Unimation Inc.

Rubber Bale Handling

At Firestone Rubber in Orange, Texas a 2000B *Unimate* is being employed to automatically palletize 75 lb. rubber bales into shipping containers. A series of
conveyors transport the bales from the manufacturing area to the packaging and
shipping area. Here, automatic machines wrap the rubber bales in plastic film
bags for placement into shipping containers. Each conveyor delivers a specific
grade of rubber to be palletized into a selected shipping container location and
configuration. The *Unimate* is adapted with a special hand mechanism to transfer
the bales from the two conveyors into the appropriate containers.

The conveyors accumulate the
bales in groups of three (one
layer) before calling for the
Unimate program to transfer
that layer of parts to the
container. Just prior to
transfer, each bale is resized
to assure that cold-flow of
the material has not distorted
its shape. Two types of containers are used; a cardboard
container with vertical
dividers holding three bales
per layer, eight layers high
in three vertical rows, and a
wooden box with no dividers.

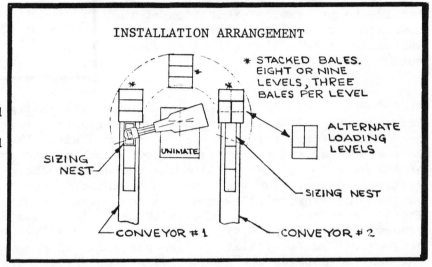

INSTALLATION ARRANGEMENT

* STACKED BALES.
EIGHT OR NINE
LEVELS, THREE
BALES PER LEVEL

ALTERNATE
LOADING
LEVELS

SIZING NEST

SIZING NEST

UNIMATE

CONVEYOR #1

CONVEYOR #2

The bales are palletized in alternate hand patterns for each layer to key them
together nine rows high. The same *Unimate* program is used for each common
pattern layer for a given container position. A special control console allows
selection of a separate program for each conveyor so that either conveyor output
can be palletized in either type container at any of the three loading stations.

Upon completion of a container, a materials handler will remove the full container and replace it with an empty one. Since there are three container locations for the two conveyors, the operator can select an alternate program to
load a second container while replacing the full container.

HAND MECHANISM - The special hand for this operation incorporates three major
features. A bale is picked up by using a large capacity vacuum system and a
special 12" x 24" vacuum cup contacting the top of the bale. Special devices
within the cup puncture the plastic film wrapper to prevent the vacuum from
rupturing the film. A pneumatically operated mechanism provides up to 56" of

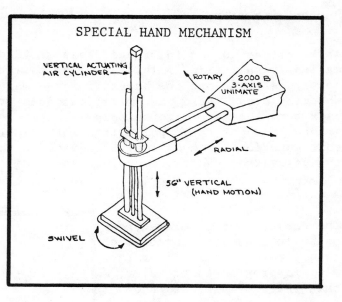

SPECIAL HAND MECHANISM

VERTICAL ACTUATING AIR CYLINDER

ROTARY

2000 B
3-AXIS
UNIMATE

RADIAL

56" VERTICAL
(HAND MOTION)

SWIVEL

vertical travel. Vertical up motion is controlled by valves and orifices and always stops at a fixed position. Downward motion is similarly controlled, but the motion is smoothly stopped by a pressure differential generated when the lowering bale contacts the bale beneath it. Two downward speeds may be called for in the program. Termination of the downward motion signals coincidence initiating a bale blow off by an upward motion of the hand.

The 2000B *Unimate* is supplied in a three axis configuration (rotary, radial and swivel). Radial and rotary motions position the vertical loading axis over the centerline of the bale for pick up and deposit. The bend drive system coupled to the special swivel carriage allows teaching the proper rotational attitude of the bale for placement into the container. There is no servo controlled vertical motion. The boom up-down actuator is replaced with a rugged adjustable positioning link to allow the boom to always move in a fixed horizontal plane. The 512 step memory is adapted with an 18x28 Random Program Selection system and a special remote operator's setup console to allow preselection of the desired programs. The flexibility and adaptability of the *Unimate* once again point up the many advantages the robot offers over special purpose automation for complex and variable manipulative operations.

APPLICATION JUSTIFICATION - Synthetic rubber manufacture is a continuous process, seven-day-a-week operation. Manual handling of the heavy bales wrapped in fragile film present untold personnel and quality problems. The *Unimate* frees two workers per shift while giving an assured production rate in excess of four bales per minute.

Convertor Palletizing

A crash program was needed. Fast delivery - ease of installation - and an automated device to keep pace with the high speed conveyor delivering up to six
hundred 45# catalytic convertors per hour - palletizing the convertors into standard industrial wire tubs. Manual loading was impractical. Hard automation and
long delivery after design - unprofitable. Solution - the *Unimate* industrial robot.

APPLICATION DESCRIPTION

A 2100B Series, 5 axis industrial robot with inexpensive hand tooling was used to
stack 58 catalytic convertors into wire cage-type transports in proper preprogrammed sequence. These are placed accurately and in the proper position in every
tote cage. The convertors are fed from a transport onto a small conveyor which
has the capability of tilting the conveyor at a 72° angle for *Unimate* pick up. It
then delivers the convertor to the *Unimate*, signals the *Unimate* to pick the convertor up and place it into its proper location. Accuracy of placement must be
maintained as convertor surfaces are critical to final finish product. The only
difficulty experienced in programming was the odd shape of the convertor having
left and right plumbing on its ends. Solution - programming technique combined
with alternate end feeding from the input conveyor.

SPECIAL FEATURES

Due to the quantity of convertors
and accuracy of placement required
a standard sequence control unit
was supplied with the *Unimate*. The
controller allowed the use of a
base routine to perform the placement of the convertors in the tilt
cage. This reduced the
number of steps required
and simplified the pro-
gramming. Since the
motions required to pick up
the convertor and to move
to a point over the tub are
the same, the base routine
repeats these steps con-
stantly. Then, the sub-
routine takes over for the
individual placement at 58
different positions of the
convertor into its proper
place in the tub. When a
tub is filled with the
proper quantity of convertors it is moved by

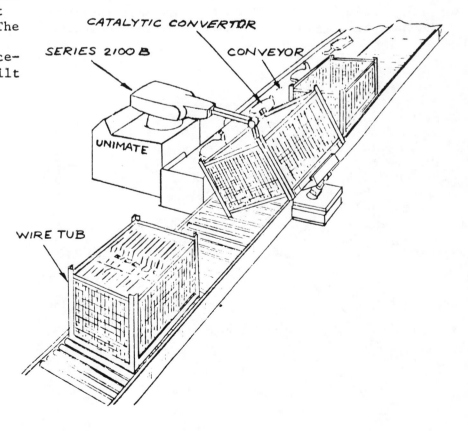

CATALYTIC CONVERTOR

SERIES 2100 B

CONVEYOR

UNIMATE

WIRE TUB

conveyor for shipment and another empty tub arrives at the loading station. The task is completed and the *Unimate* repeats the job shift after shift unattended and properly interlocked to its incoming conveyor and tilt mechanisms.

HAND DESCRIPTION

The hand which has two finger-like paddles with a total closing travel of 1/2", closes on command from the *Unimate* when it is in the proper position. Clamping action of the gripper mechanism is a parallel constant force motion generated by two linear cams; one finger motion is very small, the other approximately 1/2". Neoprene was added inside the two paddles to provide better gripping. Should convertor dimensions change, paddles may be quickly removed and new ones installed in minutes. The finished convertor has no hint that it was handled by a *Unimate*.

APPLICATION JUSTIFICATION

Since the upstream operations were all automated by special purpose automation interlocked with in-process computors, this loading station presented the bottleneck. Surely, finished convertors could have been diverted from the main convertor line by branch conveyors and palletized manually and hinder profitability of the process justification. Manual operation would surely have needed four operators per shift plus one and three quarters operators for relief. But how about product quality? Heavy separators would have to be adapted to the tote box. If this was so, larger tote boxes would be needed. Again - a redesign. As this was a new production line for catalytic convertors, justification for automating would be too difficult to manually operate, and therefore automation was the only answer and - *UNIMATE* the final solution.

Refractory Brick Loading

A leading manufacturer of refractories has installed a system for loading refractory bricks into oven cars. This turn-key system, supplied by Unimation Inc., includes a 4000 Series *Unimate* industrial robot with special tooling, conveyors, and an oven car indexer-positioner.

APPLICATION DESCRIPTION

Formed bricks are ejected out of a press and pushed onto a metal pallet on a conveyor. A single pallet and brick can weigh up to 75 lbs. Loaded pallets are accumulated in groups of three and picked up by the *Unimate* with its special tooling for loading into an oven car. The oven cars are divided in half, each of which has twelve shelves for holding six pallets at each level. Since the *Unimate* robot handles three pallets at a time, it goes to each level twice.

In the first operation, one half of the oven car is empty while the other half contains empty pallets. The *Unimate* picks up 3 full pallets from the output conveyor and loads them into the proper position in the empty half of the oven car. It then goes to the other half of the oven car, picks up three empty pallets and deposits them onto the input conveyor. Picking up three more full pallets, the *Unimate* loads them behind the first three in the oven car. The next set of full pallets will be loaded at the next lower

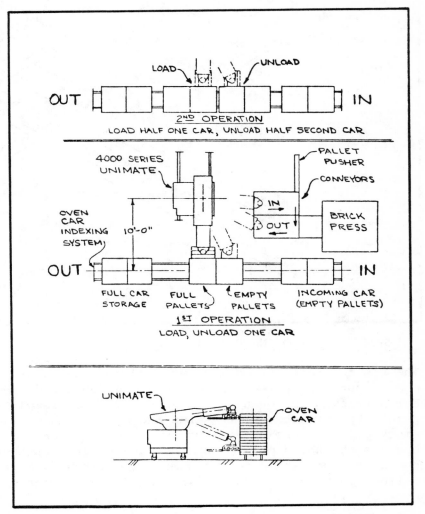

level. This sequence continues until one half of the car is loaded with full pallets while the other half is completely empty. The *Unimate* robot then proceeds to perform the second operation.

The oven car indexing system shuttles the half full, half empty car and an oven car filled with only empty pallets, into the proper position. The *Unimate* robot now loads full pallets into the second half of the first car while removing empty pallets from the first half of the second car. When the second operation is completed, the *Unimate* robot programs back to the first operation, while the fully loaded car is indexed into position for removal by a fork lift truck.

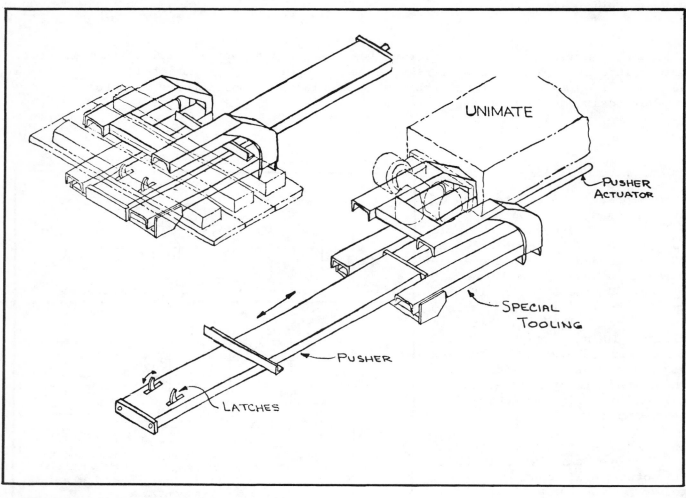

UNIMATE ROBOT TOOLING DESCRIPTION

The tooling consists of arms upon which the pallets rest. When servicing the input and output conveyors, the pallets are deposited or picked up directly by the arms. For servicing the oven cars, a pusher, which is part of the tooling, extends out and pushes the pallets off the arms and onto the shelves of the cars. In order to remove empty pallets, the pusher extends into the car and under the pallets. A latch then engages the third pallet and the pusher retracts thereby pulling three empty pallets out of the car and onto the arms. The latch is kept engaged and serves to hold the pallets in place during transfer and releases the pallets at the conveyor.

JUSTIFICATION

Two men were required to perform the task of placing a twelve pound metal pallet in front of the press and picking it up again with a brick on it - for a total weight of 75 lbs. Since the press has a ten second cycle time, the men alternated this task and picked up a loaded pallet every twenty seconds. This installation relieved the two men of this heavy, tiring task. In addition, the *Unimate* robot performs the operation tirelessly, uninterrupted 2 shifts per day 5 days per week.

The successful operation of the first system resulted in the installation of two additional robot systems, thereby effecting additional cost savings and increased productivity.

Conveyor Line Tracking

Line tracking - a new solution to the problem of accurately feeding a continuously moving conveyor. *Unimation Inc.*, working in close cooperation with customer engineers, has designed and installed a system including a Series 2000 *Unimate* robot and associated equipment to transfer 72 pound engine heads from an assembly conveyor to a continuously moving monorail conveyor at a major automotive manufacturing facility.

APPLICATION DESCRIPTION

The system consists of a powered roller accumulating conveyor equipped with a positioning shuttle to locate the cylinder head for *Unimate* transfer. Monorail carrier guides were designed to assure true position of the carrier during transfer. *Unimation* also designed and supplied a monorail pulley and mounting, together with an encoder mounting and coupling arrangement.

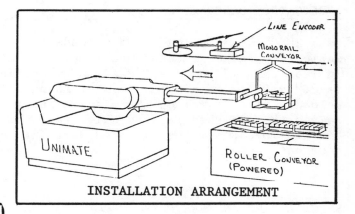

INSTALLATION ARRANGEMENT

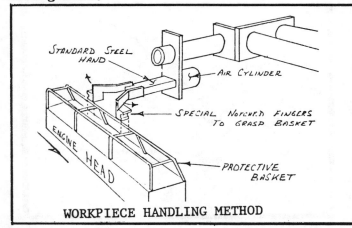

WORKPIECE HANDLING METHOD

SPECIFICATIONS

Part weight	72 lbs.
Monorail line speed	0-8"/sec.
Carriers per hour	720 max.
Carrier spacing	32 - 40"
Unimate cycle time	5 seconds

SPECIAL ACCOMMODATIONS - An encoder (positional feedback device) is coupled to the continuously moving line. This encoder is automatically set to a zero position in reference to each individual carrier as the carrier arrives at the load station. Thus *Unimate* can operate flawlessly even when the carriers are not evenly spaced. The in/out motion of the *Unimate* is slaved to the line encoder, when the *Unimate* robot is in the tracking mode. Line speed variations present no problem to the *Unimate*.

NOTE - *Unimation Inc.* offers two types of tracking systems. The system described above tracks in such a manner that it requires the equipment to be arranged so the line to be tracked travels in a straight or circular path parallel to the travel of one of the *Unimate's* articulations.

JUSTIFICATION - The *Unimate* robot displaces one operator per shift on a two shift basis, where he had been required to transfer a 72 pound part.

As part of a time-phased manufacturing development plan, the first two of a series of *Unimates* have been delivered to a major automotive manufacturer, to fully automate the edge grinding operations on flat window glass.

APPLICATION DESCRIPTION

Each 2000B *Unimate* loads and unloads two automatic glass edge grinding machines alternately in less than 17 seconds. Maximum glass size is 28" x 68", weighing up to 35#. An auxiliary unload assist device on the glass grinder raises the finished part to allow the *Unimate's* hand to move under the finished part into a nest on the grinder.

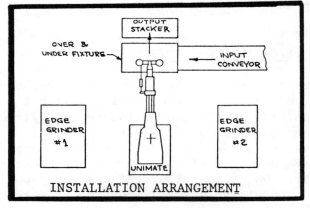

INSTALLATION ARRANGEMENT

SPECIAL HAND AND FEATURES

A special double hand allows handling two pieces of glass so that a finished part is deposited on an output conveyor as a new piece is picked up from the input conveyor. The hand consists of two sets of dual vacuum cups mounted back to back so that an unfinished glass is supported by the lower cups and a finished piece supported by the upper cups. Three clamp operators combined with special valves and timer allow independent programming of vacuum and/or blow-off on either set of vacuum cups.

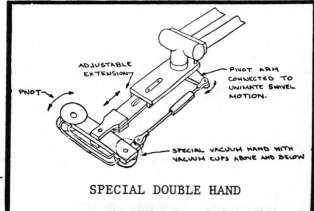

SPECIAL DOUBLE HAND

The hand is supported from the hand gear train housing and the angular position of the vacuum cups can be varied ±20% by the swivel motion. The position of the cups relative to the wrist is adjustable. The programmable swivel and adjustable cup position allow handling a large variety of part shapes. A 4x32 program option combined with Alternate Program Selection allows the operation of any 2 of 4 programs; thus *Unimate* can operate either one or both edge grinders as required.

JUSTIFICATION

The continuous process nature of the glass manufacturing industry makes this outstanding application an extremely profitable investment. Each *Unimate*, with relatively simple auxiliary equipment, replaces two workers per shift on a three shift basis. The programmability of the *Unimate's* memory in combination with the five axis flexibility allow easy setup for a wide variety of shapes and sizes of glass not possible with hard automation or less sophisticated robots.

Projectile manufa͟ture at our country's arsenals is a classic example of archaic production techniques – little significant improvement has been made since the 1940's. However, one of the major arsenals is not standing still and two projectile depalletizing systems have been installed – the heart of which is the 4000 Series *Unimate*.

APPLICATION DESCRIPTION – Empty projectiles come to the arsenals from special manufacturing facilities which forge and machine the projectiles. They arrive on standard wooden pallets and are transported to the depalletizing area for loading into tank cars. The system must remove and dispose of banding, remove plastic protective grommets and remove pallet tops. These operations are accomplished as part of a power and free palletized conveyor system which then presents the projectiles to the *Unimate*.

Four 4000A, five axes *Unimates*, two working the 155mm and 175mm line, and two working on the 8 inch line, remové the projectiles from the wooden pallets into tank cars. The projectiles weigh up to 200 lbs.

The first *Unimate* engages a lift lug which is screwed into the nose of the projectile from the six or eight position pallets to an intermediate nest. The *Unimate* unscrews the lift lug and deposits it onto a conveyor as it returns for the next projectile. The second *Unimate* in the team picks up the projectile by engaging the internal threads in the nose and transfers the part to one of fifteen positions in the tank car.

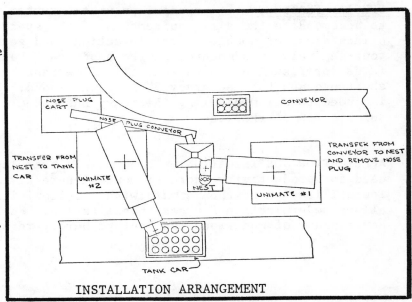

INSTALLATION ARRANGEMENT

HAND DESCRIPTION – Two types of hands are employed in this application. The first hand incorporates two features. One is a contoured cup driven by a pneumatic impact wrench to seek, find and unscrew the projectile lift lug. The second feature is a pneumatically operated lift pin to engage the lift lug eye for transport of the projectile.

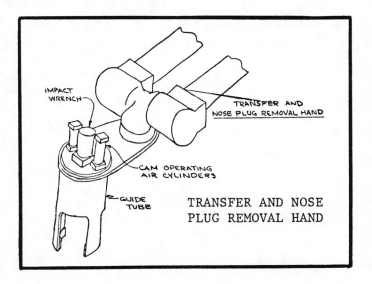

TRANSFER AND NOSE PLUG REMOVAL HAND

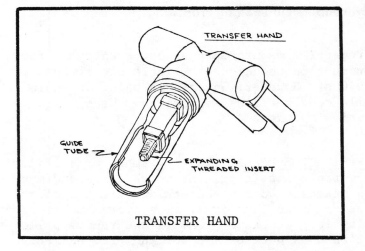

TRANSFER HAND

The second hand contains a threaded, pneumatically operated internal expanding mandred for transporting the projectile with the lug removed by engaging the internal nose threads. Both hands include long cylindrical fingers to prevent projectile sway and are inherently fail-safe on loss of air or electrical power.

SPECIAL FEATURES - A special pneumatic system was designed for the 4000 *Unimate* to accommodate the air requirements of the special hand tooling. This included a 4 inch diameter double or triple acting hand gear train cylinder. A large port four-way valve was mounted at the hand gear train housing to accommodate large air feeds for fast hand operation. To allow fast cycling the clamp time delays were shortened and the "Accuracy 3" range extended. Rotary damping cylinders were included to allow handling the heavy suspended loads. A cassette recorder allows quick setup of the line.

APPLICATION JUSTIFICATION - The arsenals are plagued with the problem that their need for productivity increase comes inherently at a time of national emergency. The utilization of the *Unimate* with its flexibility of setup and high productivity allows modernization for easy mobilization. Relieving men from the hazardous environment of munitions manufacture hardly requires mention.

When one of the nation's leading manufacturers decided to automate the grinding of their harrow discs, they contacted *Unimation Inc*. After careful study of the operation we recommended a Series 2000 *Unimate* industrial robot with 3 degrees of freedom, and equipped with a special vacuum air double hand.

APPLICATION DESCRIPTION

In operation, the *Unimate* robot's hand blows off excess dust from the blade and vacuum lifts it to a registration table where it is oriented, allowing the *Unimate* robot to regrip in a centered position. This is necessary because incoming stacks of blades can lean as much as 1" from a centered position. Having oriented the blade, the *Unimate* robot loads Grinder #1 with its left hand, and removes a finished blade with the right hand. The robot then deposits the finished blade at the nearest output station, where it picks up another from the incoming lift table. After orienting this blade and loading Grinder #2, the finished blade is removed and the cycle repeated.

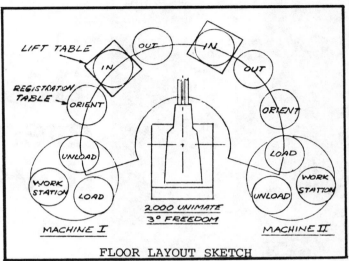

FLOOR LAYOUT SKETCH

SPECIAL HANDS AND FINGERS

The hand consists of a group of suction cups attached to spring steel fingers set on an angle compatible with the concavity range involved. Each hand has its own pneumatic system enabling it to be independently programmed. All pneumatic lines are fitted with quick disconnect couplings allowing rapid changeover from one size to another. The hands are pivoted above a self-aligning bearing allowing them to properly seat themselves on the leaning stacks of blades. Lift tables are provided, and through an electric eye indexing system, keep incoming blades at a fixed height. This eliminates the need for a wrist on the *Unimate* robot, allowing the 2000 model to lift the 90 pound discs (2 at a time), normally requiring the heavier 4000 unit.

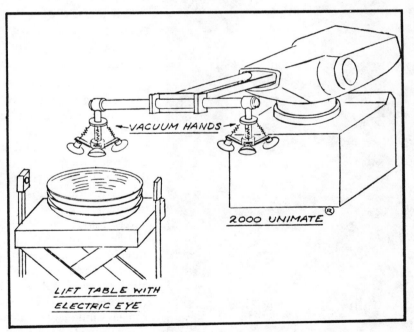

LIFT TABLE WITH ELECTRIC EYE

JUSTIFICATION

This is but another example of the *Unimate* industrial robot performing a job unsuitable for human labor. The blades, weighing as much as 90 lbs. each, are heavy and awkward; and the air is contaminated with dust and excessive noise levels. Though originally intended to satisfy OSHA regulations, the *Unimate* robot has effected substantial part quality improvement and increased productivity.

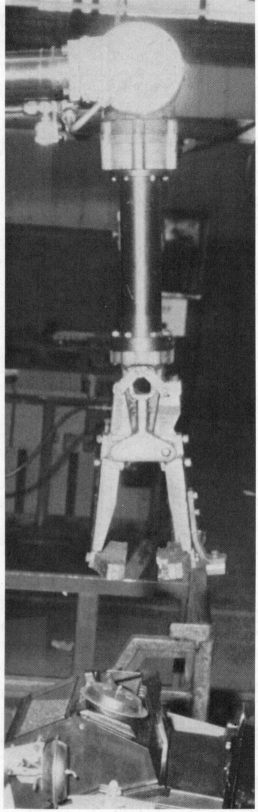

A tough but excellent job for a robot—picking up loads from a moving assembly line and placing them in shipping containers. This is how robots should be used in materials handling, performing tasks which are repetitive, boring, and physically demanding for people.

This robot 'sees' everything coming down the line!

They don't have eyes, but the robots at Harrison Radiator Div. know the exact position of each product moving down an assembly line. They need this information so that they can pick up loads on the fly and position them in shipping containers.

Very few people would enjoy performing a job that requires them to remove 30-lb air conditioners from the end of an assembly line and manually place them in shipping containers. Especially if they had to do it every 19 seconds. But, for a robot, it's an almost ideal application.

At Harrison Radiator Div., General Motors Corp., Lockport, New York, where robots are used on several assembly lines, the job is much more complicated than one of "pick and place." The assembly lines run continuously, which means the robots have to pick up the assembled air conditioner units on the fly.

Moreover, each shipping container holds 12 units, and there's a defined position for each unit. The result is, the distance through which the robot's arm has to travel from conveyor to loading position, and back again, varies from one loading cycle to the next.

"Each robot has to be able to locate an air conditioner on the moving conveyor, pick it up and determine the path to the loading position, then place the unit where it's intended, within a ¼-in. tolerance," explained Gerald Paterson, senior production engineer.

The robot has no "eyes" to see where the air conditioners are located on the moving conveyor. Instead, there's an encoder attached to the drive shaft of each conveyor. It's actuated when a fixture carrying an air conditioner breaks the light beam of a photoelectric control mounted alongside the conveyor. The encoder sends signals to the robot's microprocessor control, enabling it to track the exact position of each fixture, within 0.050 in.

A robot is mounted at the end of each assembly line, with the arm aligned with the centerline of the conveyor. This means the robot has only to extend or retract its arm to achieve precise alignment with the moving fixtures. Still, this would be

This robot 'sees'

Locating a load for pickup. The robot receives signals from a microprocessor which is linked to an encoder mounted on the drive shaft of the assembly conveyor. In this way, it always knows the exact position of each load fixture, within 0.050 inch.

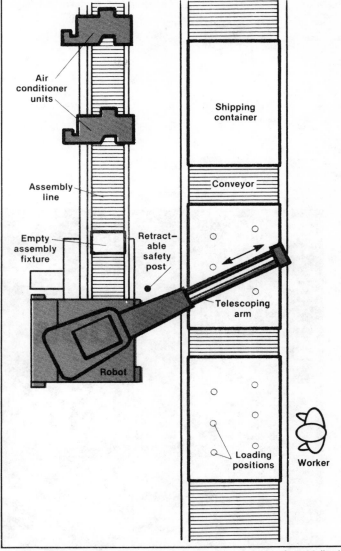

Air conditioner units

Assembly line

Empty assembly fixture

Retract-able safety post

Shipping container

Conveyor

Telescoping arm

Robot

Loading positions

Worker

Cycling automatically every 19 seconds, the robot transfers 30-lb air conditioner units from an assembly line to a shipping container. Each container holds 12 units. People are involved, but only to place dunnage in the containers and strap the final load.

impossible without tracking information from the microprocessor control.

The assembly line is a moving slat conveyor—over-under carousel design—with a series of fixtures mounted on 42-in. centers. Air conditioner units are assembled directly on the fixtures. Loaded fixtures advance to the end of the line, the robot removes the units, and the empty fixtures return on the lower flight to the head of the assembly line.

Each air conditioner is picked up by a mechanical grab at the end of the robot's arm. The grab engages a flange on the top of the air conditioner and lifts it from the

fixture. Then the arm quickly moves through a horizontal arc to the shipping container.

Workers assist in the loading process. Empty shipping containers—steel racks—are brought from storage by fork trucks and placed on a powered roller conveyor which parallels the assembly conveyor. In each container there's reusable dunnage with molded inserts which the workers arrange for loading.

First a container with a bottom sheet of molded dunnage is advanced to the first loading position. There it's automatically lifted off the conveyor bed and clamped in place.

The robot (Unimation Inc.) automatically transfers air conditioner units to the container, placing them according to a pre-programmed pattern. When six units have been put in place, the container advances to a second loading position. There, a worker adds spacers between the units, and a sheet of dunnage for the second layer of six units.

To ensure worker safety, a vertically retractable post is mounted on the worker side of the robot. There's also a movable plastic door at each of the two loading positions, which bars a worker from access to a container whenever the post is lowered.

Swinging into position. After picking up a load, the robot's arm pivots to the side and places the load in the container. It's not as easy as it sounds because the distance between the two points varies from one loading cycle to the next.

When the post is raised, physically preventing the robot arm from swinging to the side in the event of a malfunction, interlocks release the doors. The worker now can push the door to the side and reach into a container to add dunnage. The worker secures the load after the container exits from the second loading position.

Another safety device—an automatic stop switch—is provided. Although the robot's arm cycles from conveyor to container and back in much less than 19 seconds, there's always a possibility of a delay. If a fixture is not unloaded when it reaches the end of the conveyor, the air conditioner interrupts a photoelectric light beam and this deactivates the drive, causing the assembly conveyor to shut down automatically.

Loading a shipping container. The mechanical grab at the end of the arm orients the load so that it will fit snugly the proper position in the container. Then the arm lowers, and places the load on a bottom sheet of molded dunnage.

CHAPTER 2

MACHINE LOADING

Commentary

Machine tool loading and unloading is a growing robot application. The reliability and steady pace of the robot can significantly increase machine tool utilization and productivity.

In many cases, because of long machining cycle times, a single robot will load and unload several machines. The machines may be grouped about the robot or the robot may move from machine to machine on a traversing base.

Dual-hand tooling is often used to increase the robot's efficiency, or dual-function tooling may be used to load and unload a part at several different machines. Automatic gaging may be provided, with the robot loading and unloading the gage, which might be interfaced with the machine control for automotive tool adjustment.

Spindle orienting devices, chip blow-off systems, cycle timers and power-actuated splash guards are machine modifications which may be required when a robot is installed. In addition, machines may have to be relocated and part feeder/orienters must be provided. Thus, these applications can be quite complex and require careful planning.

Presented at the Robot III Conference, November 1978

Machine Loading Robots On A Planetary Pinion Machine Line

By Boris Kelly
Massey-Ferguson

Massey-Ferguson has developed an automated machining line for planetary pinion gears that utilizes three robots, nine machine tools and a transfer system to produce gears from forged blanks. The reasons for selecting robots for machine loading rather than conventional hard automation is discussed. The integration of the robots into the system with specialized hand tooling, programmable controllers, sensors, machine tool modifications and safety devices is described. Massey-Ferguson's management team approach, training of personnel and presentation of the robot-manned line to the union work force is explained.

INTRODUCTION

When Massey-Ferguson became concerned about decreasing machine productivity and utilization, it was suggested that auto-loaded machines be investigated as one means of potential improvement. The initial inquiries into automatic machine loading were directed toward conventional hard automation. These investigations coincided with the upswing of machine tool orders and increased delivery time. New Britain Machine Company proposed automating two of their vertical turners with a Unimate robot rather than conventional automation, indicating this would be less expensive and delivery time would be improved. The proposal initially met with some skepticism as to whether the application required this degree of sophisticated automation. The decision was made to go ahead with this plan after investigating the Unimate application and considering the decreased cost and improved delivery.

At this time Massey-Ferguson was still considering conventional automation for other machines in the pinion process. When the broach manufacturer was contacted about automating the broach, he responded with a moderately priced, conventional automation proposal with acceptable delivery times. In this case, it was thought that conventional automation was the way to go. Next, the manufacturers of the shaping and shaving machines used in the machining of the pinion were approached about automating their machines. Preliminary proposals from these machine manufacturers indicated long delivery schedules and high costs. The long delivery schedules were due to their current design engineering and manufacturing work loads; the high costs were due primarily to the design and engineering development costs for the specialized application.

45

Because of the high costs and extended deliveries of conventional automation, Massey-Ferguson once again turned to the possibility of automating the other machines with robots. Several robots were investigated to see if perhaps a simpler robot might be available. In the course of this search the concept of independent groups of automated machines was expanded to a line concept, in which the individual machine groups would be connected with a part transfer system. Because of the transfer system proposed by the Fab-Tec Company, it was determined that a robot with the capabilities of the Unimate was required. Based upon Unimate's willingness to undertake a system of this type and a convenient branch location in the Detroit area, Massey-Ferguson ordered two additional Unimates for the automated planetary pinion machining line. Thus, Massey-Ferguson's first venture into robot automation was launched.

SYSTEM DESCRIPTION

Process Flow

Massey-Ferguson's automated planetary pinion machining line was designed to handle five sizes of pinion gears, ranging from a 29-lb., 6.8 in. OD, 3.2 in. face width, 25 tooth, 4.0 pitch gear down to a 6-lb., 3.6 in. OD, 2.5 in. face width, 16 tooth, 5.0 pitch gear. The process consists of broaching the ID of a normalized rough forged blank, turning the faces and OD, shaping the teeth, tooth deburring and gear shaving to produce a pinion gear ready for heat treat.

The process flow is shown in Figure 1 on page 3. The sequence follows:

1. A steel tub containing approximately 200 normalized forgings is loaded with a forklift truck into the hopper. The tub is tipped, dumping the forgings into the feed section of the hopper. A slot belt with inclined cleats lifts parts from the hopper into a twist chute discharging to a roller conveyor leading to the broach. The belt feed is controlled by high and low level limit switches.

2. Parts are positioned for ID broaching by a drag link and push rod loader. After broaching they are ejected onto a roller conveyor leading to an elevator.

3. The parts are ejected from the elevator onto a roller

MASSEY FERGUSON
Automated Planetary Pinion Machining line
Process Flow

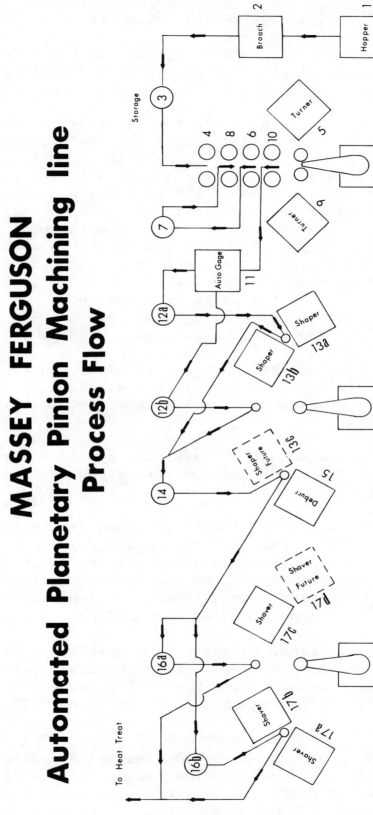

Figure 1 - Process Flow

conveyor leading to a helical storage unit with a capacity of approximately 350 parts.

4. The storage unit supplies parts, upon demand, to parts terminal #4 that positions two parts for robot pickup.

5. During the machine cycle the robot picks up two parts
& from the terminal and places them on preload table #5
6. directly in front of the New Britain dual spindle turner. When the machine cycles out, the parts on the chucks are removed by the robot and placed on the postload table #5 in front of the turner. The robot then picks up the parts from the preload table and places them on the chucks. At this operation the gear blank's OD is rough-turned and the face and hub are finish-turned. When the machine is cycling, the robot picks up the parts from postload terminal #5 and deposits them on discharge terminal #6. This area is shown in Figures 2 and 3 on page 5.

7. After leaving the discharge terminal, parts are conveyed via roller conveyors and an elevator to a helical storage unit with a capacity of approximately 50 parts.

8. The storage unit at #7 position supplies parts to parts terminal #8 that positions two parts for robot pickup.

9. The part handling at this position is the same as at #5, in that parts from the load terminal at #8 position are placed on a preposition table by the robot and then placed on the turner chucks after the chucks have been unloaded. At this operation the OD and opposite gear face and hub are turned.

10. The robot transfers parts from postload table #9 to discharge terminal #10. From the discharge terminal the parts are conveyed via roller conveyors and a lift to the gaging station at position #11.

11. The auto gage checks OD and ID concentricity, bore to hub face squareness, OD size, face and hub size. Rejected parts are shunted into a holding chute. Acceptable parts are delivered by conveyor to an elevator.

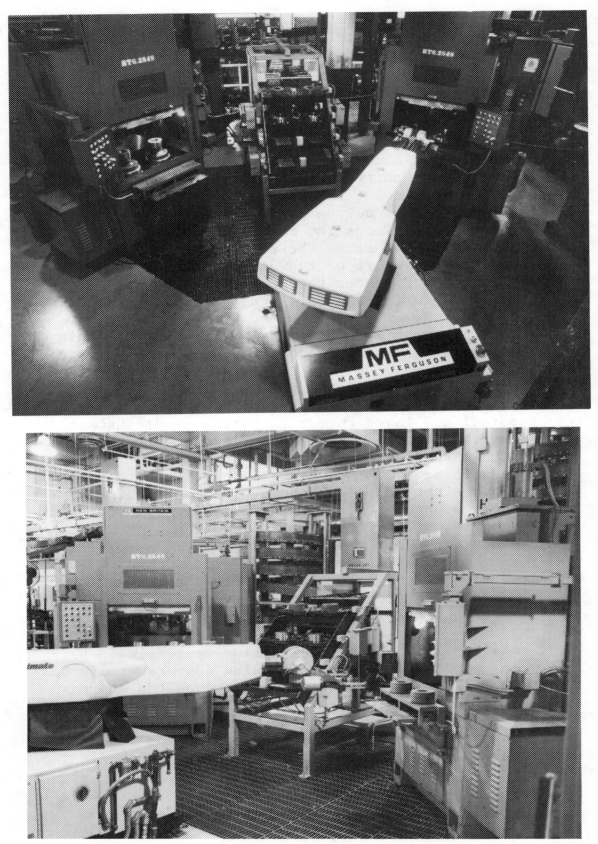

Figures 2 (top) and 3 (bottom) - Blank Turning Machine Center

12. After leaving the elevator, the parts are divided between storage units 12a or 12b, each having a capacity of 200 parts.

13. The parts flow from the storage units to terminals between the Ex-Cell-O Shear Speed shapers at 13a and 13b or 13b and 13c. The Unimate picks up parts from the terminals and places them on a preposition table at 13a, b, or c, where they are stamped to identify the machine running them. When the machine cycle is complete, the robot unloads the part from the Shear Speed, places it on the preposition table, picks up a new part and loads the machine. During the machine cycle, the robot takes the part from the preposition table and places it on the discharge conveyor. At this machining center the gear teeth are cut. This area is shown in Figures 4 and 5 on page 7.

14. After gear shaping, the parts go to a storage unit at position 14 having a capacity of 350 pieces.

15. The parts move from the storage unit to a terminal at the deburr unit. The robot takes parts from the terminal and loads them onto a two-spindle Hammond deburring machine. After the first deburr cycle, the Unimate picks up the part, turns it over and places it back on the deburr machine for the second cycle. After both sides have been deburred, the robot removes the part from the deburr machine and places it on the discharge conveyor leading to an elevator.

16. After leaving the elevator, deburred parts are divided between storage units 16a and 16b, each capable of holding approximately 325 parts.

17. Parts from the storage units flow to terminals between the National Broach Red Ring shavers at 17a and 17b or 17b and 17c. The robot removes the shaved gear from the shaver and deposits it on a discharge terminal in the same area as the load terminal. The robot then picks up a fresh part from the load terminal and loads it directly into the shaver. There are no preposition tables in this area. After shaving, parts are then sent to heat treat. This area is shown in Figures 6 and 7 on page 8.

Figures 4 (top) and 5 (bottom) - Gear Shaping
and Deburring Machine Center

Figures 6 (top) and 7 (bottom) - Gear Shaving Machine Center

Unimate Tooling and Programs

The three robots are Unimate 2105 G, 5° models. They differ primarily in hand tooling configurations and programmable controller systems.

The hand tooling for the Unimate servicing the New Britain turners is shown in Figures 8 and 9 on page 10. The two hands are air actuated with independent cylinders. Each hand has two fingers with four cats paw inserts for grippers. The fingers are of a pivoting, self-centering design to compensate for variation in rough forgings and variations between rough and finished parts while maintaining the precise positioning necessary to load parts onto the dual spindles of the turners.

The hand tooling for the robot servicing the Shear Speed shapers and Hammond deburr is shown in Figures 10 and 11 on page 11. In addition to the normal gripping motion, the hand has a seven-inch vertical stroke capability. This is necessary because the Shear Speed work spindle does not stop in the same location each time.

The hand tooling for the robot servicing the Red Ring shavers is shown in Figures 12 and 13 on page 12. Transfer of parts to and from the shavers is done with the part clamped in the hand. Positioning of the part in the machine is performed with the part unclamped to permit the part to rotate in the hand while meshing with the shaving cutter. Final positioning is done as the arbor is inserted through the bore.

The robot servicing the New Britain turners has five programs. The main program is for unloading and loading of parts for rough turning on the first machine and unloading and loading of parts for finish turning on the second machine. This is a fixed sequence, rather than random program of 46 steps with six input and two output signals.

The alternate programs are servicing only the first machine, servicing only the second machine, servicing the first machine for rough and finish cut, and servicing the second machine for rough and finish cut. These programs permit continued production in the event either turner is down for tool change or repair. Programs are manually selected.

The robot is interfaced with the machines and transfer system with an Allen Bradley Bulletin 1774 PLC, which also controls the two turners.

Figures 8 (top) and 9 (bottom) - Robot Hand
Tooling for Turning Center

Figures 10 (top) and 11 (bottom) - Robot Hand
Tooling for Shaping and Deburring Center

Figures 12 (top) and 13 (bottom) - Robot Hand
Tooling for Shaving Center

The robot servicing the Shear Speed shapers and Hammond deburring machine has ten program routines, varying in length
from 18 to 6 steps with 8 input and 8 output signals. The
programs are unload and load routines for the three machines,
load preload stations for the shapers, and input and output
conveyor routines. The unload and load programs are randomly
selected on a priority basis. If the robot is idle, it will
go to the first machine that cycles out. If the Unimate receives a signal that two machines have cycled out, it will
choose the closer shaper or either shaper over the deburr.
Loading the shapers also takes priority over loading the preload station. The robot completes the program it is in before
selecting another program, regardless of priority. In addition
to the random program selection, any program may be manually
selected or omitted from the system.

The robot servicing the shear speeds and deburr and the robot
servicing the shavers are interfaced with the machines and
transfer system with a Texas Instruments 5TI Programmable
Control System.

The robot servicing the Red Ring shavers has ten program
routines varying in length from 6 to 8 steps with 7 output and
8 input signals. The programs are unload and load routines
for the three shavers and conveyor selection routines. The
programs are randomly selected according to machine cycle
completion. All programs may be manually selected or deleted.

Machine Modifications for Robot Loading

The modifications necessary to adapt the machine tools for
robot loading were primarily in the areas of control circuitry
to enable the machines to send and receive signals from the
robots, chucking systems to facilitate loading, chip removal
and part in-position sensors.

The control circuits on all the machines were revised to indicate ready for unloading, loading, and start conditions to
the robot. The ready for unloading condition consists of
two signals that are mechanically independent but electrically
in series -- machine cycle complete and door open. Unless
both of the signals are present, the robot will not service
the machine. The ready for loading condition consists of two
signals -- spindle clear and part present on the post-load
station. This latter signal is necessary in case the part was
removed from the spindle but dropped inside the machine. The
start condition is indicated when a signal is received from
the part in postion sensor and the robot has withdrawn its

hand from the machine. At this point the robot sends a <u>start</u>
cycle signal to the machine.

The chucks on all the machines are the expanding arbor type.
They were designed to give somewhat more expansion than hand-
loaded arbors in the range of .020" to .025." The lead taper
was also increased to .100" x .125."

Chip removal was a major problem in the turning and shaping
operations. Chips on the locating surfaces result in improperly turned and shaped parts. Long, stringy chips make
loading and unloading difficult and interfere with the part
sensors.

The turning chip problems were solved by a combination of chip
breakers on the tools, tool paths, increased coolant flow, and
air blow off nozzles. The shaping chip problem was resolved
by increased coolant flow and air blow off nozzles. Although
the shaving chip problem was not as severe as the others, the
coolant flow was increased and air blow off nozzles were added.

The part position sensors have two functions: to detect if a
part is properly positioned to start the machine cycle and to
detect if a part is properly positioned for the robot to
safely load or unload the part without striking the arbor. On
the turners the sensors for <u>part in position</u> to start the
machine cycle are LED photo-sensors positioned so that the
beam will be broken if the part is not completely seated on
the arbor. The <u>part in position</u> sensors on the Shear Speed
shapers are small air ports on the locating faces of the
arbors. The air ports are connected to pressure switches
activated by back pressure when the part seats on the locating
face.

The <u>part in position for loading safely</u> sensors are LED photo
sensors mounted in front of the arbors and positioned so that
if a part is low enough to strike the arbor the beam will be
interrupted and the robot stopped. The <u>part in position for
unloading safely sensors</u> are LED photo sensors mounted above
the arbors and positioned so that the beam must be clear before
the part is removed from the machine. The sequence is such
that the robot will vertically raise the part off the arbor
but will not horizontally move the part unless the photo sensor
beam is clear. The intent here is to prevent the robot from
attempting to remove a part that is still partially on the
arbor.

LEDs in chip invies ??

Parts Transfer System

The Fab-Tec transfer system transformed the manufacturing process from a group of machining centers to an integrated machining line. Although the process is now a machining line, it retains most of the flexibility of independent machining centers as opposed to the lack of flexibility of some transfer machining lines.

The key elements in the flexibility of this system are the storage units positioned between machining centers. Maintaining these storage units in a half-filled status during normal operation permits the machines ahead of the unit or below the unit to be down for approximately half a shift and not affect the overall line production. In other words, if the machine center downstream from the storage unit goes down, the machine center upstream from the unit can continue to run and store parts. If the machine center upstream goes down, the machine center downstream can draw parts from the storage unit and continue to run.

Another important feature of the transfer system is the display console shown in Figure 14 on page 16. This console has a schematic diagram of the entire machine line with indicating lights at the various component locations. These lights indicate the status of the various machines and the locations of parts at the various terminals. There is an LED counter that is used to display the number of parts in the various storage locations as well as the number of parts run through the individual operation. This counter is also used to preset the number of cycles of the machines for tool change. When the preset tool count is reached, the robot will no longer service the machine and display lights on the console indicate which machine is ready for tool change. This function as well as the control of the entire transfer system is handled by an Allen Bradley programmable controller.

The balance of the transfer system is rather conventional in that it is composed of gravity conveyors, pump up lifts, elevators, and various metering devices. The unconventional feature of the transfer system is the was it is integrated into a transfer line that automatically transfers parts as needed. Much of this flexibility is due to the programmable controller system.

Safety Devices

The safety devices fall into two functional categories: one

Figure 14 - Transfer System Display Console

to protect the robot and machine and one to protect the operator. The first category includes devices, such as the part position sensors described in the preceding section, which prevent the robot from loading or unloading the part or prevent the machine from starting unless the part is properly positioned. Also included in this category are the serially connected signals, such as the cycle complete and door open signals, where two or more signals must be present for a robot or machine function to occur.

The devices to protect the operators are of prime importance in a robot-serviced operation. Many devices such as a perimeter barrier of photo cells and trip wires mounted on the robot boom were considered. It was decided that safety fences, safety posts and key-operated robot controls would be adequate. An outer-perimeter safety fence encloses the entire machining line and is intended to keep unauthorized people out of the area. An inner perimeter fence behind the robot console and safety chains between the machines enclose the immediate operating area of the robot. This fence has gates that are electrically interlocked with the robot control circuit. If the gate is opened when the robot is running, the robot is put into a hold mode. Closing the gate does not restart the robot; it can only be restarted from the key operated control on the console outside the inner perimeter fence.

Outside the inner perimeter fence behind each robot is a control console. It has several switches for controlling the machines and transfer system, but its prime purpose is to provide a remote location for starting and stopping the robot with a key-controlled switch. This key-controlled switch on the exterior console overrides the robot console, so that the robot cannot be restarted from the interior console if the exterior console switch is in a hold position; conversely, the robot cannot be restarted with the exterior console switch if the interior console switch is on hold. The differences between the interior and the exterior switches is that the key gives a positive lockout feature.

The safety posts are designed for use when it is necessary to have an operator working on one machine while other machines in this group are being serviced by the robot. The safety post has three safety features. When the safety pole is inserted in a receptacle in the floor to protect a particular machine area, it activates a proximity switch that prevents the programmable controller from directing the robot to service that area. The second safety feature is an electrically interlocked wand on the side of the post. If the robot, through a programming error, should try to service the blocked out area, the boom would contact the wand and electrically disconnect all power to the robot. The third safety feature is the physical barrier provided by the safety pole. If the electronic and electrical safety features fail to stop the robot, the safety post is strong enough to physically stop the robot.

SYSTEM DESIGN AND IMPLEMENTATION

System design by the machine tool, robot and transfer system manufacturers was coordinated by a Massey-Ferguson project engineer, assisted by Massey-Ferguson process and plant engineering personnel, a tool design consultant and an electrical design consultant.

Although the automated line machining process was based on existing machining techniques, the integration of the various components into a system was a sizeable undertaking. By far the largest task was design of the electronic control system that enables the robots to communicate with the machine tools

61

and transfer system. As difficult as the task was, it was made considerably easier with the programmable controllers used in the system. Second only to the control system design were the robot hand designs.

System implementation was the responsibility of the plant engineer and project engineer. They were assisted by a production general foreman and foremen who were assigned to the project full time and attended the Unimate training course. The foreman was the one who would have production responsibility for the line. Management felt it was important to have production foremen involved in the early stages of implementation. Also involved in the early stages of implementation was an electrician who later became the electronic technician responsible for system operation and maintenance. The actual installation of the system was handled primarily by outside contractors under the direction of the plant engineer.

AUTOMATED LINE CONCEPT PRESENTATION TO UNION WORK FORCE

When the automated machine line was developed there was little automation and no robots at Massey-Ferguson plants. Apprehension existed that a robot manned machining line might meet with some resistance from the union.

Early in the development of the line, the concept was presented to the union bargaining committee, with the approach that this line, with its increased productivity, would provide security for existing jobs and create additional jobs. The establishment of job classifications for the line personnel included a higher pay rate than the regular machine operator category, due to additional training and required qualifications for the positions.

The site chosen for the automated line was the main aisle in the center of the plant, with additional lighting installed. This concept openly displayed the line, thus creating interest and curiosity.

Training of the operators and maintenance personnel was important with the introduction of this automated line to the union work force. (Three electricians were sent to the programmable controller classes.) Eight hourly personnel attended the Unimate training course: three are involved with the daily operation of the line; the others might be considered backup. However, the real benefit in training more people is the familiarization of plant personnel with robots.

Although Massey-Ferguson maintains "a right to manage" provision in its contract, the company felt that the above presentation surpassed "presentation by edict." The results have been favorable with no apparent resentment from the workers.

SYSTEM RETURN ON INVESTMENT

The automated machining line is designed to produce at approximately the same cyclical rate as manually loaded machine centers, i.e., a robot takes about the same amount of time as a man to unload and load a machine for a given cycle. Without taking into account the overall system and looking only at the replacement of one operator with one robot, the payback period would be approximately 1½ years on a two-shift operation, using direct labor standard cost.

The payback period on the overall automated machining line is approximately 2½ years on a two-shift operation, using total standard cost. This is based on approximately a four-operator per shift reduction in direct labor, a reduction in rejected parts and a reduction of in-process inventory. *← DIP TAKE 7 OPERATORS*

Although the robot does not show an appreciable gain in cycle time over a manually loaded operation, it does show a significant gain in consistency. On an overall basis, this consistency results in approximately a 25% increase in productivity. This means that a four-machine robot loaded center will produce the equivalent of a five-machine manually loaded center.

CONCLUSION

Although long-term production data are not yet available, it appears that Massey-Ferguson's goals of increasing machine productivity and utilization will be obtained with the robot-loaded machining line. Reception of the robot loaded concept by the union work force has been good. Robots will definitely be considered for future auto-loading machine applications.

Presented at the Eighth International Symposium on Industrial Robots, May/June 1978. Reprinted courtesy of International Fluidics Services, Kempston, Bedford, England

Integrated Machining Systems Using Industrial Robots
By George E. Munson
Unimation, Inc.

Two machining lines are described in detail in this paper to illustrate how existing manufacturing systems can be automated by the introduction of industrial robots.

At the time of this writing, the systems were installed and in the initial stages of de-bugging. Certain phases produced unanticipated problems requiring tooling modifications. In one case, it became necessary to add assist devices in the machine tool fixturing to insure reliable work piece positioning.

In another case, the robot had to be relieved of some work piece loading manipulations in the interest of time cycle. Again, the solution to the problem was to add assist devices to the machine tool fixturing.

While these alterations were implemented with relative ease, they represent generic problem areas in machine loading applications which require close attention.

While Unimate robots have been extensively applied to loading and unloading metal cutting machines most installations to date have been limited to an isolated "work cell" in which the robot services from one to four machine tools. Work pieces coming to the work cell may have undergone operations previously and are not necessarily finished items as they leave the work cell. Without due consideration of what occurs before and after the work cell it is quite possible that work flow optimization has not been fully realized. Nevertheless, the applications to date have demonstrated the viability and benefits of this class of robot application.

For example, at the Weston Shops of Canadian Pacific Railroad, Winnipeg, Manitoba, Canada, a Unimate robot loads and unloads a vertical milling machine in the manufacture of turnout plates with dramatic results (Fig. 1). There are about 100 different plate versions ranging in weight from 7 to 32 kg. being processed in batches. Prior to installation of the 2005B series Unimate only 20% of total operating time was spent in metal-cutting. With the Unimate, cutting time was increased to over 60% and per shift output went up from about 75 to 300 pieces, a productivity increase of 400%.

At Xerox Corporation, Rochester, New York, U.S.A. three 2105B Unimates achieve similar results in the production of a family of duplicator fuser rollers (Fig. 2). Each Unimate in the system handles the input and output of two machine tools with dual hands. Two of the machine tools are CNC center-drive lathes and the entire system is under the supervision of a programmable controller. Multiple NC and robot programs facilitate fast system changeover to accommodate the family of roller designs. The system is capable of producing over 100,000 work pieces annually and consumes less floor space than previous manufacturing methods.

These are but two of a number of metal cutting installations which illustrate the benefits to be derived from combining the technologies of NC, robotics, CAM and group technology in manufacturing systems.

Diesel Engine Parts Manufacturing

In late 1975, Cummins Engine Company began to investigate methods to improve the manufacture of diesel engine parts at their Lakewood, New York U.S.A. plant. This is a modern new facility which was established to produce detail engine parts and sub-assemblies for distribution to their assembly plants.

Two of the parts produced are cam follower levers and cam follower housings. Fig. 3 shows these parts in raw casting and finished form. There are actually two different lever configurations in a set of three. Each cylinder of a diesel engine requires one set of levers and one housing (for every two cylinders) whose functions are to control fuel to the injector nozzle and to operate the intake and exhaust valves.

For reasons of economy and production volume requirements as well as the desire to relieve human operators from boring, repetitive jobs, Cummins' manufacturing engineers
realized the need to automate the existing manual systems. Initially, evaluations were
made of hard automation. It was soon established that such an approach would be too
costly, particularly since much of the existing equipment could not be utilized. Also,
special purpose automation would not have sufficient flexibility for product changes.

After evaluating existing industrial robot applications in metal cutting, Cummins asked
Unimation Inc. to study the problem and propose a solution. The resultant proposal was
to utilize four Unimate industrial robots to process the work pieces through the existing machine tools.

Fig. 4 shows the lever (cam follower line) and housing (cam box line) systems diagramat-
ically. Since the proposal met Cummins' cost, production rate and flexibility requirements, they authorized Unimation Inc. to proceed with the design and build of the line
automation. In addition to the robots, Unimation provided a central control system and
various parts feeders, conveyors and storage devices.

The design required minimal equipment relocation and some fixturing modifications.

Cam Follower Lever Line

Basic operation of the lever line is as follows (refer to Fig. 4):
1. Raw castings hopper fed to chain broach;
2. Levers, in sets of three flow to Unimate #1, 2105B, pick up station;
3. Unimate #1 alternately loads/unloads lever sets between two milling machines;
4. Milled sets are placed in transfer conveyor;
    a) Periodically milled sets are placed in the inspection station.
5. Unimate #2, 2005B, picks lever sets out of transfer conveyor and loads them
   into machining transfer line;
    a) If transfer line is not available for loading, sets are placed in storage
       pallet or
    b) If no sets are available in conveyor, sets are taken from storage pallet
       and loaded into transfer line.
6. Finished parts are automatically ejected from transfer line.

Rate of the line is 220 lever sets per hour at 100% efficiency. Fig. 5 and 6 illustrate
the process.

Cam Follower Housing Line

Basic operation of the housing line is as follows (refer to Fig. 4):
1. Raw castings are loaded into storage silo and fed to Unimate #3, 2106B;
2. Unimate #3 loads/unloads milling machine;
3. From milling machine Unimate #3 loads/unloads drilling and milling machine;
    a) If there is no casting from the storage silo or the first milling machine
       is unavailable, Unimate #3 takes housing from first mill storage pallet
       and loads drilling/milling machine or
    b) If drill/mill is unavailable Unimate #3 places housing from first mill into
       first mill storage pallet, removes a housing from the drill/mill storage
       and pallet and places it in the transfer conveyor.
4. From drill/mill machine Unimate #3 places housing in transfer conveyor;
    a) Periodically a housing is placed in the inspection station. When a housing
       is to be reinserted into the system, Unimate #3 returns to the inspection
       station after placing a previous housing into the transfer conveyor, re-
       trieves the inspected housing and places in the transfer conveyor.
    b) If the transfer conveyor is unavailable, Unimate #3 places housings from
       the drill/mill into the drill/mill storage pallet.

5. Unimate #4, 2105B, picks two housings from transfer conveyor and loads the drilling machine;
6. From the drilling station Unimate #4 loads the pair of housings into the deburr machine;
7. From the deburr machine, Unimate #4 places the pair of housings into the washer;
8. Unimate #4 removes the pair of housings from the washer and places them in the output conveyor;

Rate of the line is 112 housings per hour at 100% efficiency. Fig. 7 and 8 illustrate the process.

Central Control System

From the foregoing outlines of the functional line sequences it becomes apparent that the programmable controller in the central control plays a key role in interfacing the many different system elements.

The entire system takes into account that there will be work stoppages at random times for a variety of reasons such as jam ups, tool dressing and adjustments and component failures etc. Therefore, intermediate material storage pallets and alternate loading routines are built into the design so that, in most cases, localized trouble spots will not shut down the entire process. For example, the storage silo has a capacity for 250 housings or about 2 hours of production. The housing buffer storage pallets hold 252 parts (in 14 pallet layers) and the lever buffer storage pallet has a capacity of 1008 parts (also in 14 pallet layers). In both cases this represents about 2 hours of production.

Even if the programmable controller fails, the systems can run, but without the subroutines which provide for the other types of interrupts. Back up for Unimate failure could be manual labor but this is not too practical. Heavy dependence is placed on the Unimate robot's inherent reliability and field record of better than 98% uptime.

The programmable controller performs the basic function of program and subroutine selections and sequencing for Unimate robots #1, 2, and 3. Since no alternate actions or subroutines are required in the operations performed by Unimate #4 it is independent of the programmable controller. All interlocks between Unimate and machinery functions (parts present sensing, machine cycle complete signals and start machine cycle signals) are also arranged independent of the programmable controller.

In addition to the central control, there are local control stations at Unimate #1, #2 and #3 for mode selections, teaching (PLC) data entry and indicating counters and signal lights.

Detailed attention will now be given to each of the Unimate stations, programming and PLC control. Robot program details, point by point will not be given but it is important to remember that the start and end positions of each routine must take into consideration the possible combinations of previous and subsequent routines to avoid robot-machinery collision courses.

Unimate Robot #1

Unimate robot #1 has 4 programs:
1. Input conveyor - load/unload mill #1 - transfer conveyor - input conveyor;
2. Input conveyor - load/unload mill #2 - transfer conveyor - input conveyor;
3. Input conveyor - load/unload mill #1 - inspection station;
4. Input conveyor - load/unload mill #2 - inspection station.

All programs include go-no go interlocks.

The local control station includes a mode selection switch, and various control for inspection and frequency selection.

The mode switch has 4 positions:
1. All Equipment Operational;
2. Mill #1 out of commission;
3. Mill #2 out of commission;
4. PLC out of commission.

If mode #1 is selected, the PLC will select programs 1 and 2 alternately. If mode 2 is selected only mill #2 will be serviced. Similarly for mode 3. If mode 4 is selected the Unimate robot will alternate between programs 1 and 2 and programs 3 and 4 cannot be selected (see following paragraph).

Since the frequency of work piece inspection is required to be variable the local control station also includes thumb wheel switches, predetermining counter (1-999) and PLC data entry keys. Once the frequency of inspection is set, the number is entered into the PLC which will then select programs 3 and 4 at intervals determined by the counter setting.

Unimate Robot #2

Unimate robot #2 also has 4 programs:
1. Transfer conveyor - transfer line - transfer conveyor;
2. Transfer conveyor - buffer storage pallet;
3. Palletizing subroutines;
4. Transfer conveyor - transfer line.

All programs include go-no go interlocks.

A local control station similar to the one described for Unimate #1 provides a mode selector switch plus memory step displays, data entry keys for the PLC and a PALLETIZE selection mode switch.

The mode switch has 4 positions:
1. All equipment operational;
2. Transfer line out of commission;
3. Transfer conveyor out of commission;
4. PLC out of commission.

If mode 1 is selected, the PLC will select program 1 only.

If mode 2 is selected, indicating that the transfer line is unable to receive new lever sets the PLC will address the Unimate memory to perform program 2 followed by a sub routine of program 3 etc., etc. The first sub-routine selected will direct the robot to the last pallet position that was filled as remembered by the PLC, and so on.

If mode 3 is selected, indicating that there are no lever sets available from the transfer conveyor, the PLC will address the Unimate memory to perform 3 followed by program 4 except, in this case program 3 sub-routines will be "played backwards" to depalletize the lever sets in buffer storage. This technique simplifies program teaching and conserves robot memory. Selection of mode 4 results in playback of program 1 only.

Teach assisting display units and data entry keys at the local control station provide the means for programming the PLC for robot memory addressing.

Unimate Robot #3

Unimate robot #3 has 7 programs to accomplish its complex task:
1. Input conveyor - milling machine load/unload;

2. Milling machine - drill/mill machine load/unload;
3. Neutral position - output conveyor;
4. Palletizing subroutines, milling machine buffer storage pallet;
5. Neutral position - inspection station;
6. Inspection station - neutral position;
7. Palletizing subroutines, drill/mill buffer storage pallet.

All programs include go-no go interlocks.

As before, a local control station provides the required mode selection and Unimate/PLC programming aids and controls.

The mode switch has 4 positions:
1. All equipment operational;
2. No parts from silo or milling machine out of commission;
3. Drill/mill machine out of commission;
4. PLC out of commission.

If mode 1 is selected, programs 1, 2 and 3 will be selected in sequence by the PLC. The sequence will be interrupted by the presence of a housing in the inspection station that is to be returned to the work flow (presence of a housing here always takes precedence in a sequence). The sequence of programs then becomes 1, 2, 3, 5, 3; 1, 2, 3; 1, 2, 3 etc. Also, if the output conveyor becomes inoperative, the sequence changes to 1, 2, 7; 1, 2, 7 etc. A third interruption in the routine will occur if an inspection (count) is called for or a failure is detected in the lower station of the drill/mill machine. The sequence then becomes 1, 2, 6; 1, 2, 3; 1, 2, 3 etc.

If mode 2 is selected, the PLC selects program sequence 4, 2, 3, with interrupt sequences as described under mode 1.

Selection of mode 3 yields a sequence of programs 1, 4, 7, 3, subject again to the aforementioned interrupts.

If mode 4 is selected, the PLC is inoperative and only sequence 1, 2, 3, is available.

Programming of robot subroutines and of the PLC is done similar to the techniques described for robot #2.

Unimate Robot #4

As previously noted, the program routines of robot #4 require no PLC intervention since no alternate actions or subroutines are involved.

While description of the many routines is involved and tedious it serves to indicate the design measures taken to assure system flexibility, continuous work flow and the effectiveness of programmable control of the Unimate robot memories.

Unimate Robot Grippers and Load/Unload Requirements

A brief description of the robot grippers and machine tool work piece fixtures is in order.

Unimate #1 is a 5 axis machine employing a gripper which clamps two lever sets (6 levers) at a time in two banks. Fig. 9 shows the Unimate robot and (original) gripper arrangement. Fig. 10 shows the levers in the milling machine fixture.

The original design concept, here, was to unload levers which had been milled from one bank of the milling machine, flip the gripper over and load new levers which had been picked up from the input conveyor into the same bank. The milled parts were then deposited in the transfer conveyor and the sequence repeated for the second milling machine bank.

The motions required of the robot in loading the levers required placing the lower ends into the fixture and then pushing them into a vertical position, allowing them to pivot in the grippers as they were pushed. Essentially, the reverse of this sequence was required to unload the levers.

This technique resulted in rapid wear of the gripper pads which led to unreliable positioning of the levers in the fixtures. After considering various alternatives it was decided to add a) additional guides for the levers in the milling machine fixtures and b) to introduce "flip out" pins in the fixtures. In so doing it was then possible to load the fixtures, release the robot grippers and then rotate the work pieces into position with fixed push pins located between the gripper pads. For unloading, the "flip out" pins are actuated when the fixture clamps are released, placing the work pieces in the required position for unloading. This arrangement eliminates the need to rotate the levers in the gripper, hence eliminates pad wear.

In addition to these changes, it was decided to alter the gripper arrangement so that both banks of lever sets are unloaded simultaneously followed by loading of two new sets. (It is expected that this will reduce the overall robot cycle time and, by eliminating several gross motions of the robot's wrist, reduce long term wear.)

Unimate #2 is also a 5 axis machine with a simple design which grips three (1 set) levers at a time. Fig. 11 shows the arrangement just after pick up from the transfer conveyor.

Dual grippers which operate independently are used with Unimate #3. With a housing ready to be loaded in one gripper, the other is used to unload a housing. Since the first milling machine is continuously rotating (at a slow rate), the robot is required to load and unload "on the fly". Sychronizing signals derived from the mill trigger the robots load/unload sequences.

Loading and unloading of the drill/mill machine is also somewhat complex since the housing goes through two machining cycles. With a part in one gripper which has come from the first mill, the robot unloads a housing from the upper section of the drill/mill and immediately reloads it. The robot then drops to the lower section, unloads it and reloads with the housing that was just removed from the top section. Loading of both sections requires a sliding action which was originally to be performed by the robot and required a 6 axis machine. However, because the time cycle became prohibitive this function was removed from the robot's routine and accomplished by a powered pusher in the machine tool. Fig. 12 shows loading at the first mill and Fig. 13 shows loading at the drill/mill machine.

Unimate robot #4 also employs a dual gripper as shown in Fig. 14 at the deburring station. The drilling machine, deburring machine and washer are all load/unloaded two work pieces at a time. However, pick up at the transfer conveyor is done one at a time. Therefore, each of the grippers are independently operated.

Conclusions

In the preface to this paper, it was stated that unanticipated problems occurred in the start up phase of the lines, relative to Unimate robot work stations #1 and #3. These difficulties are elaborated on in the text. The preface also suggests that these problems are of a generic nature.

This is particularly true when introducing robots into existing (manually operated) systems. The tendency often times, is to attempt to duplicate the human operators manipulations without benefit of his vastly superior (to the robot) abilities.

In the case of Unimate #1, the difficulties inherent in the simple operation of pivoting work pieces into position were underestimated and a high risk of equipment damage introduced.

In the case of Unimate #3, while the manipulations to load and unload the work pieces were well within the robot's capabilities, the time to perform the required motions was prohibitive.

The solutions to both problems being implemented at the time of this writing, are straight forward and relatively simple to implement. But lost time and additional expense has been incurred.

The roboticist (and user) will do well to critically analyze application approaches and not over estimate the robot's capabilities. Pratical "assists" for the robot will generally increase reliability and, where well thought out in advance, can usually be introduced within the economic contraints.

Acknowledgments

The author wishes to acknowledge the kind assistance and cooperation of the management of Cummins Engine Company and in particular of Mr. Thomas K. Knapp, Manufacturing Engineering Advisor at the Jamestown plant in preparing this paper.

Acknowledgment is also made of the invaluable assistance provided by Mr. John E. Mattox, Manager, Applications Engineering and his staff at Unimation Inc.

Figure 1. Unimate Loading & Unloading Vertical Milling Machine

Figure 2.   One of Three Unimate Robots in Machining System at
            Xerox Corporation.

Figure 3.   Cam Follower Levers (left) and Cam Follower Housings
            (right). Raw Castings and Finished Parts (Cummins
            Engine Company)

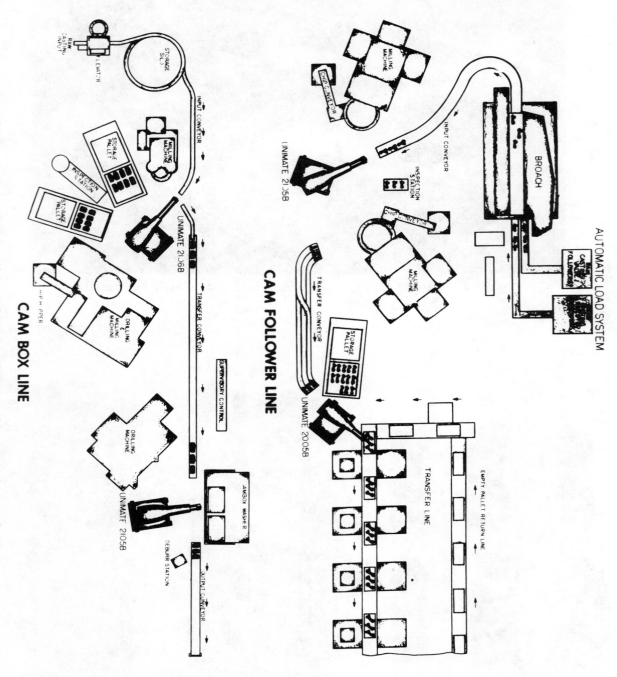

CAM BOX LINE

CAM FOLLOWER LINE

Figure 4. Cummins Engine, Diagram of Cam Follower Lever and Cam Follower Housing Machining Systems.

Figure 5.   Unimate Robot #1 in First Work Center, Cam Follower
Lever Line.

Figure 6.   Unimate Robot #2 in Second Work Center, Cam Follower
Lever Line.

Figure 7.   Unimate Robot #3 in First Work Center, Cam Follower
            Housing Line.

Figure 8.   Unimate Robot #4 in Second Work Center, Cam Follower
            Housing Line.

Figure 9.   Unimate Robot #1 Gripper; Designed to Handle 6 Cam
Follower Levers.

Figure 10. Cam Follower Levers in Milling Machine Fixture.

Figure 11. Unimate Robot #2 with Cam Follower Lever Set, at Pick Up Point of Transfer Conveyor.

Figure 12. Unimate #3 Loading Milling Machine on Cam Follower Housing Line.

Figure 13. Unimate Robot #3 Loading Drill/Mill Machine on Cam Follower Housing Line.

Figure 14. Unimate Robot #4 at Deburr Station on Cam Follower Housing Line.

Presented at the Robot V Conference, October 1980

Unimate Application at Cummins Engine Company

By William R. Riche and James G. Boerger, Jr.
Cummins Engine Company

In 1974 Cummins purchased a Model 4000 Unimate to be used in load and unload of a Cincinnati Horizontal Broach. This application was not as successful as originally hoped and the Unimate was subsequently removed from production and placed in storage.

Approximately 36 months later the Walesboro Components Plant of Cummins began a search for an appropriate re-application of the idle Unimate. We selected a bank of four machines that are used in the progressive machining of our high volume Rear Cover Plate. This application would necessitate the rework of each of the four machines, new hand tooling and a sixth degree of motion on the Unimate.

We decided to sequence the Unimate into production starting with loading and unloading a twin spindle New Britain Model 66 Vertical Lathe. The New Britain would require all new electrics, spindle orient and minor fixture modifications.

Obstacles included the design and purchase of new automated incoming and outgoing conveyors incorporating part orientation and sensing devices, finding a suitable machine rebuilder for the turn-key operation, building a sufficient parts bank and encouraging employee acceptance by putting away past negative attitudes towards this type of automation.

Complete benefits of this system have not yet been fully realized as we are still progressing towards adding more machines to this system. In the interim we have gained a good knowledge of pitfalls that may be encountered in this type of system and therefore can react in anticipation rather than in retrospect.

CUMMINS ENGINE COMPANY

Cummins, founded in 1919, is the world's largest independent manufacturer of diesel engines. We manufacture and sell a diversified line of in-line and V-type heavy-duty diesel engines, components and replacement parts in worldwide markets. Substantially all of the Company's sales are attributed to engines, parts and related products. The Company's principal market is the U. S. On-Highway Truck Industry, with every major U. S. truck manufacturer offering Cummins Engines as standard or optional equipment.

Major off-highway customers are the construction, mining, agricultural, oil and gas, logging, marine, industrial locomotive, electrical generator, compressor pump and other special purpose machinery industries.

Cummins also manufacturers and markets crankshafts, turbochargers and related components for diesel engines and industrial equipment; oil, fuel, air and water filters for truck and passenger car engines, and reconditioned diesel engines and parts. [1]

ORIGINAL APPLICATION

Cummins" first attempt in applying industrial robots to production equipment was in July of 1974, when Cummins purchased a Unimate Model 4000 Industrial Robot. The robot

[1]   Cummins Annual Report - 1979

was to be used in the automatic loading and unloading of a
Cincinnati Horizontal Broach.

For several reasons, including:

. Poor casting quality.

. Inadequate part presentation capabilities.

. Non-compensating fixturing.

. Overall system complexity.

the system soon proved to be unsuccessful. The Unimate was
subsequently removed from production and placed in storage.

Its failure to perform to original expectations created a
feeling among many employees that this type of technology
would never be successfully applied at Cummins.

REAPPLICATION

For approximately three years the Unimate remained in
storage and available for reapplication. During this period
the Jamestown, New York Plant of Cummins, purchased and
installed three Unimate Industrial Robots. This installation
proved to be very successful and prompted us to give closer
attention to reapplying our idle robot.

Relying on the expertise of those previously associated
with the Unimate and our Jamestown personnel, the Walesboro
Components Plant began to examine areas where we could reapply
the Unimate.

Our goal was to select an operation that would be cost
effective showing an attractive return on investment (R.O.I.)
while allowing us to pursue this type of technology so as to

reduce the associated risk of development so that it could be comfortably applied to other areas.

After exploring several possibilities we decided upon using the Unimate to load and unload a bank of four machines used in the progressive machining of a Rear Cover Plate. The Rear Cover Plate (as shown at right is a component measuring approximately 1" X 8.5" X 9.5" and weighs about 6 pounds. Its function is to house a seal which surrounds the crankshaft to prevent oil leakage.

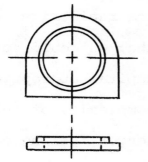

The Unimate would load and unload each of the four machines including a Kearney & Trecker Mill, New Britain Twin Spindle Vertical Boring Machine, Natco Multiple Spindle Drill and a Multiple Spindle Tapper.

This bank of machines was chosen for several reasons including:

1. We would achieve a $45,000 annual cost reduction representing a 21% return on investment with a 3.7 year payback.

2. The New Britain was running internal to a Warner & Swasey 2AC Lathe. The 2AC would be eliminated and reapplied to another area needing more lathe capacity.

3. We were producing approximately 420 pieces per day on two shifts. This would allow us some flexibility in developing this technology while minimizing production interruptions.

We were given appropriations of $155,500 for the project implementation.

As the project began to develop it became obvious that all four machines involved would require much more extensive rework and upfitting than originally anticipated. This coupled with the rapid rise in costs due to inflation made it very apparent that we could not implement this project within the approved capital appropriations. Additionally, the timing and complexity of this installation became paramount and we were no longer feeling comfortable with the success of this operation.

Upon further examination we determined that we should still pursue this application but that we should revise our approach. We decided that rather than apply the Unimate to all of the machines simultaneously we would sequence the application, adding only one machine at a time to the Unimate. We decided to start with the New Britain for several reasons:

1. It would be the "key" machine, with the addition of all new electrics, spindle orient, etc. Its success would determine the eventual outcome of the entire system.

2. We could still develop this type of technology to provide for its comfortable application to other areas.

3. We would be able to apply the Unimate with minimum disruption in production.

4. We would make a Warner & Swasey 2AC available to another area.

5. By using the sequencing approach of adding only one machine at a time we felt our chances of a successful operation were greatly improved. This was important in gaining employee acceptance to this type of automation. The fact that this same unit had once before been unsuccessfully applied made employee acceptance all the more difficult but also all the more important to achieve.

We could now focus our attention on the upfitting of the Unimate, the rebuilding and upfitting of the New Britain and the purchase of new incoming and outgoing powered conveyors.

Some of the specific requirements for this installation included:

1. Unimate Robot - 4000 Series.

    a. Required the conversion from five to six axis and the addition of 3-OX and 3-WX channels.

    b. General maintenance upfitting.

    c. The design and build of new hand tooling.

2. New Britain Model 66 Twin Spindle Vertical Boring Machine - each chuck completes one piece.

    a. Because of part configuration, spindle orient would be required. This meant new pulleys, belts, all necessary interlocks and mounting brackets to orient the spindles within $\pm$ 1/2 degrees.

    b. New automatic sliding guards and controls.

    c. New electrical panel incorporating controls for

doors, spindles, chip removal and signals to and from the Unimate.

d. Modifications to the operators station to include new functions such as a key lock selector switch to convert the machine from Unimate to manual control.

e. Adding automatic door opening and closing devices.

f. The addition of high volume flushing capabilities.

g. New ladder diagram & panel layout indicating points of interaction between the New Britain and Unimate.

3. Incoming conveyor - from the Kearney & Trecker Mill to the New Britain.

a. This is where the initial part orientation would take place. It was necessary that the conveyor design be such that parts could only be loaded in one position.

b. The conveyor would be required to sequence the parts so that two pieces on 18" centers would always be available to the Unimate.

c. The conveyor would be 25' long and would require (3) 1.5" rotating air cylinders, (2) scan-a-matic reflective eyes and positive block type location to assure proper part orientation.

d. We used Xenerol powered conveyors, capable of traveling approximately 60 feet per minute.

4. Outgoing conveyor - from the New Britain to the Natco Multiple Spindle Drill.

   a. This also would be a Xenerol powered conveyor. Approximately twenty feet long it would contain one scan-a-matic electric eye to detect an open position that the Unimate may deposit two pieces.

   b. All rollers would be plastic coated so as to reduce noise and prevent marring the finished part.

SEQUENCE OF OPERATIONS

Assuming the system is full we developed the following sequence for our process:

1. Unload New Britain - two pieces.

   a. Completes machining cycle.

   b. Tool Coolant stops.

   c. Spindles orient - Coolant flood begins.

   d. Door raises - clamps de-energize.

   e. Unimate removes two pieces.

2. Outgoing conveyor.

   a. If the electric eyes detects no open position the Unimate will wait, otherwise it will deposit two pieces.

3. Incoming conveyor.

   a. The scan-a-matic reflective eyes will determine if two pieces are available on 18" centers, if not the Unimate will wait. If two pieces are present and in proper orientation the Unimate will proceed to load the New Britain.

4. Load New Britain.

    a. Loads, two pieces.

    b. Clamps energize.

    c. Door closes.

    d. Coolant starts.

    e. Machining cycle starts.

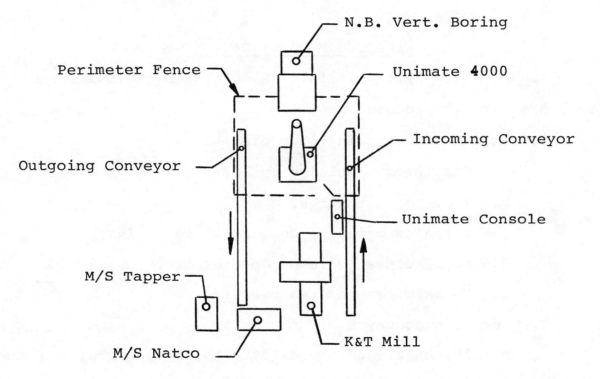

REBUILDING AND UPFITTING

The Unimate would be sent to Unimation for all rework and the design and building of new hand tooling. The hand tooling would have the ability to pick up two pieces simultaneously on 18" centers, accommodating the centerline spread on the New Britain chucks.

Unimation also had the responsibility for the design and building of the incoming and outgoing conveyors, incorporating all necessary part sequencing and orientation devices.

All hardware necessary for the New Britain conversion (excluding new electrical panel) was supplied by New Britain.

Remaining was the responsibility for the marriage of all components in the system. While we have the expertise within our own organization to efficiently bring all of the components together we felt that it was best to pursue this aspect with an independent machine rebuilder for a variety of reasons including:

1. Because of our prior unsuccessful attempt to de-bug this type of system on our floor we felt it was very important that a substantial amount of the de-bug process should be accomplished at a remote site.
2. Our maintenance backlog was extremely high and would have made implementation a much slower process.

Dixie Machine Rebuilders of Nashville, Tennessee was chosen to perform all rebuilding, upfitting and integration of the system. We have done a great deal of rebuilding with Dixie and have been very impressed with the quality of their work and their thoroughness.

We had the rebuilt Unimate sent directly from Unimation to Dixie along with the new powered conveyors. All that remained to simulate the entire system at Dixie would be to send the New Britain.

Before we could release the New Britain from production it was necessary that we build a 12,800 piece bank of parts. At a 400 piece per day requirement this would allow 32 days for rebuild and de-bug. The bank was built by temporarily adding third shift and by working any necessary overtime.

Dixie's responsibility would be:

1. Replace existing panel and components with new, incorporating controls for doors, spindles, chip removal and signals to and from Unimate.

2. Modify operator's station to include new functions including key lock selector switch to convert the New Britain from Unimate to manual control.

3. Install automatic door opening using mechanical and pneumatic parts supplied by Cummins from New Britain.

4. Add high volume flushing capabilities to the New Britain.

5. Mount spindle orient (including mechanical), pneumatic and motor parts) as supplied by Cummins from New Britain.

6. Provide new ladder diagram and panel layout indicating points of interaction between the Unimate, New Britain and conveyors.

7. Participate in a complete system runoff.

SYSTEM RUN-OFF AT DIXIE

Several problems occurred during our run-off including:

1. Several wrong components were shipped from New Britain to Dixie.

2. Parts of the conveyor system were temporarily lost/
    misplaced.

3. Some of the electric eyes were broken in transit.

4. We felt that the New Britain flushing capabilities
    weren't sufficient. A new pump was installed.

5. Numerous electrical failures.

Several trips were taken to Dixie until we were confident
of a sound system.

IMPLEMENTATION AT CUMMINS

As a result of our remote site de-bug process, when the
system was brought to Cummins implementation was a simple task.
Within a matter of a few days all components of the system were
completely installed and operating efficiently.

The fact that we had very few problems with the system on
our floor was instrumental, I believe, in promoting credibility
and employee acceptance of this type of technology.

A SAFE INSTALLATION

Foremost in our minds was insuring the safety of those
employees working in and around this system. Several pre-
cautions were taken:

1. A five foot high fence was placed around the entire
    system. The gate was equipped with an interlock that
    would immediately stop the Unimate should it be
    opened. Also the gate was locked and only authorized
    personnel given a key.

2. Chips produced at the New Britain are flushed to the rear of the machine which is outside of the fenced area. Thus, their removal can be accomplished without interrupting production.

3. The Unimate console was located outside of the fenced area.

4. A remote kill switch was located at the mill operator's station. Should he see any malfunction he could immediately shut down the system.

5. The layout of the system provided the mill operator the capability to visually monitor the system.

6. We purchased a cassette recorder that the Unimate program may be saved in any eventuality. One copy of the program is kept in Maintenance while another is kept in Manufacturing Engineering.

7. Three employees were sent to Unimations facilities in Danbury, Connecticut for a course in maintenance and programming of the Unimate.

8. After installation, coolant dripping from parts quickly proved to be a greater problem than anticipated. This was corrected by removing a section of the wood block flooring surrounding the New Britain and replacing it with grating.

FURTHER APPLICATION

The Unimate program was deemed a learning process since conception. This line of thought has been altered as the boring machine application has grown in acceptance.

Although the Unimate Robot currently sits idle eighty percent of the time as it waits for the boring machine cycle to complete, this available time has not been used to pursue automation blindly. The continuing program has been pursued not in the name of automation, but hoping to optimize the usage of both human resources and automation. The issue brought to question was whether greater return and efficiency could be realized by further automating machinery or by combining operator job responsibilities. It was a question of whether to continue the original plan to fully automate the Rear Cover Plate Line or to require one person to operate two or more machine tools.

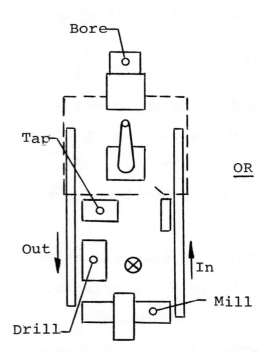

Combined operator responsibilities
(One person operates three machines)

OR

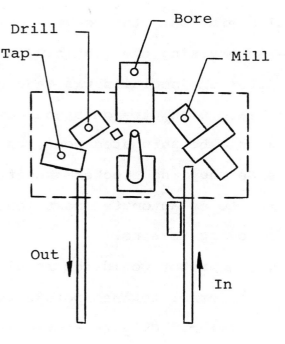

Fully automated line
(Robot loads four
machines)

After reviewing the feasibility of automation versus combined operator responsibilities, it became evident that with both efficiency and potential in mind, that pursuing further automation on the Rear Cover Plate Line would create the greater return. Not only would the potential exist to automate some or all of the machinery on the line, the robot was also deemed more capable than a human of enduring a heavy workload and more efficient at maintaining the chip making process. This review of automation versus human resources utilization did not prove futile, however. As a result of the review, it was determined that automating the Warner and Swasey Tapper was unnecessary. The tapper's cycle was short enough to allow manual operation; the robot's operator could run the tapper while supervising the machining center.

After the decision was made to further automate the Rear Cover Plate Line, the question remained as to which machine tool should be automated next, the Natco Multi-Spindle Drill or the Kearney and Trecker knee type mill. The mill was selected to be automated next instead of the drill because of the following reasons.

- The robot would not require any physical atlerations in order to incorporate the mill into the existing system. Only reprogramming would be necessary.
- The mill's tool change frequency was low, which would create a minimum of interference with the existing system, promoting efficiency.

. Automating the mill would require a relatively low
    investment, the major costs were in the mill's
    electrical redesign and a new mill fixture. Thus,
    this low risk project would also provide an excellent
    opportunity to design and implement a robotic system
    entirely using Cummins labor.

Thus, this Kearney and Trecker Mill automation project evolved from, not only the continuing original program, but from a
cost effective opportunity, producing a payback period of less
than one year. Overall, the following three phase program was
developed.

PHASE	STATUS
I. AUTOMATE BORING MACHINE	IMPLEMENTED
II. AUTOMATE MILL	IN PROCESS
III. AUTOMATE DRILL	FUTURE PLANS

After determining that the mill would be automated next,
a floor layout was devised which would accommodate a smooth,
efficient flow of materials. An intermediate table was added
to reduce the total amount of travel required by the robot and
the incoming and outgoing conveyors were rearranged to reduce
the amount of floor space required. The machine tools were
rearranged to allow adequate space for tool changes, while
also keeping the required robot travel to a minimum to reduce
the load/unload time.

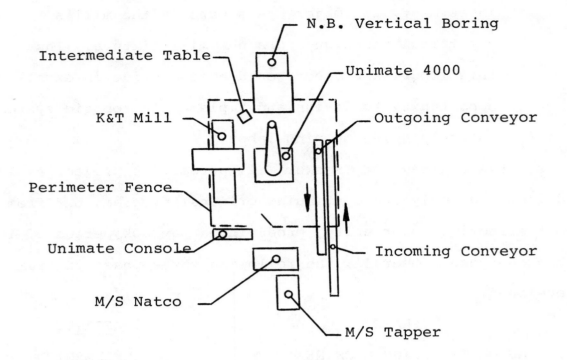

N.B. Vertical Boring

Intermediate Table

Unimate 4000

K&T Mill

Outgoing Conveyor

Perimeter Fence

Unimate Console

Incoming Conveyor

M/S Natco

M/S Tapper

The major expense involved in automating the Kearney and Trecker Mill was due to the new mill fixture. This fixture would require hydaulic clamping, interface with the mill for automatic clamping, and a workpiece orientation device to position the part once it is placed in the fixture. The following concepts were used in designing the mill fixture; concepts that were developed as a result of experiences gained from the boring machine application.

. Maximum clearance between the fixture and workpiece was allowed, not only to accommodate normal robot inaccuracies, but also to allow for casting variances and fixture repositioning inaccuracies.

- The fixture was designed to position the part against the locators, requiring the robot to only place the part in the fixture.

- The locators were designed to be totally covered by the workpiece, thus the chips can't fall on the locators, facilitating chip removal.

- Chip removal is crucial; all nooks and cranneys which could allow chips to build up were avoided.

With these concepts in mind, a mill fixture was designed to facilitate automation, aiding both in workpiece location and chip removal.

The mill itself required redesign in order to send and receive electrical signals from both the robot and the fixture, providing electrical interface between the system's components. The electrical redesign was completed by a Cummins Electrical Engineer who had attended the Unimate Training School. By redesigning the electrical package in-house, as well as designing the mill fixture utilizing Cummins personnel, complete control over the project was realized as well as total responsibility, at a significant cost savings. The following electrical signals were provided in design to enable "communication" between the various components of the system.

- A signal indicating whether the fixture clamped or unclamped was provided by a limit switch.

- A signal indicating that the machining cycle is complete and the mill table has returned to the home position was provided also by limit switches.

These signals provide a medium by which the coordination and interaction between the various members of the system are possible. These signals also prevent costly wrecks; for example the fixture clamped signal tells the robot not to unload the parts until they have been released, protecting the fixture, robot, and workpiece.

The electrical controls of the robot system have had a significant impact on the success of the WCP Rear Cover Plate installation. These controls include electric eyes, part present sensor, or simply a limit switch. In any event, the Cummins philosophy has been to only include enough controls to insure a sound application; feeling that there is a saturation point where too many controls will hinder a system rather than protect it. The major control required in automating the mill was a limit switch which indicates when the mill table is out of position. Part present sensors in the mill fixture were omitted when the risk versus cost was evaluated.

Placing the automated mill into production will not be done at a remote site as the boring machine was at Dixie Machine Rebuilders. The implementation will be done on the machining line, by coordinating the sequence of required events in such a manner that production interference will be minimal. By proving out the mill fixture and building the mill's electrical panel before the actual implementation, the production downtime, due to machinery relocation, electrical rework, and reprogramming the robot, will be less than ten days.

This in-plant implementation has been greatly aided through the acceptance of the boring machine application.

Future plans involve incorporating the multi-spindle drill into the automated Rear Cover Plate Line, completing the original project. Automating the drill will be pursued in the same manner as the mill and boring machine was; by designing a fixture that will facilitate automation, redesigning the drill's electrical components, and by redesigning and installing newly automated drill entirely using Cummins personnel.

SUMMARY

The Unimate Robot as applied to the Rear Cover Plate Line was initiated as a learning process, and has served its purpose well. Since initiation of the boring machine application, robotics has become accepted widely enough at Cummins to allow the purchase of two additional robots at the Walesboro Plant. In addition, the fore mentioned expansion on the Rear Cover Plate Line is underway. But, the robot on the Rear Cover Plate Line has provided more than educational benefits. After the total project has been completed, the robot will be doing the work once required of three men, and has provided efficiency gains as well.

Presented at the Robot IV Conference, October/November 1979

Robots As An Extension To Skilled Labor

By Dennis Mudge
Lunkenheimer

Skilled help is scarce and expensive. Changes in todays manufacturing environment allows many tedious, fatiguing, and repetitive tasks to be mechanically performed by robots as manually controlled work is transitioned to auto-cycle control equipment. Such changes tend to take advantage of, and expand upon, an operator's process knowledge. As this trend continues, the operator's contribution becomes more technically oriented and less physically oriented with significantly greater product output being realized.

Processing needs for greater productivity and reliability are not new, however; the contribution of individual machine operators is made more significant today than in past years.

INTRODUCTION

During the years since the second world war, both economic and technical changes have taken their toll on the availability and cost of skilled labor. Inflation and negotiated wage increases have driven up the cost of established skills. Changes in basic processes tend to increase job requirements and further increase the cost of labor. As well, the variety and number of skilled and technical positions required in today's industrial machine is substantially larger than its World War II predecessor. In short, an individual with technical talent has more, and in many cases, more interesting career opportunities to choose from than previous generations did.

This competitive and high cost labor environment is a multi-faceted problem. It combines with the ever-present need for cost effectiveness, market service, and product revisions to create the greatest management challenge yet faced. This paper discusses a manufacturing dimension that utilizes today's technology to solve some of today's problems - ROBOTS AS AN EXTENSION TO SKILLED LABOR.

A LOOK BEHIND US

Since the start of the industrial revolution (Circa-1825) manufacturing people have been seeking processes with two basic characteristics:

1. Cost Effectiveness

2. Repeatability

In early ventures into the production of firearms in this country, manufacturers had little in the line of quality, repetitive equipment with which to produce rifle components. As a result, gunsmiths were required either to produce a weapon completely from hand-crafted methods or finish the final fitting and assembly of components, which had been produced by apprentice or journeyman labor . As standards, (primarily set by the government) became more exacting, machinery was developed which took the control of the process from the hands of the craftsman, and placed it in the repetitive cycle of substantially built pieces of machinery which, of course, were controlled by lesser skilled operators. The use of in-process gaging also emerged which moved the control of product integrity further from the puts and takes of the skilled craftsman and into the realm of lower cost process engineered capability.

This trend continued into the manufacture of every mass-produced item from shoes to bicycles to automobiles. Even in the manufacture of non-mass-produced items such as machinery and special equipment, cost effectiveness and process reliability were major objectives. It is, however; in this segment of industry where the non-repetitive nature of the product or production item precludes investment in automated processes that individual skills and a great deal of hours are relied upon for manufacture and assembly.

CYCLE CONTROL

As division of labor and economies of scale are realized by reliance on machinery with auto cycle control, competitive dependence on this equipment grows. Below is a graphic comparison of various modes of machine control as related to volume manufacturing, along with explanations of each mode/machine value.

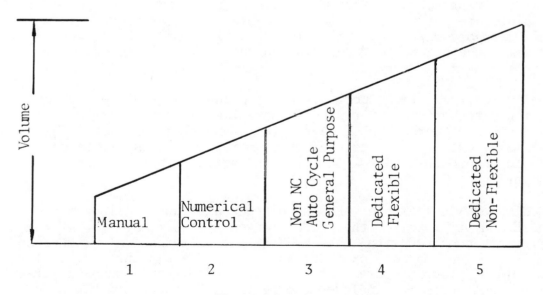

Figure 1.

Figure 1

Machine/Mode Graphic Representation

1. Manual: This group includes machines such as turret lathes, engine lathes, conventional manual controlled milling machines, drilling machines, etc. Equipment in this group is characterized by the need for an operator to perform a work sequence of two or more elements on the same workpiece. Its utilization is particularly suited to small and one-off lot sizes.

2. Numerical Control: Machinery and equipment with full automatic cycle, governed by numeric programming, through the use of a numerical control console, stand-alone computerized numerical control console, or direct numerical control from a central computer which controls two or more pieces of equipment. This mode of cycle control contributes greatly in the production of low to intermediate volume production where a sequence of operations are performed on a workpiece with either single or multiple workpiece handlings required to complete work which the machine can perform. Because of flexibility in cycle control and the relative simplicity of tool requirements, the NC group has been segregated from other conventional auto cycle equipment. As well, in terms of volume capability, it has been placed between manual and auto cycle because conventional auto cycle equipment can, generally speaking, out perform NC on cycle time comparisons.

3. Auto Cycle: Conventional non-numerical control metal cutting and forming equipment which has full auto cycle enabling it to perform one or more workpiece operations in a specific cycle sequence. Generally, sophisticated tool arrangements are required to produce parts of different configurations across the same machine. Available speeds, and feeds somewhat limit auto cycle machines in adapting to a wide variety of workpiece materials and configurations. These constraints are not as narrow in comparable N.C. equipment. Equipment in this category may be economically set-up for runs extending from three times the set-up to (depending on the job) the life of the machine.

4. Dedicated Flexible: Specially built production equipment which is designed to perform a variety of operations across a narrowly defined group of work pieces. Although set-up times on such machines may normally extend to fifteen hours or more, lot sizes are large enough to absorb this cost with little negative impact on piece costs per run.

5. Dedicated Non-Flexible: Equipment such as transfer lines that is designed and utilized for the production of a specified item. Volume requirements combined with costs of alternative methods of production must dictate the enormous expenditures for such equipment. Typically, all machining/process operations are performed without manual intervention. Sequences may include loading, shuttle/index, machining,

gaging, unload, etc. This type of equipment arrives at the highest possible economies of scale, but is dependent on extremely high volume for justification.

INTERMEDIATE VOLUME MANUFACTURING

For the purposes of this paper, intermediate volume manufacturing will be defined as the production of machined parts utilizing either numerical control or conventional general purpose auto-cycle control machinery in volume great enough to yield a run time which is between four and fifteen times the set-up time. This level of production demand, combined with the use of auto-cycle equipment, offers the manufacturer the best opportunity to incorporate production tasks into cells.

Once again, it is of particular interest to note the set-up cost differences between conventional auto-cycle equipment and numerical control equipment utilizing standard tooling. The conventional piece of equipment may require from two to five times as much set-up as its NC cousin. Also, NC equipment offers capability and flexibility not available in traditional auto-cycle equipment. These differences compile to permit lower volume manufacturing requirements to fall within the reach of the set-up to run ratios described above.

As the control and regulation of processes increase, all indications seem to point to a diminished role for the hitherto indispensable operator. While his attendance may yet be required, job content diminishes as education, experience, dexterity, work hazards, responsibility for material or product, and even physical effort are gobbled up by mylar, wires, and steel. But, let a system failure, material variation, process omission, or any other shortcoming occur and the first question asked is, "Where was the operator?", or "Why didn't he recognize the shortcoming or understand the situation?" Perceptive managers foresee the need to keep high cost equipment operating at its optimum. They also recognize when job evaluations point to labor grade ranges which are less than suitable for their interpretation of the job. Often managers will respond to this problem by bending the job description enough to attract the "right" man.

Frequently the gerrymandering of job descriptions leads to a particular group or individual that has demonstrated not only the technical competence to perform the work, but also a propriety interest in the job. Just as frequently this particular group of individuals or individual will ultimately become bored by the lack of involvement and challenge once the start up and debugging phase of the installation is complete. When this happens, proprietary attitudes have a tendency to deteriorate which can lead to adversity by design or default. The best way to avoid such a quandary is to structure the operating evnironment to utilize the required skill at a rate that is sufficient to maintain the skill.

One popular way to fully utilize specific skills is to break the cell

set-up responsibilities away from the operating responsibilities. This
is commonly done with a set-up classification which is responsible
for set-up and initial quality of the production run, while the actual
run is performed by another individual working under a loader-operator
classification. In this capacity, the loader-operator is required to
physically handle the part from machine to machine within the cell and
may also be required to perform in-process gaging and make necessary
machine adjustments. While the loader-operator completes the production run, the set-up man usually moves to another set of machines
to begin set-up for different production requirements. This method
of cell operation takes advantage of division of labor economies; still,
much can be gained by the introduction of robot control over the loader-
operator functions.

A CASE HISTORY OF ROBOT
CONTROL IN CELL OPERATION

Once a cell is set-up and turned over to a loader-operator, the cell
may be viewed as a single manufacturing unit. Manufacturing people
can readily see the advantages (cost, control, through-put) in
substituting a robot for the loader-operator in a production cell.
An example of this is shown in Figure 2, a cell presently in operation
at Lunkenheimer. This cell takes advantage of various control elements
(manual, NC, auto-cycle) and links NC and auto-cycle controlled units
with a robot to create a cell with both robot and operator participation.
Set-up to run ratios for work within this cell fall within the parameters established in the preceding Intermediate Volume Manufacturing
Section.

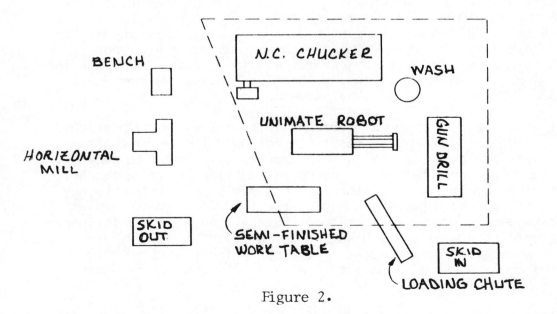

Figure 2.

Lunkenheimer valve body manufacturing cell

Equipment within the dotted line area is robot controlled, equipment outside this area is under the control of the operator. Part progress through the cell is as follows:

ITEM	DESCRIPTION	CONTROL
1.	Fill loading chute	Manual
2.	Pick part from loading chute	Robot
3.	Load/Unload N.C. Chucker	Robot
4.	Wash part	Robot
5.	Load/Unload Gun Drill	Robot
6.	Place piece on semi-finished bench	Robot
7.	Remove part from semi-finished bench	Manual
8.	Load/Unlaod Milling Machine	Manual
9.	Place part on out-station skid	Manual

All inspection and in-process adjustments are performed by the operator.

The establishment of this cell with the use of a robot permitted the joining of various auto-cycle elements into a controlled process, whereas the previous manufacturing employed manual control only, with process control being established by production scheduling between operations. Volume of any individual workpiece was not sufficient to support the investment required for this arrangement. However, flexibility in each machine within the cell, along with the robot's adaptability to various workpiece configurations, did allow overall volume to be pumped up to justifiable levels without set-up time adversly affecting lot size and unit cost objectives.

Using the cycle control diagram in Figure 1., the old process appears as shown on the next page:

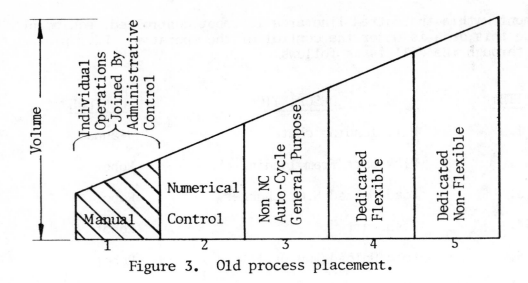

Figure 3. Old process placement.

The current process introduces new dimensions in control and operator involvement.

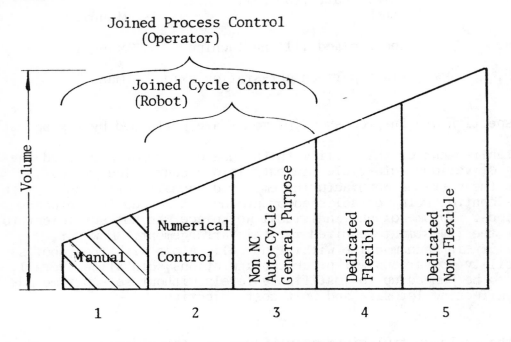

Figure 4. New process placement.

The following operational advantages of this cell contribute to a favorable return on its required investment:

1. Fewer Manual Handlings

2. Fewer Operations Overall

3. Reduced Planned Production Costs

4. Higher Operating Efficiencies

5. Increased Control Over the Process

6. Shorter Factory Through-Put

7. Reduced Scrap

From the operators standpoint, the following changes in his functional role occurred:

1. New technology was brought to the manufacturing environment.

2. Assembly of this technology into a single work unit allowed the operator's skills to grow, covering set-up and trouble shooting of equipment and machining tasks which either did not exist in the past or were covered under several separate job descriptions.

3. The operator gained more control and responsibility over equipment and processes than he enjoyed in the past.

4. A robot handled much of the tedious and repetitive machine loading, leaving the operator free to perform other operations which required his talent and increased his productivity.

5. Because changes in technology and assembly of that technology brought more content to the job description, the base earning rate was proportionally increased.

CONTINUED CHANGE

As a result of the success of this layout, other elements of work on the parts, processed through the cell, have been addressed and are to be incorporated into this cell in coming months. Among them are milling, drilling, and burring operations. To accommodate this work, the cell will be changed as shown in Figure 5.

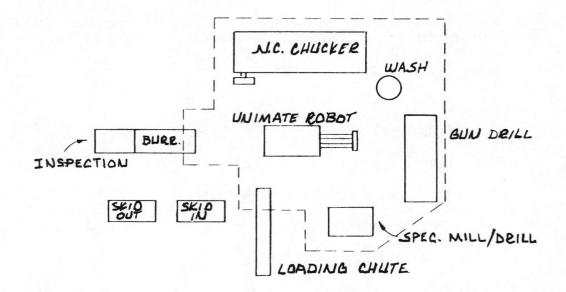

Figure 5. Lunkenheimer valve manufacturing cell with changes planned for November, 1979.

The changes from the current cell layout and the one shown above include an additional special purpose milling/drilling machine which will complete machining of the part family and re-location of the operator's section to facilitate his movements. When this re-arrangement is complete, the operator will be trained to "teach" the Unimate 2000B robot. Robot "teaching" is presently done by the process engineer.

SUMMARY

Lunkenheimer has made a significant cost reduction on a group of manufactured items by combining their similarities to take advantage of economies of scale which were out of reach of individual items. This was not done with high cost special purpose equipment. Instead, it was done with lower cost flexible machines which lend themselves to the variety of demand which exists for the particular product. Complementing the operator's effort, a robot performs part transfers tasks which would otherwise require another operator in the cell. This unique blend of human resources and machinery is an example of today's technology complementing a limited availability of skilled labor. It is interesting to note that this trend began with the Industrial Revolution over 150 years ago!!

Figure 6. View of cell showing Unimate robot, Lodge & Shipley 12/25 NC chucker and De-Hoff gun drill

107

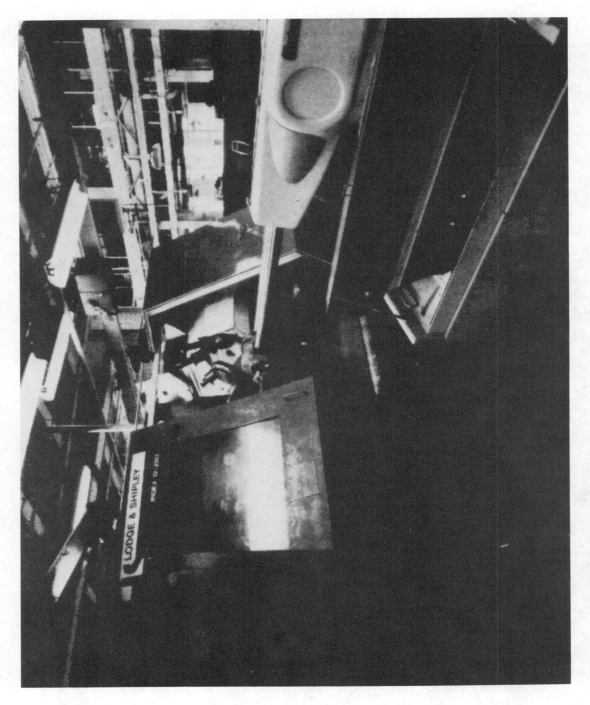

Figure 7. Robot loading NC chucker

Figure 8. Robot loading gun drill

Presented at the Robot V Conference, October 1980

Robot Utilization in Automatic Handling Systems and Transfer Machines

By Robert G. Fish
GMC/Hydra-matic Division

INTRODUCTION

Automatic transfer machines of the "Inline-Lift & Carry" type have been machining transmission cases at Hydra-matic Division, G.M.C., Willow Run since the 1950's. Some of these transfer machines have as many as 170 stations per line. To keep a machine with so many stations running at one time has been a major concern to both machine builder and machine user. Over the years these lines have been broken down into smaller sections with interlocking mechanisms manned by employes loading or unloading parts on demand; using overhead monorail systems as storage or floating devices so that better "uptime" could be maintained. This method greatly increased efficiency except it did not allow for human inefficiencies and other contingencies.

In early 1970, Hydra-matic Division began its first endeavor into automatic loading and unloading of transmission cases on one of its inline-lift & carry machine lines.

Our first "Robot" was a hydraulically operated pick and place unit which unloaded a transmission case from a section "A" transfer machine and loaded a section "B" transfer machine. Depending on existing conditions, it would also place parts into or take out of a twenty (20) part capacity storage unit automatically on demand.

This unit being a prototype design encountered many changes in the debugging stage and unusually high production schedules made it extremely difficult to finalize the project. In July, 1973, before we could completely make it fully operational we had to remove the robot and storage unit to make area available for a new transmission program.

Then in 1978, a machine supplier showed us their newly developed handling system. This system automatically loads and unloads parts from their inline-lift & carry transfer machine lines utilizing a "CYCLOIDAL" drive indexing elevator to elevate parts out of the transfer machine and place them in a "carrier" which when elevated gravitates by means of blue steel rail to the next transfer machine to be loaded by reversing the before mentioned sequence.

After seeing the potential of this system, we contracted with the supplier to research and develop this system to autoload and unload their "PALLET" type transfer machine which they were building for Hydra-matic to machine a new transmission case at our Three Rivers Michigan facility.

SYSTEM DEVELOPMENT

After designing a "Mechanical Manipulator" to load and unload transmission cases from handling system, then to transfer machine pallets, it was decided that:

1. There was too much mechanical gadgetry.

2. Each unit would have to have its own
    peculiar design. (too costly)

3. We did not have enough machine cycle
    time to do all the required motions
    which included a blowoff of the part
    during the unload cycle.

At this point, it was decided to look into the possibility of using a robot to load and unload the parts. Analysis of all parameters were involved such as

1. Capability of performing desired motions
    within the transfer machine cycle time of
    17 seconds.

2. Capable of handling approximately a 100
    lb. load with .060 repeatability.

3. Safety

4. Flexibility and ease of maintenance.

5. Performance record

After preliminary analysis of available robots, we asked
Cincinnati Milacron to set up a simulation test at their
facility using a T3 Robot. The machine supplier sent a transfer machine base and pallet along with an experimental robot

"HAND" for tryout. These two suppliers, coordinating with
Hydra-matic, were able to successfully meet every requirement,
including the articulative movements necessary to blowoff
coolant on the part during the unload cycle.

We then approved the use of the T3 Robot and authorized
the build of one (1) handling system unit with intent to purchase four (4) units if successful.

The handling system consists of the following:

1. A T3 robot to unload part from a "carrier"
    and load a transfer machine pallet.

2. An accumulating power conveyor, elevator
    and blue steel ride rail to convey empty
    "carriers" to the transfer machine unload
    station.

3. A T3 robot to unload transfer machine pallets
    and load the handling system carriers.

4. An accumulating power conveyor, elevator
    and blue steel ride rail to convey parts
    in carriers to the transfer machine load
    station.

5. Eighty (80) "carriers" to convey parts
    from machine to machine (float).

After successfully building and trying out a complete unit, we authorized the purchase of the entire automatic handling system for one (1) complete pallet transfer machine line which includes eight (8) T3 Robots; four (4) to load and four (4) to unload four (4) pallet transfer machine sections.

Now that we had purchased industrial robots for the first time, we had new considerations confront us; most important being the upgrading and training of personnel to operate and maintain this equipment. Training sessions were immediately set up for both production and maintenance personnel as well as meetings on a regular basis to keep everyone involved informed as to the status of the project.

Other considerations during the planning stage was the location of the equipment so that we could load or unload parts manually if necessary. A "flag" system was installed on all machine pallets to denote good or bad pallets (This flag when manually positioned will tell the robot whether to load or not to load parts into machine pallets.).

R O B O T   S E Q U E N C E

The robot sequence for the four (4) "LOAD" robots is as follows: (See Figure #1)

1. Home position

2. Intermediate position (ready or holding)

3. Grasp part and remove from the handling system carrier, (Empty carrier is then automatically returned to be reloaded by unload robot.)

4. Load part into transfer machine pallet and ungrasp (Pallet part present switch must be made up before the robot will load part into the machine pallet.)

5. Return to intermediate position. (ready or holding)

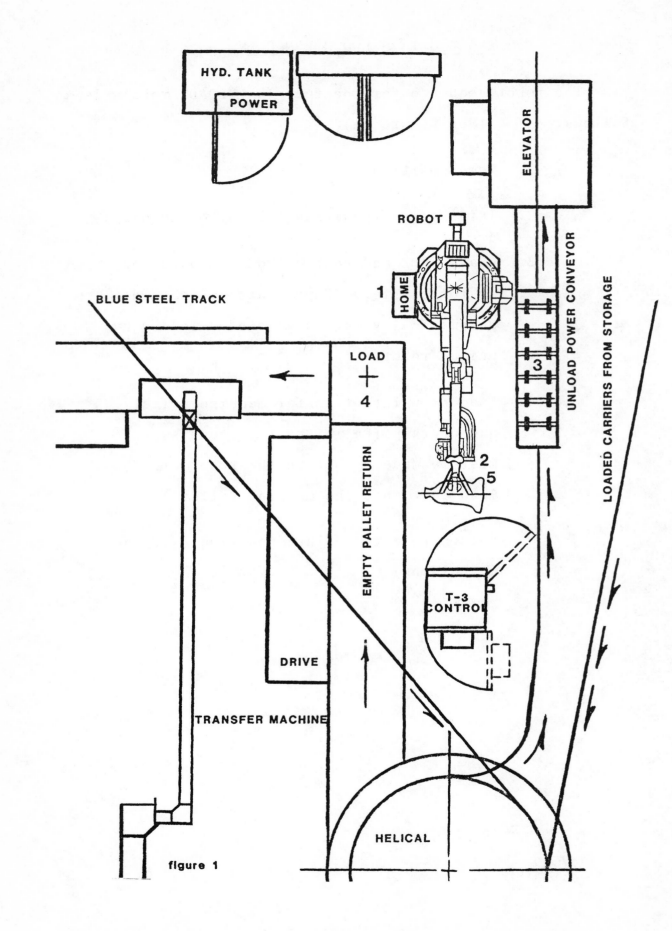

HYD. TANK

POWER

ELEVATOR

ROBOT

HOME

1

BLUE STEEL TRACK

UNLOAD POWER CONVEYOR

LOADED CARRIERS FROM STORAGE

LOAD

4

3

EMPTY PALLET RETURN

2

5

T-3 CONTROL

DRIVE

TRANSFER MACHINE

HELICAL

figure 1

The robot sequence for the four (4) "UNLOAD" robots is as follows: (See Figure #2)

1. Home position

2. Intermediate position (ready or holding)

3. Grasp part and unload transfer machine pallet. (Robot lifts part up thru a blowoff cabinet for external blowoff while special compresses air lines in robot hand blowoff "pan face".)

4. Load the handling system carrier. (Loaded carrier is then automatically conveyed by the handling system to the next transfer machine load position.)

5. Return to intermediate position. (ready or holding)

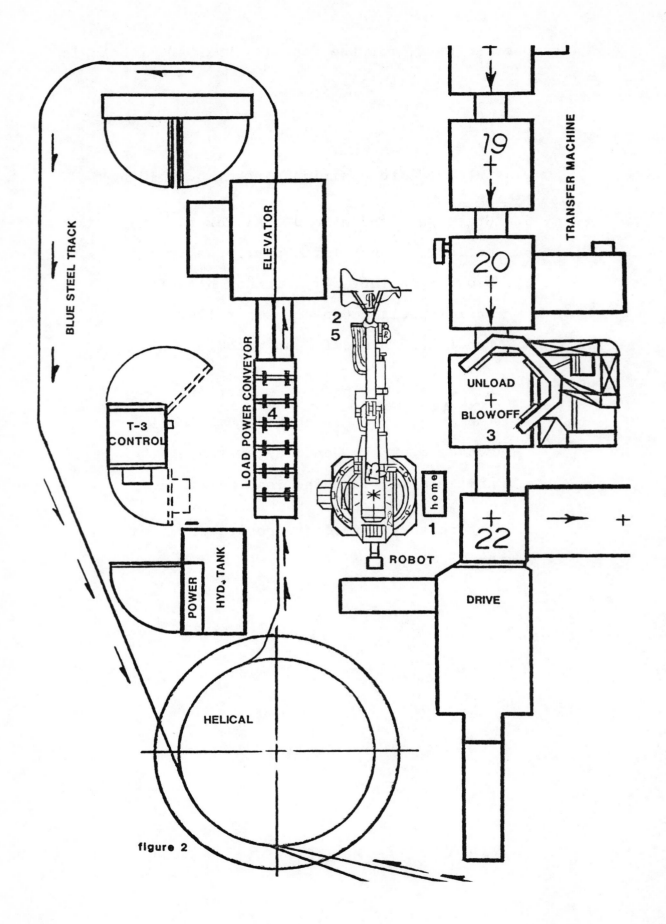

figure 2

OTHER APPLICATIONS

Another robot application presently being installed is at our Windsor, Ontario, Canada plant. This application is also a system handling transmission cases. This system uses only one (1) robot to both load and unload transfer machine; thus reducing by one half the number of robots previously required to do the same job.

Also at the present time we are engineering a system for our Warren, Mich. plant, which will automatically load and unload five (5) complete pallet transfer machine lines (each line has five (5) sections) which will utilize twenty-five (25) robots. The unique thing about this system is that it will allow parts to be conveyed from any transfer machine section to any subsequent transfer machine section (from any of the five (5) "A" machines to any of the five (5) "B" machines and etcetera).

This will increase production by allowing parts to by pass machine sections that are "down".

S U M M A R Y

This system was installed in the spring of 1980 and is presently on line and running. Both production and maintenance personnel are enthused with its implementation.

The eight (8) T3 Robots installed as part of the Handling System have made this method a success and in turn, the handling system has made the robot a success. We feel the complimentary action of the two has many future applications and thereby, many future successes.

PHOTO #1

Empty "carrier" locked in position for robot to load.

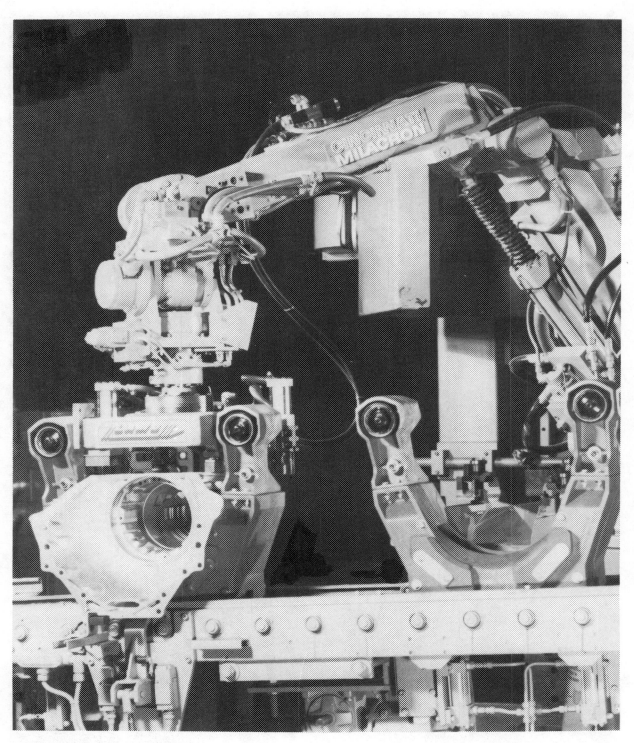

PHOTO #2

Represents robot loading part into "carrier" or unloading part out of "carrier".

PHOTO #3

Represents robot loading part into machine pallet or
unloading part out of machine pallet.

PHOTO #4

Robot removing part from machine pallet, up thru blow
off cabinet.

PHOTO #5
Overall view of one (1) complete handling unit with robots.

CHAPTER 3

DIE CASTING

Commentary

Unloading die casting machines and handling the castings and scrap are operations to which robots were first applied and, today, remain one of the more important robot applications areas.

The robot may simply remove a casting from the die casting machine, placing it on a conveyor or into a container, or the robot may unload the casting, quench it, place it in a trimming press and dispose of the sprue and runners after trimming.

The robot may also load inserts into the dies, spray die lubricant onto the dies or ladle molten metal into the machine. A single robot may tend two die casting machines, unloading each one alternately.

Robots unloading die casting machines can provide an excellent economic return by reducing or eliminating labor cost, decreasing scrap and increasing the production rate of the machine.

Presented at the Eighth International Symposium on Industrial Robots, May/June 1978. Reprinted courtesy of International Fluidics Services, Kempston, Bedford, England

A Systems Approach To Robot Use In Die Casting

By W. M. Goldhamer
President Superior Die Casting

Earliest efforts to achieve operator-free automatic cycling of die casting machines started many year's ago and achieved a high degree of efficiency in the production of relatively small, simple parts, cast from the lower melting temperature alloys on small machines. Very fast cycling speeds have been achieved, using simple self-contained unloading means, and such machines are therefore totally outside the scope of interest regarding use of robots in die casting.

More recently larger self-contained die casting machines, in the range of 100 to 500 tons locking capacity have been introduced which cast, quench, trim and eject castings automatically, from within the machine's platens. Production experience with these machines is still quite limited. However it appears that their relatively limited internal available space does limit casting complexity, and their higher cost does not yet make them competitive with a conventional die casting machine integrated with a robot and adjacent trim press.

Therefore, this paper is confined to discussion of the use of robots with conventional die casting machines of more than 100 tons locking capacity.

In the seventies, robot use in die casting has spread to most of the western industrialized countries and Japan. More than a million hours of robotized production have been logged in both captive and custom operations.

Robots are being used in die casting for a wide variety of operations. These include casting extraction, die lubrication, casting inspection, molten metal ladling into the injection chamber, loading inserts into the casting die, casting quenching, loading into and removing castings from trimming dies and other secondary operation equipment, and loading castings onto various types of conveyors. Also, with the right interfacing circuitry, robots are also being used as system controllers.

To the writer's knowledge, not all of these robot operations are being employed in any single plant, and, in many plants, only a few of these operations are being robot performed. This wide variety in usage mode stems from substantial differences in purpose and concept for robot use in die casting, and in widely varying individual plant operations and conditions.

In the sixties, most early robot applications in die casting were simply to replace casting machine manual operation. This was because of unavailability of skilled operators and because the nature of the work and the environment in which it was performed were fatiguing.

Although this simple substitution of robot for manual operator was intended to merely assure production, it was also recognized that the manual operator's function involved far more than just extracting castings from the machine.

Most of the operators mechanical functions could be performed by a programmable robot with 6 axes of freedom and/or continuous path capability. However, other functions performed by a skilled manual operator involved complex sensing and judgment.

The human brain's ability to observe by sight, sound and even smell and feel, and respond with operational and machine adjustments for observed variations in the process were, and still are, beyond the commercially available robot's capability.

Varying ability to offset the lack of this human "closed loop response capability" has been one of the three significant causes for the development of different modes of robot use in die casting.

The other two factors causing different use modes are the particular plant's processing economics, and the varying proficiency of its tools, machines and equipment. Some or all three of these factors have governed the selection of robot functions in each plant, and in some cases the original selection has precluded or made difficult a subsequent expansion of robot usefulness.

Thus, an examination of the reasoning behind the selection of robot functions is important to understanding their widely divergent use patterns. It also can provide a basis for appraising their future place in the die casting industry.

In order to operate a zinc die casting machine without a manual operator, the following functions, previously performed by the operator, must be performed automatically:

1. The die faces must be lubricated.
2. If the means for lubricating moves between the open die halves, its removal before die closure must be assured and detected as functioning.
3. Safety means to prevent human intrusion into the open die space must be provided and detected as functioning.
4. Machine cycle must be started and its elements timed automatically.
5. Upon die opening, the casting must be removed from the ejectors and then moved out from between the die halves.
6. The parting faces must be cleared of any remaining flash.
7. The casting "shot" must be inspected to assure that no portion remained within the die.
8. If the casting must be quenched, means for quenching must be provided and actuated.
9. The casting shot must then be deposited onto means for conveying it away from the machine.
10. Automatic metal temperature control must be provided.

In the case of an aluminum die casting machine, the same functions as for zinc plus these additional must be provided automatically:

1. The metal injection piston must be lubricated.
2. The molten metal must be transferred from the holding furnace to the injection chamber.

Although it is theoretically possible for a robot plus sensor's, controllers and interfacing circuitry, to perform all of the casting machine operator's manual functions, it is rare for this to be the selected course of automation.

The reasons for this are the difficulties the individual producer faces in accurately prejudging his own achievable cost effectiveness and the practical problems anticipated within his own operation.

Of course the die caster knows that his larger size machines are slower cycling than smaller machines. He is aware that with the smaller machines a single robot's time may be totally occupied with merely unloading and transferring castings. Whereas, with machines of 500 or more tons clamping force, the cycle may be slow enough to possibly permit the robot enough time to function with several post-casting operations.

However, even though presented with case history evidence of success

130

elsewhere, the individual die caster must judge factors unique to his own operation: the condition and suitability to automation of his casting dies, machines and equipment; the length of continuous production runs available to him; the complexity and size range of castings produced; the degree of uniformity of operations required; his existing methodology of secondary operations; suitability of plant lay-out; availability of required skills for design and maintenance of the more sophisticated equipment and electrical systems required.

Frequently, the die caster is unsure of his ability to move to full automatic cycling in one step. He will decide to gradually progress towards automatic cycling by replacing only some individual manual operations, such as metal feeding or die lubricating. For this he will buy already proven special purpose equipment such as an automatic ladle or automatic die sprayers which are less costly than a robot. His reasoning being that he will leave the manual operator on the machine both as a safeguard against die or machine damage and as a "time hedge" until he is convinced that other elements of automation are well within his capability.

Generally speaking, it is for these reasons, and recognizing the individual plant's differences in both circumstances and judgments, that initial robot installations in die casting have varied as to concept for use.

In some cases where the robot was purchased initially, die lubrication was considered that plant's paramount problem in automation. The robot was used immediately to duplicate manual spray patterns, which can constitute a significant time element in the casting cycle. Later, automated lubrication using other means has proved practical, freeing the robot for more cost effective functions.

In some instances, initial success with the robot for simple casting extraction has encouraged later consideration of its use for other functions.

In many cases, however, where there was lack of an original "system concept" for automating the entire process, there have developed serious problems in later achieving optimum effective robot use.

The "system concept" of die casting automation should be based on a few, very fundamental considerations. First, the system must include a complete process flow plan from raw material to finished product. Second, all operations, which can normally and economically be done in continuous flow, should be so integrated to minimize work-in-process inventory and handling costs. Third, starting with the casting machine as the higher cost center, the system must permit optimum cycling speed and continuous utilization of the casting machine and then of other equipment in descending order of capital cost wherever possible. Fourth, each piece of equipment, in relation to its cost, should be kept occupied continuously performing only those functions not capable of being performed by less costly equipment. Fifth, orientation of parts should be maintained as far as possible in the process flow to minimize handling and reorientation cost. Sixth, ultimate closed loop control of each element of processing should be contemplated even though not presently possible.

This "system concept" we consider to be the essence of any plan for the ultimate "totally automated" die casting facility. In such a concept programmable robots will be essential, particularly for custom die casting production.

Unfortunately, only the well advised among purchasers of automation equipment for die casting have started their efforts with a good "system concept". Too often, unfortunately, they have been sold by the easier, less arduous approach of "try one, you'll like it!", and too often this easier, initial approach has proved ultimately to be difficult and costly to the die caster. To that extent, it has somewhat impeded the more rapid and wide-spread use of robots in die casting.

The "system concept" is not an easy approach for the die caster. For example, its first essential is a process flow plan which must translate into plant layout of machines and equipment. The typical die casting machine requires a multitude of below grade service connections for air, water, power, fuel, lubricants, etc. The overhead space is defined by need for crane access and ventilation means. When one thinks in terms of an automated center around the casting machine to establish good process flow, one must think of access for molten metal supply, means for conveyorized removal of gates and other metal to be remelted, space for robots, trim press, quenching means, other secondary operation equipment, controls conveyors and access room for maintenance and service. And, to be cost competitive, one cannot squander floor space.

Yet, automation is the inescapable future in die casting, and consequently each step taken without a "system concept" plan can prove disastrously expensive if it necessitates repeated moving of equipment and services.

The space problem is in some respects even easier than the operation sequence problem for a custom producer. Some castings require only one secondary operation after casting, while others may require several. Thus, flexibility versus specialization of automated centers must be resolved as best suited to the particular die caster's product mix.

Probably he will elect to have some of each type, and here the use of additional robots or robot modules will come into question. Already we are seeing instances where castings made in multiple impression dies may be most economically tapped, drilled, broached, and reamed using multiple spindles, before the castings are trimmed from the gate. Depending on the casting machine cycle speed, a second robot or robot module then may be required to transfer the gate into the trim press after casting machining.

In other instances, a supplementary robot module, with only a three motion freedom, transfers an already trimmed casting into a lathe for a turning operation. This is because the first robot, used for casting extraction, inspection and positioning in the trim die, is already being completely utilized.

CONCLUSION

The writer is aware of a number of instances where the initial installation of robots never has been expanded because no ultimate plan for real automation was created. In these cases full economic justification has never been achieved, as the robots are either under utilized, or good reliability of the other components in the system is lacking, or the robots are performing functions far below their capability.

In these cases, the die casting industry's expanding technology will pass them by, as the industry's use of programmable controllers, closed loop control of process variables, and more extensive and cost effective use of robots in a "system approach" achieves complete die casting automation.

8th SDCE INTERNATIONAL DIE CASTING EXPOSITION & CONGRESS

Presented by the Society of Die Casting Engineers

Cobo Hall · Detroit, Michigan
March 17-20, 1975

PAPER NO. G-T75-134

Comparative Costs of Automation in Die Casting Using Robots

WALTER K. WEISEL

Unimation, Inc., Farmington, MI

Presented at the 8th International Die Casting Exposition & Congress. Reprinted courtesy of the Society of Die Casting Engineers

Application

Economics of automation is of particular significance when words like cash flow, cost savings, productivity, inflation and payback period are used.

Die casters should be thinking in terms of industrial robots which have logged millions of hours of field experience. Robots are debugged and proven tools which figure heavily in future manufacturing operations. This paper provides examples of key areas based upon reviews of hundreds of proven (and cost justified) robot installations in the die casting industry.

INTRODUCTION

The die casting process represents one of the oldest means of forming metal, however, the basic process has remained unchanged for many years. New innovations have been few and far between, thus creating keen competitiveness between suppliers. Skyrocketing cost increases for material, capital equipment, labor and facilities coupled with two-digit inflation makes the economics of automation of the utmost importance.

Today's robots are more skillful and faster than those of only a few years back. There also has been a significant change in their reliability which now enables them to be more dependable than the blue collar worker. The economics of automation discussed in this paper are those related to the industrial robot.

Hundreds of robots are currently used in the die casting industry to do much more than simple part extraction. High-speed servo controlled arms can span over 20 ft., which usually is enough to unload two machines. Their speed not only paces the die casting machines that they serve, but allows them enough extra time to perform trimming operations as well. Other robots are not only unloading, quenching and trimming, but are also loading inserts into the die casting die

cavities. A typical set-up is seen in Fig. 1. The most obvious place to start looking for economic savings with robot automation is in the area of labor displacement.

Labor Savings

The blue collar worker with his wages, fringes, supplies and facilities represents one of the most costly commodities within the plant operation. Wages have increased dramatically over the past several years and those who are looking ahead know that more escalation is coming.

Fig. 1. Typical set-up with a robot serving two die casting machines, with inspection stations and a quench tank.

Fringe benefits now account for an additional 20 to 30% of the worker's base salary and new contracts have been awarded in some industries with medical, dental and retirement benefits which will increase the percentages even further.

In the die casting industry, wages plus fringe benefit pay are generally calculated at $14,000 to $16,000 per year, per man. In some parts of the country the amounts are less, but they are expected to catch up soon.

A robot designed to unload, quench and trim two machines costs approximately $35,000. In most instances, labor displacement alone is sufficient for payback on a straight return on investment formula. The robot can pay for itself if it only operates one die casting machine on a two-shift operation. When figures are added for savings in trim press labor, the payback period is shortened even more as shown in Table 1.

Table. 1. Typical examples of payback period of robot installations.

FORMULA: $$P = \frac{I}{L - E}$$

P = Payback period in years

I = Total investment of robot and accessories

L = Total annual labor savings

E = Expense of robot upkeep

TYPICAL
INPUTS: I = $35,000

L = $14,000 (die cast industry, includes fringes)
E = $1,700 one shift use
    $2,500 two shift use

EXAMPLE 1: One shift - double machine unload
(can also reflect two shift single machine)

$$\text{Payback} = \frac{35,000}{28,000 - 1,700} = 1.3 \text{ years}$$

EXAMPLE 2: Two shift - double machine unload

$$\text{Payback} = \frac{35,000}{56,000 - 2,500} = 0.65 \text{ years}$$

(Note: the labor figure includes $7,000 for trim press labor and fringes, less than a full shift. Use the full labor figure if trimming for two machines.)

EXAMPLE 3: One shift - single machine unload, quench and trim
(note: the labor figure includes $7,000 for trim press labor and fringes, less than a full shift. Use the full labor figure if trimming for two machines.)

$$\text{Payback} = \frac{35,000}{(14,000 + 7,000 - 1,700)} = 1.8 \text{ years}$$

Additional economic savings are frequently overlooked when the displaced labor figure is used and yet there are several other areas which should be taken into consideration. These may be savings from such things as the operators' company-furnished tools, gloves, rags, toilet facilities and showers. One medium-sized die casting shop running two shifts is known to have spent over $20,000 for gloves and rags alone for the die casting operators. And, gloves are said to be getting harder to obtain and prices for them are increasing dramatically. Locker rooms, toilet facilities and cafeteria facilities all represent valuable plant space which will not require enlarging if company growth is achieved with automation. All these "overhead" savings should be considered when computing savings that result from automation.

Increased Production

In one test it was proven that when the operator paced the machine, rather than the machine pacing the operator, an increase in production of approximately 30% could be achieved. Some situations may have less potential and some may have more.

However, this example is in line with production increases which have been achieved by automating die casting machines with the Unimate robot. It isn't difficult to assign a dollar value to an extra 30% of production. These increases might have enough effect on production to eliminate the need to buy another die casting machine, or even cut back on overtime or additional shifts.

Reduced Scrap and Better Quality

In the test cited above less scrap was produced during the test than on the normal shift operation. This reduction in scrap was attributed to the die casting machine cycle being more consistent and dies being maintained at more constant temperatures. The shot size becomes more uniform and dies operate better, longer and with less maintenance when run with consistent cycles. Die lubrication patterns are preset or programmed since they too are part of the full automation process. The result of automation is less scrap, better quality, greater die and machine life. Each of these factors has a dollar value.

One user of robots claims that he can justify them on quality and reduced buffing operations savings alone. That user casts a 30-lb. automotive trim part which is too large to come straight out from between the tie bars. When run manually, operator fatigue caused nicking of the parts since it was difficult to avoid hitting them against die clearance points and the tie bars while they were being extracted. The robot was programmed to extract the part, clearing all obstructions, and since it repeats its programmed path within 0.050 in., it never damaged a single part. That robot installation resulted in labor savings (three shift operation), elimination of secondary buffing operations, 20% less scrap, increased production output, automated trimming and better quality. Based on the experience of that first installation, nine more robots were added in a one-year period.

Die Maintenance

For initial economic analysis of automation, it is unlikely that die maintenance will be, or even should be, considered. However, after a robot has been used successfully to automate an operation, there will be a definite increase in die life due to the more consistent die operating temperatures. The high speed, stiff, servo-controlled arm of a robot can get in and out of the die much faster than the human operator when unloading. The result is that die open time is minimized, die temperature is more constant, and there is less wear and more parts are made each hour.

Material Handling and Floor Space

Savings in material handling is another factor that is not generally considered in the initial justification formula. The robot's ability to trim parts, once extracted, eliminates the need to stack the parts in tote bins to be moved to the trimming department. The robot also eliminates the alternate method where parts are carried off via conveyors to be trimmed. Such conveyors require floor space which becomes available for other uses when the conveyors are replaced by robots. If conveyors were not the means for moving parts, the cost for manual or forklift truck moving of the parts should be included in the calculations.

When a robot is used to unload, quench and trim, it is usually possible for the robot to feed the casting to the trimming operation without changing its grip or the part orientation. The robot is easily programmed to return the scrap directly to the machine's holding furnace. Each operation that can be added to the robot's work load eliminates a handling operation which has a cost.

Scrap Remelt Costs

Melting costs are an area of economics rarely considered for justification of robot labor. Until recently these costs have not been significant. However, melting costs now play a major part in economic justification and will be even more significant in years to come. At today's energy prices, it costs approximately three cents to remelt each lb. of metal. That three cents is a raw energy cost and does not take into consideration costs for flux or metal losses.

If one die casting machine, running manually, produces 160 5-lb. shots per hour the total output per shift is 6,400 lb., half of which might be castings and the remainder remelt material such as sprues. Experience shows that proper application of an industrial robot will normally reduce scrap by 20% or 640 lb. for the example quoted. At remelt costs of three cents per lb. that 20% is a total dollar savings of $19.20

per shift, per day, per machine. Therefore, if a robot operated one machine on a one-shift basis for 250 work days per year, the energy savings alone would be $4800 per year per machine.

Safety

Although safety has always been a consideration when applying automation, it has never received as much attention as in recent years. OSHA requirements are still unclear in many areas and some experts feel they will become stricter in years to come. When operator aids are used, they are often costly both in dollars and lost production. The robot offers a safe method of insert loading and casting removal. The benefits and cost savings make it very attractive for automating with safety in mind. No one can assign a value to a lost human arm. However, if a machine or control failure should happen to cause a robot arm to get caught between the dies, the only costs are time and money.

Leasing Economics

Many companies are putting off plans to automate equipment because of inflation and almost daily increases in operating costs. When such conditions exist, leasing becomes a very economic means of obtaining capital equipment.

Although it is totally impossible to even consider a five year labor contract which carries with it no increases in benefits or salaries, a fully-equipped robot can be leased for five years at a fixed monthly payment that is hundreds of dollars less than today's wages for a die casting machine operator. Lease plans become even more attractive when overtime or multi-shift work is involved since no additional charges are involved in a lease.

Leasing frees-up working capital and eliminates the need for large outlays of cash. Available capital can be used for profit making production. Leasing should be looked upon as another tool for the justifying of automation in today's economy.

Basic Payback

Table 1 shows calculations for the payback period when considering labor displacement only. Since all repair parts are normally covered under a warranty plan, the per shift maintenance costs E only includes labor. Based on the fairly typical values in the Table, one robot running two die casting machines on one shift will have a full payback in 1.3 years. A two-shift operation with one robot running one die casting machine will have a payback time of 1.37 years. However, it is more accurate to use a formula that reflects the increased production and what effect that has on capital equipment (die casting machine) write-off.

Table 2 shows the effects of increases in production for one robot operating two shifts and tending only one die casting machine. The value of the die casting machine is assumed to be $100,000. The first example in the Table shows that a 20-percent increase in the production rate results in a one year payback period when the yearly write-off value of the machine is considered. Example two in Table 2 shows the payback period with labor displacement only. The period is 1.37 years. These examples show the importance of considering all cast factors. The analysis can be expanded by including values for remelt cost savings, space, quality, etc.

SUMMARY

Many devices on today's market are called robots. In actual operation, many prove to be slow and rather unreliable extractors with limited capability. The full benefits of automation can best be realized with the easily programmed, high speed types which can be quickly re-programmed as needs change.

Hydraulically actuated and electronically controlled robots are the best. Hydraulic actuation gives high speeds and rigidity needed to operate with large castings. Electronic controls provide the accuracies required. The electronic reprogrammable machines are easily set up for new tasks as jobs come along. In addition to being fast and accurate the robot must have built-in mobility so it can be moved to other job sites when work loads demand.

The industrial robot is not difficult to justify, particularly if all the cost factors listed in Table 3 are considered. Everyday inflation is adding to the formulas for economic justification of industrial robots. Ironically, widespread use of robots could be a very positive factor in checking that inflation.

The die caster has little control over the prices paid for the zinc, magnesium and aluminum alloys that are cast. But, he can control how he uses these metals in the casting process. Smart application of automation today is the answer to increasing productivity and staying competitive in the future.

Table 2. Payback period for robot when production rate is affected.

FORMULA:
$$P = \frac{I}{L - E + Q(L + Z)}$$

P = Payback period in years

I = Total investment, robot and accessories

L = Total annual labor savings

E = Expense of robot upkeep

Z = Capital value of associated equipment (typically 15% of acquisition costs)

Q = Production rate coefficient

EXAMPLE 1: Two shift operation - single machine unload, $100,000 die casting machine getting 20% increase in production rate.

$$\text{Payback} = \frac{35,000}{\$28,000 - 2,500 + 0.2 (28,000 + 15,000)} = 1.03 \text{ yrs.}$$

EXAMPLE 2: Two shift operation - single machine unload, labor consideration only as in Figure 2.

Payback = costs of robot and accessories
labor – equipment maintenance

$$\text{Payback} = \frac{35,000}{28,000 - 2,500} = 1.37 \text{ years}$$

Table 3. Points of economic justification.

1. Labor savings and supporting tools and service

2. Increased production

3. Better quality

4. Reduced scrap

5. Increased die life

6. Better utilization of capital equipment

7. Material handling

8. Floor space

9. Safety

Reprinted from Die Casting Engineer, May/June 1978

Robots are good workers

Chuck Symmonds
*General Manager, Dowagiac Division
Du-Wel Products, Inc.
Dowagiac, Mich.*

Du-Wel's Dowagiac plant has developed a special fondness for its Unimate robots, which are on the job every day, produce parts for three full shifts, take no breaks for lunch or coffee, and no time off. Employees have come to see the robots as their slaves; with high temperature ranges near the die casting machines and with other production conditions, operators would rather supervise robot-automated systems. Management sees the Unimate robot not in terms of job elimination, but in terms of increased efficiency of operation.

Since 1946, Du-Wel Products has been producing parts for the automotive and appliance industries, for communications equipment, calculators, office machines, photocopiers, office furniture, architectural accessories, farm machinery, and garden tractors.

Du-Wel engineering and design have been recognized for excellence of parts produced in both zinc and aluminum. The company operates five production plants with a total of 30 zinc and aluminum die casting machines ranging in capacity from 12 to 2400 tons. The Dowagiac, Mich. plant is the aluminum division.

Continuing study of die casting production procedures for cost reduction and superior parts quality and consistency has resulted in several manufacturing improvements, including use of industrial robots. Reprogramming the robots for different parts is done by two specially-trained workers. The set-up man sets the die in place and takes a shot. The electrician/programmer trains the robot to remove the completed casting and makes any changes required in the sensor panel. In general, the robots are used to run tougher parts and longer runs. The increase in parts production is very satisfying.

Several of the parts produced could not be run efficiently without a robot. These parts involve intricate dies and hard-to-handle configurations. In such cases, the scrap rate had run as high as 50%; these rates were drastically reduced by robot-assisted production.

Parts now being cast include one for an automatic washing machine for Westinghouse and a shift lever housing for GMC's Saginaw Div. The robot unloads two 800-ton machines, cycling and putting both parts in a quench tank, readying them for the trim presses.

The operation starts with the closing of the die casting machine to the robot's left, which is casting four parts for a washing machine. When the die casting machine opens, the robot extracts the part from the die. Next, the die is sprayed with a lubricant, and the part is moved past a sensor into the quench tank. The die casting machine is then recycled, as the robot moves to the die casting machine on its right, removes a transmission housing, and readies for recycling. Should both units open simultaneously, the robot automatically services the faster die casting machine. Although normally the robot serves one and then the other, it can run either separately, if required.

It's a real pleasure to watch the system make parts. Recently, the set-up man turned on the machine Tuesday morning and the operation kept going until Friday evening without interruption. Such increased uptime plays a big role in making Du-Wel more competitive. After evaluating robots and other unloading devices and automating techniques, the company selected the Unimate, produced by Unimation Inc., Danbury, Conn., because of its programming ease, versatility, and ability to expand into trimming. When looking for productivity increases, the shots-per-hour figure isn't as important as the overall increase in part-production uptime.

Use of the robots provides improved part quality, lengthened die life, high utilization of equipment, more competitive pricing, better conservation of energy and control over cycle timing, reduced scrap rates (by 15 to 20%), and longer die life. The robot does its part in energy conservation by substantially reducing scrap. As a result, we reduce dross losses and save fuel because we don't have to remelt. Also, the robot's consistency enables the prediction of production every shift, every day, so the company can be more accurate in scheduling and delivery. The increased uptime made possible by the robots provides the basis for quoting prices, purchases of new equipment, and more profit.

Reprinted from PRODUCTION Magazine, March 1978, Bramson Publishing Co.

The Tireless Worker

The consistent performance of a Unimation robot allowed Doehler-Jarvis to realize a 10 percent boost in productivity and to cut the labor requirements of its casting machines by 60 percent

Although a productivity increase of up to 10 percent is a principle benefit, it is the combination of benefits that proves the cost effectiveness of the robots for Doehler-Jarvis in Toledo. Other advantages the company reports include:

—consistency and precision of performance
—improved product quality, therefore less scrap
—reduced labor requirement—one technician for each three-robot casting machine installation
—less employee exposure to hazards of the diecasting environment
—automation of die care tasks and upgrading the operation by including automatic trimming in the production routine
—reliability of the units—a reported 95 percent uptime makes it the most reliable part of the automated system.

The parts handled by the Unimates are aluminum automatic transmission cases ranging in weight from 25 to 75 lb. They are supplied to Ford Motor Co. and to the Hydra-matic and Detroit Diesel Allison divisions of General Motors. Doehler-Jarvis Castings Div. of NL Industries Inc. is one of the only three domestic outfits with the capacity to produce these large, heavy transmission castings.

Following a management decision to evaluate the opportunities for automation with robots, several managers attended a Unimation Inc., seminar in Danbury, CT. From what they learned they were able to develop a program to integrate Unimate robots into parts production operations. The design was to increase output and reduce scheduling problems by automating equipment.

Consistent Production. Robot control of the casting operation results in a well regulated, consistent production rate. The robot's pace is steady. An operator's work pace fluctuates considerably.

And the quality of Unimate's product is always the same. According to Dave Neville, production supervisor at Toledo plant #2: "The (manual) product will not have the same uniform quality because the operator can't or won't always service the die properly. You can bank on the Unimate's reliability.

"The big advantage of the robot is that it lubricates a die (with a parting agent) more consistently than does a man. The lubrication sequence is essential for die life and quality castings. And the industrial robot performs this sequence correctly every time."

Robot use also helps minimize such people problems as the training and retaining of employees. It is increasingly difficult in the industry to hire operators for diecasting machines because the job environment is least preferred. Therefore the turnover of help in casting departments is fairly high. Robots solve this problem by reducing the labor requirements by about 60 percent.

Another advantage to be gained by the use of robots is less employee exposure to hazards common in casting operations. Consequently, robots reduce the safety-related risks in the diecasting environment highlighted by the OSHA regulations, such as "no hands in die."

Doehler-Jarvis began to look into robots, though, because of the productivity increases the units could spark.

A Boost for Productivity. "The finances and engineering time we put into improving the casting machines and the dies for automation was reflected directly in the productivity increase achieved with the Unimates," according to Neville. Productivity climbed from an average of 33.3 castings an hour to 38.6 for a 16 percent improvement.

While putting out more castings an hour, the robots also reduce scrap and therefore save energy in remelt-

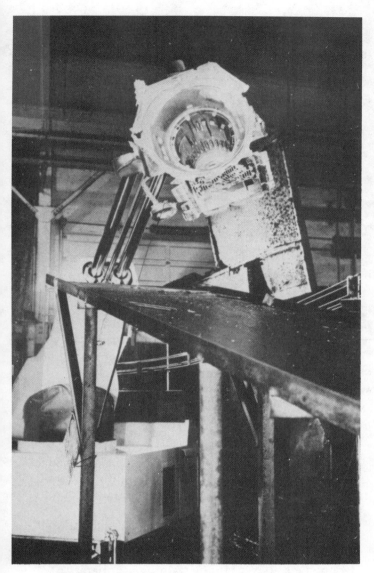

After unloading the diecasting machine, the Unimate places the transmission case on a chute to an inspection table

Flash control also is important in automated diecasting operations. Control of the shot size helps to control flash, but dies must be designed and built to eliminate unwanted flash. Operators can observe the flash and remove it. A Unimate lacks vision. Consequently, dies must be of higher quality for automatic operation.

In the event excessive flash does occur, sensors in the die slides stop the cycle.

A die redesign program to make dies less troublesome included sealing them tighter, using desurgers to stop the water hammer-type noise, and closer-fitting die components.

Operation Sequence. The robot's function begins with it in a "ready" position, waiting for the 2000-ton diecasting machine to open. If the machine does not open, the robot doesn't move, but an alarm is sounded. When the machine opens, the robot enters the die and signals the machine to eject the casting. It then grips the casting and places it on a chute to an inspection table. As safety precaution, the part must touch an electrical limit switch in the chute before it is released by the robot. After removal of the casting, the Unimate activates the timer-controlled water sprays used to cool the casting die.

In its next operation, the Unimate grasps a spray gun, enters the die, activates the gun, and lubricates all die surfaces. It sprays in a precise, programmed pattern reaching all areas of the die, withdraws, and signals the machine controls to close for the next cycle.

At the cycle end, it puts down the gun, returns to the start position, and waits for the machine to open again.

Many of the automatic diecasting machines at Doehler-Jarvis because of part size and shape, use the "drop-through" method for discharging castings. Some jobs, however, do not lend themselves to that sort of automation. Nor is that method flexible enough to permit automation of die care tasks or upgrading the operation by including automatic trimming in the production routine.

So, the company decided to phase a Unimate into the trim department to replace a robot of another make which lacks sufficient flexibility: it uses positive stops in two directions, is accurate to within about 0.060, and cannot be programmed as quickly and easily as the Unimate. It is a simple, electrically motivated device.

Much of Unimate's value derives from its ability to imitate the movement of the human arm, including the wrist. Those movements are used in the lubrication of the dies—"an operation we could not do with other types of automation," reports Neville. "We made several efforts, but the dies were either under-

ing. "We are realizing about a 4.5 percent improvement in our casting quality performance," Neville said.

Setup. Operational consistencies are critical to Unimate's success. It demands better dies and more restricted operating conditions. Because a Unimate cannot make progressive adjustments as a man can, it forces more precise tooling and top-notch maintenance of machines.

Whereas an operator can compensate for changes in shot size in the event of metal fluctuation in the furnace, a robot cannot. So, Doehler-Jarvis changed ladles on the diecasting machines from gravity-filled to vacuum-filled Lindberg ladles.

about Unimate

139

Here the robot sprays a parting agent in a precise pattern to lubricate all sections of the die. Next it triggers an air blow-off to clear and dry the die

or overlubricated. Some of the mechanical systems actually tore themselves apart."

The wear area on the Unimate centers on the wrist because of the weight of the mechanical parts. Spare parts are stocked and installed as needed. Certain components may be fabricated in-house: Doehler-Jarvis engineers modified the design of the casting grippers to suit specific casting configurations.

Taped Programming. Easily programmed, the Unimate is led through an operation and its memory records each movement. This operation is done at a slow mode and can be completed between 15 minutes and four hours, depending on program complexity.

Doehler-Jarvis uses 4-channel tape cassettes which permit faster program changeovers and rapid reprogramming. Tape changeovers usually require some fine tuning unless the program is being used on the original Unimate and on the same diecasting machine. But if the die from an earlier automated production run is put on another Unimate-operated machine, the cassette allows speedy adjustment, integrating those steps into the robot's memory.

The company maintains a Unimation Program Storage and Verification Unit to record established robot programs on the cassettes for all of its Unimate robots. Although the program capability for the units is 240 steps, a total of about 50 steps are programmed into the Unimates used in the production of most transmission castings.

Once a Unimate is installed and running, there may be a need for adjustment of its operation. The Unimate program may be revised to account for changes in casting process variables. The flexibility of a pro-

grammable robot is evident: it can be quickly taught program changes or an entirely new program.

Installation. Center line distances between diecasting machines were adequate for the Unimates' effective operation. No major equipment relocation was necessary, and Doehler-Jarvis personnel did all the work, including modifying the electrical circuits.

Additional flexibility is gained by the maintenance of a Unimate as a spare unit. Installing it wherever it may be needed is simplified by identical floor mounting plates at each location.

To operate and maintain the robots, Doehler-Jarvis employs about 20 technicians trained by Unimate. Maintenance department electricians and casting technicians are sent to the training seminars.

Evaluating the Economics. To evaluate the performance of the Unimate, Doehler-Jarvis set up two equivalent machine installations, one was operated manually and the other used the robot. Outputs from each of the diecasting machines after long production runs showed that for the same number of productive hours, the robot yielded a 10-percent greater output.

As a result of the proved productivity increase, the Doehler-Jarvis Toledo operations uses 11 Unimate industrial robots with its diecasting machines. Four others are nearly installed and five more have been requested.

Doehler-Jarvis began automating its diecasting machines with Unimates 11 years ago at its Pottstown, PA, plant. The division now has about 50 Unimates operating in three of its plants, but has only recently used them in production of large, complex transmission cases parts. ▫

Presented at the Robot III Conference, November 1978

Automated Aluminum Die Casting

By Marshall R. Oakland
Honeywell Inc.

With the advent of more sophisticated industrial extractors the possibility of fully automating the die casting process become a reality. Integration of the industrial automatic extractor with an aluminum die casting machine to achieve total automation, in our case, has proven successful. Labor costs have been substantially reduced by implementing multiple machine loading and as a result an extractor amortization period of less than one (1) year has been achieved. Since the extractors were placed into production they have proven to be reliable and have not caused major machine downtime. At Honeywell, automation has proven to be beneficial. We hope our experience will contribute to the technological progress of the die casting industry.

Introduction

The decision to acquire automatic extractors was based on potential cost savings. At Honeywell, our die casting machines are all equipped with automatic ladles and automatic lubricators so it was not an extremely difficult task to add the automatic extractors. By fully automating the die casting process a more consistent casting cycle has been realized because the die cast operator no longer controls the casting cycle. As a result we have achieved better quality castings at a lower cost.

Automatic Process

Previously our die casting process could be considered a semi-automatic operation. That is, the die cast operator removed the parts from the machine, initiated the die cast machine cycle, degated the parts from the runner and then disposed the runner into the melt furnace. Operator chore time was all conducted within the die cast machine cycle time except parts removal. Currently our die casting process is fully automated. Automatic extractors now perform most all functions previously performed by the machine operator. Operator responsibilities now consist of monitoring machine operation, correcting machine malfunctions, tending the melt furnaces and parts removal.

Equipment Protection

Automation has created a need for additional features designed to protect the machines and tooling from accidental damage. Safety interlocks have been added to the die cast machine and to the automatic extractor that prevents their operation until both units are in their proper position to begin the casting cycle. Safety features have also been added to the automatic lubricator that detects the presence of a runner and parts that have not been removed from the mold. This feature prevents machine closure until the runner and parts have been removed from the mold and the operator physically actuates a cycle reset circuit. When the runner and parts are removed from the mold they pass through a sensing station that is designed to detect missing parts. If a part is missing the sensing station will not initiate the casting cycle and the automatic extractor stops at a reset position and will remain in that position until the machine operator corrects the problem and restarts the casting cycle. The added detection devices, to the machines, and the tooling are a worthwhile feature because to date not one incident of damage has resulted from a machine malfunction.

Cost Reduction

Automating our die casting process reduced direct labor by approximately 50% since one machine operator now performs the tasks that formerly required two operators.

Production Performance

Initial installation and debug of the first automatic extractor proved to be the most time consuming and the most troublesome. Most of the initial problems can be attributed to lack of familiarity with the new equipment. After the initial installation the following extractor installations were accomplished with very few debugging problems. Since the automatic extractors were placed into production they have proven to be reliable and relatively trouble free.

Conclusion

Thanks to the excellent cooperation I have received from all individuals involved I feel this application is a good one and could be used by other manufacturers with a similar application.

See Attachment I
Sequential/Visual Presentation

Attachment I
Sequential/Visual Presentation

Casting Process

Figure 1 shows the die casting equipment necessary to perform this operation. In the foreground is the extractor. Immediately behind that you see the aluminum die casting machine and then moving to the right you see the sensing station, the melt furnace, and the degating press. Our die casting process consists of removing the parts and runner from the machine, swinging clockwise through the sensing station and into position to place the parts and runner into the degating press. After the parts are degated the runner is removed from the press and returned to the melt furnace after which the extractor proceeds to the home position ready to repeat the process.

Parts Removal
Figure 2

Here you see the casting cycle has been completed and the extractor arm has entered the die to remove the runner and the parts.

Sensing Station
Figure 3

After removing the parts and runner from the die the parts are then moved through a sensing station. The sensing station serves two functions: 1) It insures that all parts have been removed from the die, 2) it signals the die casting machine to recycle.

Reset Position
Figure 4

After the parts are passed through the sensing station the gripper arm then rotates the runner and parts from a vertical position to a horizontal position. Here you see the parts and runner in position ready to be inserted into the degating die. This position also functions as a reset position if a part is missing. The extractor will remain in this position until the problem is corrected.

Degating Station
Figure 5

As the parts are being placed into the degator die you see the tapered
locating pins used here to ensure that the parts are properly located.
This helps prevent damage to the parts during the degating operation.
The manifold and pipes that you see pictured here are for the purpose
of keeping the die clean and free of flash. This helps insure that
the parts are seated properly prior to degating. After the parts
have been properly located in the degating die the parts are then
degated from the runner.

Runner Removal
Figure 6

Once the parts have been degated from the runner, the runner is then
removed from the degating die.

Runner Disposal
Figure 7

The extractor arm then moves counter-clockwise to a position over
the melt furnace, where the runner is disposed into the furnace
for remelt, the extractor arm then moves back to the home position.

FIGURE 1

FIGURE 2

FIGURE 3

FIGURE 4

FIGURE 5

FIGURE 6

FIGURE 7

CHAPTER 4

INVESTMENT CASTING

Commentary

In the investment casting or lost-wax casting process, industrial robots are used in the manufacture of shell molds. The shell molds are formed by dipping wax patterns into ceramic slurry and coating the wet slurry with sand. As many as six or more layers of slurry and sand may be applied to build up the shell to the desired thickness.

Robots offer the advantages of carefully and consistently handling the shells, thereby promoting uniform shell thickness and a higher yield of good castings. Robots also significantly increase productivity through their ability to handle heavy loads. Some robots used in the production of investment casting shells are handling payloads in the range of 2,000 pounds.

Reprinted from Precision Metal, May 1974

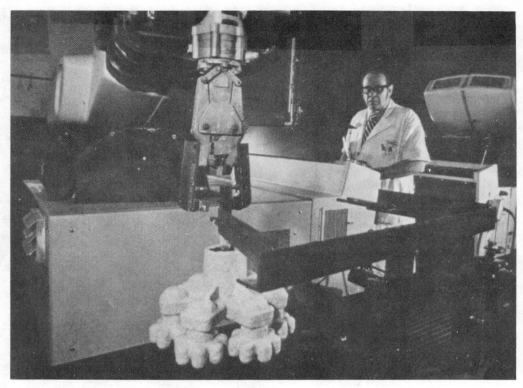

In the first phase of robot automation, wax pattern trees are manually loaded on a pick-up stand. The robot removes the patterns and performs the dipping operations.

Investment casting takes a step toward automation

Investment casting is a labor intensive process. Any labor saving device should, therefore, lower costs. One foundry has taken a major step in this direction by replacing men with robots to make shell molds.

Generally, investment castings require little machining, little finishing, and may be made in an infinite number of configurations. But preparation of the shell molds takes experience, judgment, and a critical eye. The job is tiring and tedious. An industrial robot, however, is ideally suited for the work.

Mold weight and silica coverage invariably fluctuate when a man does the job. What's more, there is insufficient assurance that the re-sultant castings will all meet specifications. A poorly filled out mold, a cold shut, or a number of other faults can mean a scrapped casting or, at best, a high-cost, time consuming salvage job.

In contrast, once a robot is taught the sequence of dipping, rotating, and swirling each tree-like cluster of patterns, the robot will repeat the sequence without any variation. The result is uniform shell molds and castings of consistent quality.

Evinrude Motors Div. of Outboard Marine Corp. has one of the most advanced foundry operations in the country. To improve casting quality and increase output, the foundry now uses Unimate industrial robots for making ceramic shell molds. These castings are used in a variety of leisure-time products ranging from outboard motors to snowmobiles.

As designed by Unimation, Inc., these robots quickly adapt to changeovers from one type of casting to another. Complete reprogramming is usually unnecessary. A few changes in the prior program will generally suffice and are made on-line.

Typical changes include: programming a different pick-up point for grasping the tree of patterns; adjustments of the spin time speed; correction in the dip depth to keep slurry out of the sprue cup.

If the difference between two runs of castings is weight alone, there is rarely a need for a program change. Additionally, the robot can handle trees which are much too heavy for a man to maneuver, hour after hour.

Varying weights and metals cast. Investment castings made in Evinrude's foundry range in weight from ¼ to 8½ pounds. The ceramic shell molds usually require six coats. Coats one and two are flour-fine slurries of colloidal silica followed by fine-grain stucco sand refractory. Coats three, four, and five employ coarser-grain slurries and stucco sand. The sixth coat is a slurry only and serves as a sealer to bind the previous stucco coats.

After a drying period, the wax pattern is melted out in a steam autoclave, leaving an empty shell. The shell molds are fired and metal is poured into the hot shell. Alloys such as 8617 steel, 410 and 413 stainless steel, aluminum, and bronze are routinely poured.

Automation is in two phases. In the first phase, now underway, robots will make the molds. Workmen will supply trees to be coated and remove the coated trees.

During processing, racks containing as many as 30 trees are wheeled to the robot's working area. Trees are hand-loaded individually onto a pick-up stand to which the robot reaches when ready to coat the next tree. The newly coated trees are put on a similar set-down stand to a positioning accuracy of ±0.05 inches. The stand is unloaded manually at this time.

Following the last coat, the trees are dewaxed in a pressurized steam autoclave and fired at 1700°F to fuse the grains of silica.

As advanced as this mold-making is, a planned second phase will

Between coats of slurry, the pattern trees are manually loaded on a drying rack. When ready for the next coat, the trees are loaded on the pick-up stand for robot dipping.

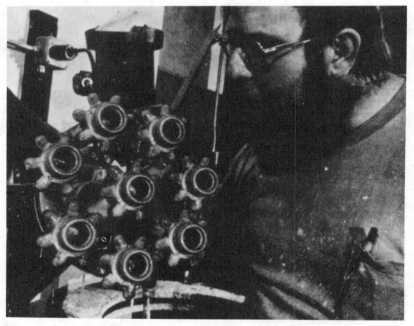

Man made shell molds are less uniform than robot made molds. They also take longer. The use of robots in mold making improves quality and lowers cost, and allows the handling of large, heavy molds.

increase the degree of automation. A conveyor will carry the trees automatically to the pick-up point. Another conveyor will carry coated trees away from the coating area. Operation of both conveyors will be controlled and monitored by each individual robot served by the conveyors. The result will be a major step forward in foundry mold making.

The industrial robot now controls the entire first phase of the operation, including: slurry mixer motors; air valves for fluidizing the bed of stucco; and gate valves in the fluidized bed dust control system.

Control and monitoring of the outbound conveyor will also be carried out by a robot. Determining that the space where the robot intends to place the coated tree is no longer occupied by the previous tree will be a simple matter of making the additional electrical connections and programming the additional signals.

After removal from the slurry, the trees are swung in an arc while rotating to evenly distribute the coating. The robot is about to start the cycle in this photo.

These investment cast parts have been made by robots. Used in engines, these parts are called ball joint assemblies.

Programs are straight-forward. Programming a Unimate consists of actuating push-buttons which drive the robot's arm, hand, and fingers through the desired motions. After each motion is satisfactorily completed, it is recorded in a magnetic memory. Thereafter, the movements will be repeated automatically, sequentially, and precisely.

The table summarizes the three basic programs which are sufficient for the usual coating procedure.

Several programs can be carried simultaneously in the robot's memory. Changes from program to program may be manual or automatic.

The many advantages of investment cast parts have, at times, been outweighed by cost considerations. Automating the shelling process with its resultant labor and cost savings advantages will, however, serve to increase the competitiveness of this important foundry process. **pm**

BASIC PROGRAMS FOR AUTOMATED MOLD-MAKING

Program 1
Coat 1:
Pick up tree from stand
Turn off slurry mix motor
Dip tree in slurry (fine grain)
Raise tree to obtain even flow of slurry over surface
Restart slurry mix motor
Lower tree and spin out excess slurry
Raise tree to obtain even flow of remaining slurry
Move tree to fluidized bed of stucco (fine grain)
Turn on air to fluidized bed
Insert and withdraw tree from bed
Turn off air
Pause to permit excess stucco to drain from tree
Place tree on stand

Program 2
Coat 2:
Pick up tree from stand
Pre-dip tree into binder
Raise and spin tree
Turn off slurry mix motor
Dip tree in slurry (fine grain)
Raise tree to obtain even flow of slurry over surface
Restart slurry mix motor
Move tree to fluidized bed of stucco (fine grain)
Turn on air to fluidize bed
Insert and withdraw tree from bed
Turn off air
Pause to permit excess stucco to drain from tree
Place tree on stand

Program 3
Coat 3, 4, 5:
Pick up tree from stand
Turn off slurry mix motor
Dip tree in slurry (coarse grain)
Raise tree to obtain even flow of slurry over surface
Restart slurry mix motor
Move tree to fluidized bed of stucco (coarse grain)
Turn on air to fluidize bed
Insert and withdraw tree from bed
Turn off air
Pause to permit excess stucco to drain from tree
Place tree on stand
Note: The sixth coat uses the first portion of Program 3 but omits the subsequent steps for applying the stucco coat.

Reprinted from Precision Metal, February 1980

Robots are working in investment casting

Integrated robot cuts shell production cost, improves quality

A robot integrated with a conveyor and transfer station by a programmable controller offers high productivity for investment casting operations.

Since investment casting is essentially a manual operation, efforts to mechanize or automate shell production have generally proven cost-effective. To obtain a ceramic shell of proper and uniform thickness, however, the pattern must go through a series of different motions.

A robot is ideally suited to shell production due to the repetitive nature of the operation, but to make it cost-effective, a robot must meet certain criteria:

1. It must be capable of carrying heavy loads.
2. Motions must be very smooth to handle delicate wax patterns.
3. Shell making must be interfaced with a conveyor to eliminate manual handling.
4. The robot must withstand foundry environment and be serviceable by plant maintenance personnel.

A unique robot system, intended specifically for investment casting, has been designed by Shell-O-Matic in conjunction with Cercast, Inc.

The design concept of the robot shown here is to mechanically duplicate manual dipping. With a capacity of 400 lbs (180 kg) several clusters can be dipped together. The result is the production of ceramic shell molds with maximum consistency at minimum cost.

The robot features an electromechanical drive system rather than the hydraulics usually used in robots. Its programmable controller is available with a memory capacity of 40 switch-selectable dipping programs. Eight timer settings can be dialed-in independent of the controller for maximum versatility.

Operation, including transfer from and to the conveyor, is fully automatic. The conveyor advances while the robot goes through a dipping cycle, eliminating time loss. The transfer system can also accept unbalanced loads, further increasing versatility.

The Shell-O-Matic robot dipping system comprises a dipping center (manipulator with a variety of tanks, fluidized beds, rainsander and auxiliary equipment) and a conveyor interfaced with the manipulator.

Productivity of the automated dipping system is affected by the size and structure of the wax patterns and the type of slurry and stucco. Typical cycle times range from two to four

The Shell-O-Matic robot arm, holding clusters of patterns coated with ceramic shell material, is interfaced with a conveyor by a programmable controller. Electromechanical drive permits robot to handle 400 lbs. smoothly.

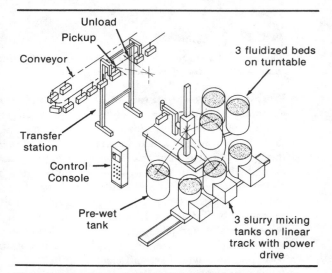

Diagram shows how the dipping robot/conveyor can be set up. Robot picks up patterns at the conveyor, completes the dipping cycle and returns patterns/shells to the conveyor.

Robots are working in investment casting

minutes, including transfer from and to the conveyor. The more clusters the robot can dip together, the higher the output per hour, so multi-cluster dipping is essential to maximizing productivity.

The benefits of this type of operation are consistent ceramic shells, which reduce scrap rate and finishing costs; reduced labor costs; the ability to produce large and heavy castings or clusters of castings; and freeing the operator from tedious, repetitive work.

Reprinted from Precision Metal, February 1980

A programmable robot for investment casting

Programmable industrial robots can be cost-effective additions to investment casting, if they are suited to the foundry environment and have sufficient programming flexibility.

The first programmable industrial robot was used in the investment casting industry in the late 1960's. Since then, the relatively small number of robots applied to investment casting are providing a good return on investment in many areas of the investment casting process. One reason for the slow application of robots is that they have not been designed to cope completely with the foundry environment and the particular needs of the investment caster.

Tight floor space has ruled out many robot applications. The inability of some robots to cope with the foundry environment (such as abrasive dust in the air) has added to maintenance cost, making the robot less attractive economically. As investment casting production volume goes up, so does casting tree weight. But if trees are kept smaller because of robot payload capacity, productivity suffers.

In the area of control, robot manu-

facturers have adopted state-of-the-art electronics, which improve reliability and simplify maintenance. Robot controls should have adequate memory to store programs for a variety of parts on-line, so that auxiliary, off-line storage devices are not necessary. Permanent program libraries can be stored off-line, however.

A tough challenge for robot manufacturers has been to provide tracking capability to follow a moving object. Costly stop-and-go systems can thus be eliminated.

A robot meeting these criteria can provide a good return on investment, and improve casting quality and productivity. One example is the Versatran robot, which was manufactured by AMF until Prab Conveyors, Inc. acquired the line in February 1979. Versatran robots are now produced in Prab's Kalamazoo, Michigan facility. They offer features and characteristics important to the cost-effective automation of the investment casting process.

Versatran robots employ direct-drive motions at the end of the arm, eliminating gearing and other drive mechanisms. A direct-drive, hydraulic motor provides gripper as well as

Editor's note: This article was prepared from a paper given by Walter Weisel, Prab Conveyors, Inc. at the 27th annual meeting of the Investment Casting Institute.

The FC Series Versatran, with a payload of 2000 lbs. and programmable motions, can handle both heavy and delicate pattern trees.

simple yaw motion. A major advantage is the robot's cylindrical coordinate system. The XYZ coordinate system can be combined with an additional axis for traverse and one, two or three axes of wrist motion, combining simple straight line motions which are important when tracking conveyor lines or placing parts into close openings.

The robot drives and mechanical construction allows the Versatran to handle up to a 2000-lb (900 kg) payload, enabling it to handle heavy castings or trees with many smaller castings.

Computer control is a significant feature of the Versatran robot. The 600 series control system operates with a Digital Equipment LSI-11 microcomputer, which directs all activities of the system. The computer is capable of directing other pieces of equipment and receiving signals to complete system operation. It can store up to 64 different part programs, which can be selected by remote control. The programs can perform repetitive operations with a minimum of memory.

The control has the ability to program velocity and acceleration/deceleration for robot motions. This is extremely important when handling delicate wax patterns at maximum speed. Eight acceleration/deceleration ramps can be programmed directly from the input keyboard.

The controller can track a moving conveyor with the robot mounted in either a floor or overhead position. The robot arm is placed in the tracking mode at the end of the dipping or core-handling cycle. Conveyor position and speed signals from a linear potentiometer provide data to synchronize the robot traverse axis motion with the conveyor, and the normal pick-up/drop-off program is resumed. The tracking axis, designated the seventh axis, operates independently of conveyor speed, and the robot "thinks" it is looking at a stationary programmed point.

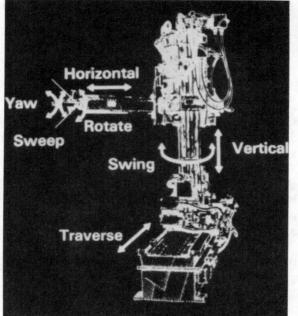

The Versatran robot has a cylindrical coordinate system. Drawing shows the seven motions possible with the robot.

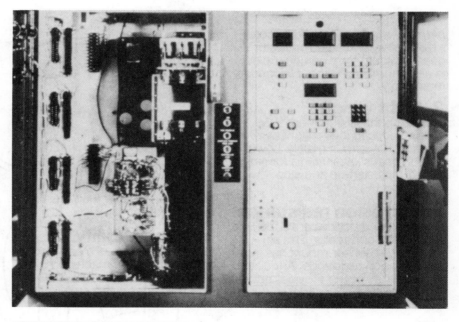

▲ The 600 Series control system can be applied to all Versatran robots. The microcomputer-based control can be interfaced with other equipment in the production system.

◄ A permanent library of robot programs can be stored in this magnetic cassette recorder. The tape provides a backup and an efficient means of storing production programs.

157

Presented at the   Robot IV Conference, October/November 1979

Robot Controlled Mold-Making System

By Frank A. Moegling
TRW Metals Division

This paper presents the operational aspects of a robot-controlled mold making system as used in the investment casting industry.

The system was designed to utilize a very specific shell process; however, after installation it became quite apparent that the system would accommodate nearly any shell process, regardless of complexity or drying requirements.

In this installation, the robot controls all ancillary equipment, including conveyors, sanders and dip tanks. Future expansion of the system will include automatic mold drying control.

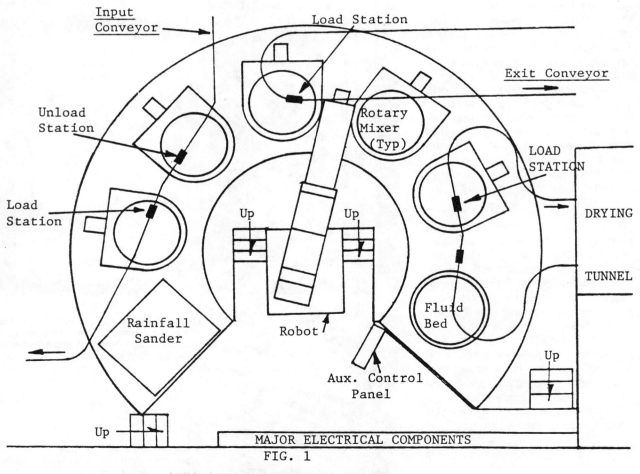

FIG. 1

AUTOMATIC MOLD-MAKING SYSTEM

GENERAL

This robot-controlled mold-making system was designed to produce ceramic molds as required in an investment casting foundry which utilizes the lost wax casting technique. The product mixture consists of blades, vanes, and some structural components as used in airborn and land-based gas turbine engines.

Basically, the lost wax casting technique involves the following operations:

1. Wax Injection - wax, in a semi-molten state is injected into a die under high pressure to form a wax pattern that is the exact duplicate of the desired finished metal part.

2. Wax Assembly - The wax patterns (above) and other necessary components, such as sprues, gates and down-poles are assembled into a cluster. This configuration allows many parts of the same configuration to be poured simultaneously once the finished mold reaches the foundry.

3. Mold Dipping - The above clusters are then immersed into a ceramic slurry. This slurry coats all sur-

faces and exactly conforms to the now enclosed wax
cluster. Immediately following this slurry coating,
the cluster is then sanded in a fluidized bed or a
rainfall sander. In this operation, the individual
grains of sand cling to the wet surfaces of the
cluster. The ceramic coated cluster is now permitted
to dry out before application of subsequent coats of
slurry and sand. This drying is usually done in a
separate chamber or tunnel where temperature, humidity and air flow are carefully controlled. After
multiple applications of slurry and sand the wall
thickness of the mold reaches a point where it will
be strong enough to contain the molten metal that
will eventually be poured into it. At this point
the mold is given additional final drying time and
proceeds to the next operation.

4. Dewaxing & Mold Firing - The ceramic encased wax
   cluster is inserted into a steam autoclave where the
   wax is melted out under controlled steam pressure
   and temperature. This pressure meltout allows the
   wax, which has a higher coefficient of thermal
   expansion than the surrounding ceramic to melt and
   run out without cracking the shell.

   Once the wax has been removed, the shell mold is now
   fired in a kiln to obtain maximum strength, cleaned
   out, inspected, and finally sent to the foundry for eventual
   pouring.

DESIGN CONSIDERATIONS

An industrial robot lends itself well to the redundancy
of the mold-making operation. Because it is physically stronger than most people and can repeat programs without variation,
mold quality can be accurately controlled. This capability
allows great flexibility while still operating under strict
process procedures and quality control requirements. Further,
the robot's ability to control ancillary equipment and to safeguard itself against extraneous equipment malfunction makes it
a prime candidate as the "system controller" in a mold-making
facility where one robot is to be employed.

Preliminary design criteria include the number of molds
to be processed per day, site constraints such as clearances
from existing structures, input of wax patterns into the
system, output of completed molds to subsequent operations and
overall envelope size and weight limitations. Large envelope
size and maximized robot lifting capability allows multiple
cluster dipping which results in high efficiencies and maximum
return on investment.

Once equipment manufacturers have been selected and site
preparation design has been completed, the single most important consideration is the transport conveyor and the robot
"hand" tooling design. This interface must be coordinated
through both the robot and conveyor suppliers to assure a
smooth working and trouble-free system.

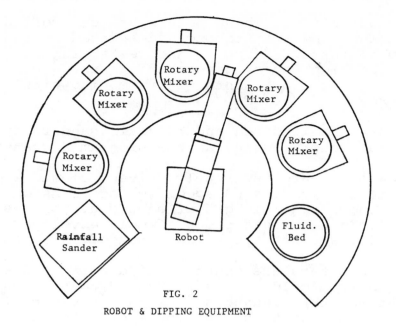

FIG. 2

ROBOT & DIPPING EQUIPMENT

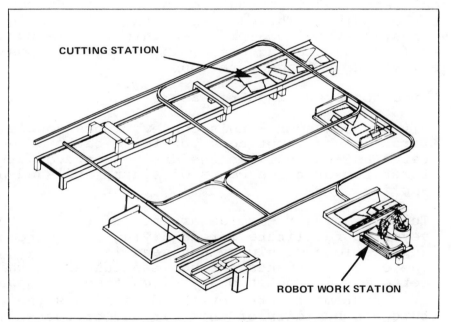

2. This schematic of the mold-making system shows the robot and ancillary equipment it controls, including conveyors, sanders, and dip tanks.

INSTALLATION SPECIFICS

The system under discussion is comprised of a robot surrounded by five rotary slurry mixers, one fluidized bed, one rainfall sander, and three conveyors. All of the above have been placed within the reach of the robot in order to operate well under the maximum physical limitations imposed by the robot manufacturer.

The 42" rotary slurry mixers contain 300 gallons each of various slurries that are applied at different times during the

mold-making operation. These mixers are controlled by variable
frequency drives and are stopped and started by the robot as
the program dictates. Stopping and restarting is necessary
because of the high forces generated against a wax airfoil
pattern when it is dipped into slurry moving with peripheral
speeds in excess of 300 SFM.

Both the fluidized bed and rainfall sanders are turned on
and off by the robot, again, as the program dictates.

The conveyors are separately driven, closed loop conveyors
that index mold envelopes to specific locations on command from
the robot. The overall length of each conveyor has been carefully calculated so that a group of mold envelopes enter, are
processed and leave the system as a batch.

The wax clusters are mounted on "dipping plates" which in
turn are hung from mold carriers affixed to the conveyors on
48" center-to-center distances. As the conveyors index toward
the load or unload station, the conveyor takes a path that
brings the carrier to a tangent point within the robots rotary
and vertical reach and is normal to the arm of the robot when
rotated to that point. The carrier is then clamped in place
with a cylinder-operated clamping mechanism. This positions
the carrier (and the dipping plate) in the optimum position for
repeatable load and unload operations.

The three conveyors involved are:

1. Air dry conveyor - this conveyor does double duty,
    first as the input conveyor for a batch of wax pat-
    terns and secondly as the drying conveyor after the
    first two or three coats of slurry and sand have been
    applied.

2. Tunnel conveyor - molds are transferred to this con-
    veyor to facilitate forced drying of the remaining
    applications of slurry and sand in the drying tunnel.
    The conveyor snakes through the tunnel in which tem-
    pered air is continuously circulated. A separate
    HVAC system, located overhead, monitors the tempera-
    ture and humidity of this air and corrects devia-
    tions as they occur. Over all system accuracy is
    plus or minus 2 degrees F. from setpoint from no-
    load through full-load conditions.

3. Download conveyor - once the molds are completely
    dipped and dried in the tunnel conveyor, the robot
    places the envelopes on the download conveyor. This
    conveyor removes the molds to an area where clusters
    are removed from the dipping plate and sent to the
    dewaxing area.

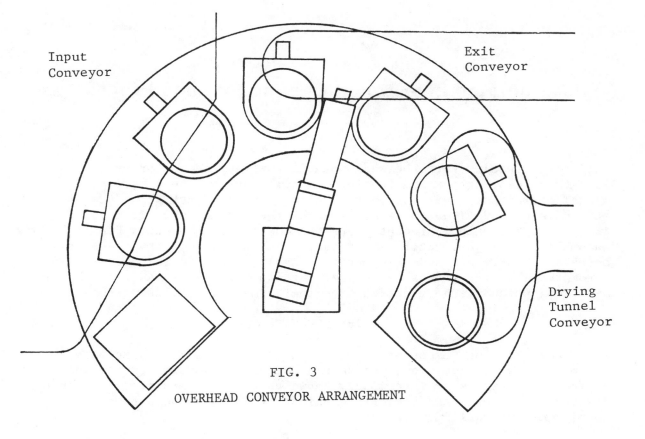

Input Conveyor

Exit Conveyor

Drying Tunnel Conveyor

FIG. 3

OVERHEAD CONVEYOR ARRANGEMENT

The robot stores programs in its memory and these programs are automatically accessed under normal operating conditions. This is accomplished utilizing an optional present counter on board the robot and bar readers that decode patterns of bars mounted on each dipping plate. The bar coding identifies the particular wax patterns mounted to the dipping plate and selects the program for successive dips. As the batch cycles through the various programs the counter modifies the program selection technique and actually "lies" to the robot even though the bar reader reads the same bar code over and over again. The circuits that encode the bar reader output signals and modify the program selection with respect to cycles accomplished was developed utilizing in-house capability.

The system as described has exceeded our most optomistic projections and is currently being modified with even more features and additional material handling systems. These will provide even greater outputs with lower mold scrap losses, and lighter stronger molds.

Presented at the Robot III Conference, November 1978

Automated Investment Casting Shelling Process

By Edward G. Laux
Howmet Turbine Components Corporation

Until very recent years the shell manufacturing process for producing
investment castings, using the lost wax process, has been basically a
manual operation. Over the years, various means of assisting the operator have been devised to remove some of the manual labor from the
operation, but only in recent years has the shelling process been fully
automated. A major factor making automation possible in this type of
operation has been the development of highly maneuverable robots. The
automated shelling process at Howmet Turbine Components, Austenal Dover
Division, which was developed around the Unimate robot, is discussed.

INTRODUCTION

The Howmet Turbine Components Corporation, the Dover N. J. facility,
are producers of precision investment castings primarily for turbine
engines employing what is commonly known as the lost wax process. This
discussion concerns the use of the Unimate robot for the automated production of our shelling process.

For those who may not be familiar with the lost wax process, I feel that
a brief description of the process is necessary for a better understanding of the automated shelling process. The lost wax process begins by
making a wax pattern of the object to be cast. A number of these wax
patterns are assembled into a cluster around which a monolithic ceramic
shell is formed. The shell consists of a number of successive coats of
ceramic slurry followed each time with a stucco sand until the desired
shell thickness is obtained. The wax is then removed from the shell and
metal cast into the mold thus producing the desired object.

The manufacture of the ceramic shell is a very important part of the investment casting process. In fact, it is probably the most important
factor in determining the quality of the casting, therefore, a consistently high quality shell is essential. The Unimate is the central figure in the automated manufacturing of this quality shell.

DISCUSSION OF EQUIPMENT & OPERATION

First of all, let's look at the general layout of the automated shelling
system (Illustration #1). The model 4000 Unimate is located in the center to permit excess to all the auxiliary equipment. To provide the required versatility to do the complete shelling operation, the Unimate
has been mounted on a shuttle mechanism so that it operates in two positions as seen in the diagram. Around the Unimate in the front area, the
ceramic slurry pots, the stucco sanders and a loading station are
arranged in a circular pattern. Also, in this area is the logic control
center and the Unimate control console and the program control console.

In the rear area are three conveyors; (1) a two tier conveyor for 1st and 2nd dip coats, (2) a conveyor for back-up dip coats, and (3) a output conveyor for final drying. In the rear to the left of the conveyors is the environmental control equipment for the conveyor rooms.

Due to the environmental requirements of our shelling process, the conveyors have been enclosed in separate rooms. The shelling area and the first conveyor room which handle the first and second dip coats, are maintained at 75°F and 45% relative humidity. The back-up conveyor room which handles the 3rd through the 12th dip coats, is maintained at 75°F and 25-30% R.H. with the output conveyor room where final drying of the shell takes place, also at 75°F but with a maximum R.H. of 20%.

The Unimates that are used at Austenal Dover, as well as the other Howmet Divisions, are the 4000 series models which have a number of special adaptations to meet our requirements. The Unimate on this automated line has 16 programs of 64 steps in each program. Fourteen of the programs are consumed by the various shelling cycles with two programs remaining open for any special programs and for trouble shooting. They are equipped with a heavy duty wrist gear train for two reasons; (1) it allows greater load capacity, which with our adapter arrangement allows a maximum load of about 250 lbs., and (2) due to the very abrasive atmosphere in the area wear becomes a problem and the heavy duty gear train gives longer life.

Another adaptation which we have incorporated because of the dusty abrasive atmosphere is the pressurization of the entire robot. A slightly positive pressure is maintained internally so that air leaks out all the openings minimizing the infultration of dust into the unit. Since it is necessary to provide clean air for this pressurization we duct air from an air conditioning unit or from an adjacent area into the Unimate.

Our Unimates have five programmed articulation with the fifth axis of movement, which is wrist yaw, converted to constant spin. This permits constant rotation of the clusters during any mode of the robot, which is vital during the dipping and draining operations for producing a consistently uniform shell.

The Unimate console has been modified to include extra time delays and a random program selector (RPS). The RPS permits the selection of any program in the robots memory as indicated by the logic control center. It also allows manual selection of any program when the Unimate is separated from the logic control center. The extra time delays are required to supply a sufficient number of stationary positions of the clusters for proper draining during a dipping program.

The hand which has been adapted to the wrist and is used to clamp the cluster adapter is Howmet designed. It is operated pneumatically by the clamping air cylinder that is built into the Unimate wrist.

Since there are a number of different shelling systems it is necessary
for the Unimate to know which programs to use on each cluster adapter
as it proceeds through the shelling operation sequence. The Unimate is
dependent on the photo eye scanners, proximity switches and solid state
counters of the logic control system to compare information and conclude
which program is to be run. The scanners read the program information
from a binary code card attached to each adapter handle. At the time the
operator fastens the clusters to the adapter, he selects the proper binary
code card for the shelling operation required by those particular clusters
and affixes the card to the adapter handle. The loading station is
equipped with a scanner which reads the code card, records the information in the control memory and signals the Unimate which program is to
used for face dipping. The robot then picks up the adapter, carries
out the dipping and sanding operations (Illustration #2) and deposits
the adapter on the lower tier of the first conveyor.

The 2nd and 3rd dip coats are the same for all our shelling systems,
therefore, the face coat conveyor is not equipped with photo eye scanners. When the Unimate picks up an adapter from the upper tier of the
first conveyor for 3rd dip, it shuttles to the back-up position, then
proceeds with the dipping and sanding operations and deposits the adapter
on the upper tier of the second conveyor. At the end of each shuttle it
is essential that the Unimate be in the exact position before proceeding
with the program, therefore, the shuttle mechanism has been equipped
with micro switches and magnetic locking devices to assure proper positioning.

The back-up conveyor is equipped with a scanner on each tier for reading
code cards and informing the Unimate which program is to be run. The
majority of the clusters receive a maximum of seven dip coats at which
time they are deposited on the drying conveyor. If clusters are to receive more then seven coats they are deposited on the bottom tier of the
back-up conveyor for further dipping and then onto the drying conveyor
at the completion of the proper number of dip coats.

The operating sequence of one complete cycle of the shelling operation
is shown in Illustration #3. The program control console indicates which
cycle step and cycle half is in operation along with the program that is
being run on the Unimate.

The Unimate is electrically interfaced with the other equipment so that
it controls the operation of the equipment at the proper time in the
programs. In this way the indexing of the conveyors, the operations of
the stucco sanders and rotation of the dip pots are operated at the
appropriate times in the programs. The Unimate also receives signals
from the equipment that everything is in readiness to continue or start
a program; i.e. when indexing of the conveyor is completed a signal is
sent to the Unimate that the conveyor is ready for a pick-up. Proximity
switches are located at the conveyor pick-up points to signal that if an
adapter is not present the Unimate will not perform that program but will

wait for the signal from the logic control center to proceed with the next program. The incoming and outgoing signals that go directly to the Unimate are programmed into the Unimate memory during the teaching process using the OX and WX signals on the control console.

When designing conveyors for use with the robot, there are several important factors which must be taken into consideration. The hangers on the conveyors must be spaced accurately. Each hanger as it is indexed into the Unimate pickup position must be within 1/8", in both horizontal and vertical position, of the programmed position of the Unimate at the pick-up step. If the hanger positioning is not accurate, the pick-up or deposit of the adapter will not be performed properly and will result in destroyed clusters. The conveyor driving mechanism must be consistently accurate each time it indexes and must not allow any drift of the conveyor after the indexing is completed. We found it necessary to install detents on the conveyor to prevent drifting of the conveyor after indexing. The number of hangers on the conveyor will depend upon the required drying time and the estimated cycle time.

One thing which I am sure that is of prime interest to everyone is the production rate of this automated shelling equipment. But first, before discussing the production rate of this automatic line, let's consider the factors that effect the production rates. The whole system, of course, is paced by how fast the Unimate performs the operations for which it has been programmed to meet the requirements of your particular operation.

The production rate is dependent upon a number of factors peculiar to each individual shelling system, mainly such as:

(1) Cluster size, which determines the number of clusters per adapter.
(2) The number of dip coats required by the various clusters. The number of dip coats in our system varies from seven minimum to twelve maximum.
(3) The drying time required between dip coats.
(4) Required drain time of the clusters after removal from the dip pot and spinning time to remove excess dip.

It is important to have operators who are well trained in the operation of the Unimate and how to program. Once the line is in operation, an operator, depending upon how well he understands the shelling process and the operation of the Unimate, can do a number of things to fine tune the programs and reduce cycle time. For instance, one way we have reduced cycle time is to program as many Unimate positions as possible in accuracy two instead of accuracy one. This of course, is not possible where the Unimate has to perform an operation such as a pick-up or deposit, but the majority of the steps do not require this accuracy and can be programmed in accuracy two. This reduces the time for the Unimate to move from one position to the next by slightly altering its path. By fine tuning the programs we have been able to reduce overall cycle time by 2 to 3 minutes from the original start up cycle.

This automated shelling line was designed based on an estimated 21 minute overall cycle time for a total of seven dips and an average of four molds per hanger. These figures were arrived at after considerable analysis of cluster size, required drying times, shelling cycles to be used, etc. and some experience from other Howmet automated systems. At a 90% efficiency rate, this would produce 247 molds in 24 hours. Due to added shelling process requirements, which have been inserted in the programs since the original planning, and clusters requiring more than seven dips, the average complete cycle time is now at 27 minutes. This means that every 27 minutes the Unimate completes the shelling operation on an average of four molds. Based on a 90% efficiency factor this is a production rate of 191 molds per 24 hour period.

These production rates, of course, are peculiar to our operation for producing our shell and for our particular products, and only gives an indication of what can be produced with other systems. Production rates of a Unimate automated line, for any other shelling system, and another product line, would have to be evaluated on an individual system basis.

This automated shelling operation was designed for future expansion with increased productivity. It was designed so that a second Unimate can be added to the line (Illustration #4). The shuttle mechanism will be removed and the present Unimate located in a fixed position to do the first three dip coats. The second Unimate will be located adjacent to the present one and will do the back-up dip coats. A transfer station will be added for transferring the clusters from one Unimate to the other after third dip coat. The back-up conveyor and the output conveyor will be expanded sufficiently to maintain proper shell drying times. With the addition of the second Unimate, we expect an increase in productivity of 75 to 80% over the present system.

CONCLUSIONS

The use of a fully automated shelling process for producing investment castings has a number of advantages over castings produced by the manual shelling process. These advantages are not just the results of the automated line experience at the Dover facility but have also been reported by other Howmet Divisions where automated shelling lines are in operation:

(1) More Consistently Uniform Shell.
    The automated shelling process produces a more consistently
    uniform shell which results in higher quality castings.
    The robot does not get tired and is consistent in the
    motions of dipping, draining and stucco sanding 24 hours
    a day just as it was taught. A man tires as the day
    wears on and his motions are not consistent, thus causing
    the shell thickness to vary. This has been confirmed by
    weight comparison of completed clusters from automated
    line vs. those produced manually.

(2) Cast Savings.
     The cost savings for reduction in labor using an auto-
     mated system, based on a manual shelling rate of 2.3
     molds/man-hour, is as follows:

Manual Labor Costs

$$\frac{191 \text{ Molds}}{2.3 \text{ molds/man-hr.}} \times \$7.50/\text{hr.} = \$623/\text{Day}$$

Automated Labor Costs

1-1/4 men/shift x 3 shifts x 8 hrs./man
   x $7.50/hr. = $225/Day

Annual Labor Cost Savings

($623/Day – $225/Day) x 240 Days/Yr. = $95,520/Yr.

This results in a payback period of approximately 3 years
on labor saving alone. However, an additional cost savings is also experience in improved casting quality.
Another division of Howmet, with more experience using an
automated line, has reported cost savings due to improved
quality almost equal to the savings in labor costs.

(3) Better Material Control.
     The more uniform shell also results in better control of
     material costs. With proper programming of the robot,
     the shell can be controlled at an optimum thickness re-
     sulting in better material control.

(4) Reduced Scrap Rates.
     As a result of the better quality shell, the casting
     quality is improved, thus reducing the scrap rate.

(5) Better Casting Surfaces.
     The consistent face coating operation results in better
     casting surfaces which reduces the amount of mechanical
     finishing required on the castings. With the incorpo-
     ration of automated shelling in one Howmet plant, they
     were able to reduce the size of their finishing depart-
     ment by better than 50% because of the improved casting
     surfaces.

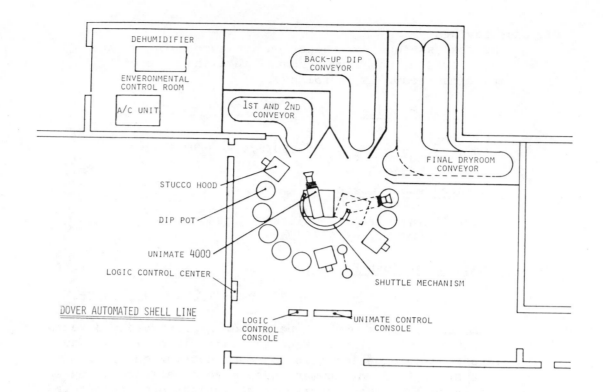

DEHUMIDIFIER

ENVERONMENTAL
CONTROL ROOM

A/C UNIT

BACK-UP DIP
CONVEYOR

1ST AND 2ND
CONVEYOR

FINAL DRYROOM
CONVEYOR

STUCCO HOOD

DIP POT

UNIMATE 4000

LOGIC CONTROL CENTER

SHUTTLE MECHANISM

DOVER AUTOMATED SHELL LINE

LOGIC
CONTROL
CONSOLE

UNIMATE CONTROL
CONSOLE

Illustration 2

SEQUENCE OF ONE CYCLE

CYCLE STEP	CYCLE HALF	DIP	DESCRIPTION	INDEX
0	2	3	REMOVE MOLD FROM FACE COAT CONVEYOR (TOP), RUN PROGRAM #7 1st HALF.	SHUTTLE UNIMATE TO BACKUP
1	2	3	RUN UNIMATE PROGRAM #7 (2nd HALF), PLACE MOLD ON BACKUP DIP CONVEYOR (TOP).	BACKUP DIP CONVEYOR
2	1	12	IF MOLD ON BOTTOM OF BACKUP CONVEYOR, RUN PROGRAM #14 AND PLACE ON OUTPUT CONVEYOR.	OUTPUT CONVEYOR
	2	7	REMOVE MOLD FROM BACKUP CONVEYOR (TOP), RUN PROGRAM 10, 11 OR 12, PLACE ON BACK-UP (BOTTOM) OR OUTPUT.	BACKUP DIP OUTPUT IF #10 OR #11
3	1	11	REMOVE MOLD FROM BACKUP (BOTTOM), RUN PROGRAM #13 OR #14, PLACE ON BACKUP (BOTTOM) OR OUTPUT.	OUTPUT IF #14
	2	6	REMOVE MOLD FROM BACKUP CONVEYOR (TOP), RUN PROGRAM 8, 9, 10 OR 11, PLACE ON BACKUP (TOP).	BACKUP CONVEYOR + OUTPUT IF 10 OR 11
4	1	10	REMOVE MOLD FROM BACKUP (BOTTOM), RUN PROGRAM 13 OR 14, PLACE ON BACKUP (BOTTOM) OR OUTPUT.	OUTPUT IF #14
	2	5	REMOVE MOLD FROM BACKUP (TOP), RUN PROGRAM 8 OR 9, PLACE ON BACKUP (TOP).	BACKUP CONVEYOR
5	1	9	REMOVE MOLD FROM BACKUP (BOTTOM), RUN UNIMATE PROGRAM 13 OR 14, PLACE ON BACKUP (BOTTOM) OR OUTPUT.	OUTPUT IF #14
	2	4	REMOVE MOLD FROM BACKUP (TOP), RUN UNIMATE PROGRAM 8 OR 9, PLACE ON BACKUP (TOP).	BACKUP CONVEYOR
6	1	8	REMOVE MOLD FROM BACKUP (BOTTOM), RUN PROGRAM 13 OR 14, PLACE ON BACKUP (BOTTOM) OR OUTPUT.	OUTPUT IF #14 AND SHUTTLE
7	1	2	REMOVE MOLD FROM FACECOAT (BOTTOM), RUN PROGRAM 6, PLACE MOLD ON FACECOAT (TOP).	NONE
	2	1	REMOVE MOLD FROM LOAD STA. RUN UNIMATE PROGRAM 1, 2, 3, 4, 5, PLACE MOLD ON FACE COAT CONVEYOR (BOTTOM).	FACE COAT CONVEYOR

ILLUSTRATION #3

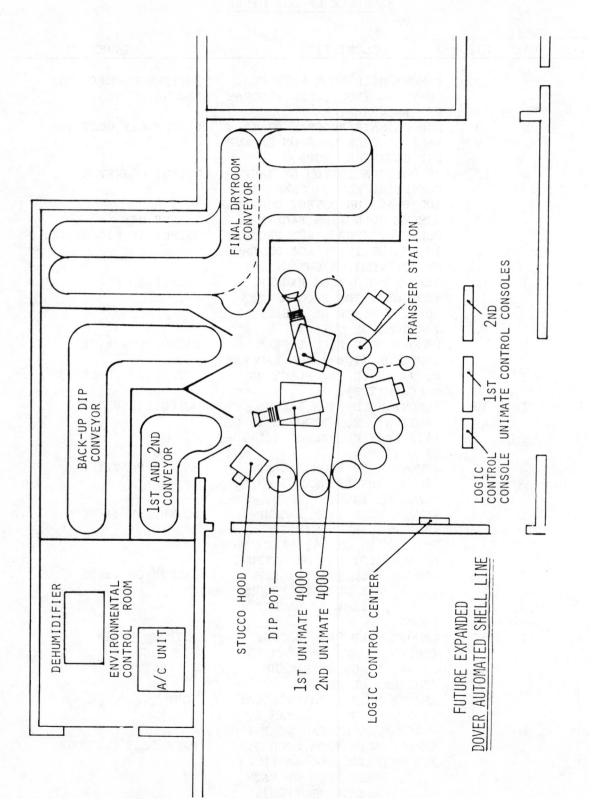

DEHUMIDIFIER

ENVIRONMENTAL CONTROL ROOM

A/C UNIT

BACK-UP DIP CONVEYOR

1ST AND 2ND CONVEYOR

FINAL DRYROOM CONVEYOR

TRANSFER STATION

STUCCO HOOD

DIP POT

1ST UNIMATE 4000

2ND UNIMATE 4000

LOGIC CONTROL CENTER

LOGIC CONTROL CONSOLE

1ST

2ND

UNIMATE CONTROL CONSOLES

FUTURE EXPANDED
DOVER AUTOMATED SHELL LINE

Illustration 4

CHAPTER 5

PRESS LOADING

Commentary

One of the earliest robot application areas, loading and unloading of stamping and forming presses, is being considered with renewed interest. Robots offer a means of complying with safety requirements and still have the flexibility to handle batch-run operations.

Robots may not meet speed and accuracy requirements for some press operations; however, for press-to-press transfer and for handling large or heavy stampings, robots merit serious consideration.

Vacuum systems are often used to handle parts and dual-hand tooling may be employed to increase efficiency.

The major advantages of robots for press loading are related to removing people from potentially hazardous, noisy, monotonous oeprations. As an alternative to other forms of press automation, robots may significantly reduce change-over time for batch-run parts.

Robot-Loaded Stamping Presses Keep Pace with Production

Many stampers say robots are too slow to efficiently load and unload presses. Well, it's simply not true. Given a chance, robots can keep pace with production, improve safety, and reduce costs

We've heard it said more than once at PRODUCTION Round Tables that industrial robots aren't fast enough to feed stamping presses . . . that they can't even keep pace with an operator feeding a machine at 12 strokes per minute.

To determine how much substance there is in this argument, we broached the issue to A. O. Smith Corp. Since 1972, its Milwaukee works has used four robots to unload and load two press lines; each comprised of three 800 ton Verson presses. The adjusted cycle on these lines is 340 parts per hour. Another robot, operating alone between two 800 ton presses, works at 390 parts per hour.

The application of robots by A. O. Smith is unique from several

standpoints: Not too many companies use robots to load and unload presses; all three lines operate at high production rates; parts handled are relatively heavy; the robots are specially-designed with extended booms and six degrees of movement. Many of them are utilized to their full travel. Each robot is controlled by programs stored in a memory device which controls all movements, optimum speed, and acceleration of servo loops.

Unimation Inc., Danbury, Conn., supplied A. O. Smith with a total of six Unimate 2000 robots. Five are used in production. The sixth is a backup substituted for units pulled off the line for periodic maintenance.

Large Stampings. The Unimates serving the two three-press lines are loading and unloading left and right side front inner rails used in the A-style frame for intermediate-size automobiles. The robot working alone handles rear upper control arm cross bars for A-style frames. Both frame members weigh about 24 lb, a load which puts no strain on the 60-lb capacity the Unimates have at full extension.

A. O. Smith specified Unimates with six programmable degrees of movement so that the robots could be used with off-center dies and have the flexibility to accommodate, at some later date, presses arranged in a different fashion.

Surprisingly the latter situation

The Unimate at the right extracts an inner rail initially formed from a blank in a press (not shown in picture) and places it into the press in the center for secondary forming. The robot at left then takes the part from that press and places it into the press at the left which completes the forming

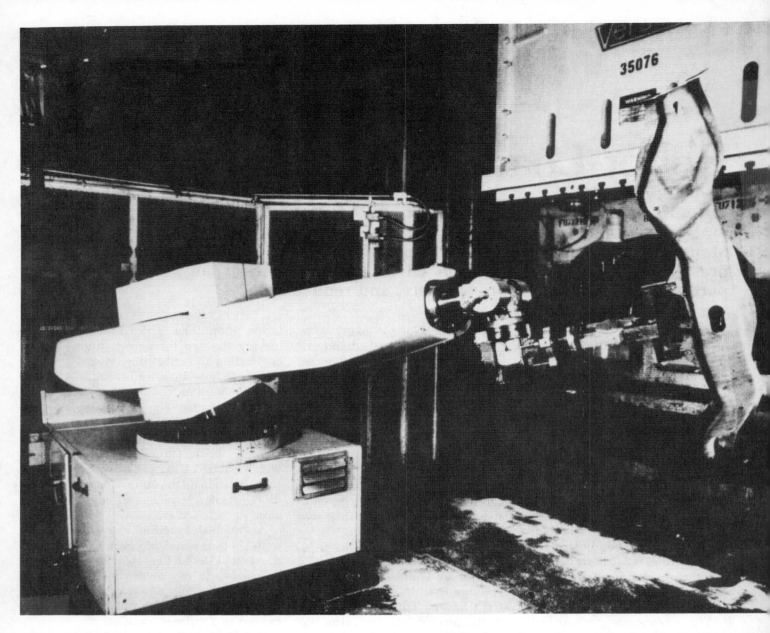

has never come up in Milwaukee. The presses served by robots have been dedicated to producing the same parts since the lines were set up for robot feed. Dies in most other presses at A. O. Smith are pulled two to four times a week.

Robots and Safety. A. O. Smith became involved with robots as part of a drive to keep every operator's hands away from the point of action on a press.

According to Robert Allen, director of safety for the Milwaukee works: "The Unimates were purchased strictly as safety equipment, not for cost reduction purposes." Although OSHA officials liked the Unimate approach to keeping hands out of the die, Allen believes they represent a very ex-

pensive solution to the safety problem. Other senior plant managers at A. O. Smith share Allen's skepticism.

The young, but experienced, engineer responsible for robots at A. O. Smith, R. R. Lefebvre, supervisor-Electronic Systems, sees the Unimates in a different light.

"The Unimates represented what we thought was the very best equipment on the market at the time for unloading a press and loading the next press in a line without going to the expense of less flexible transfer-type press lines."

Cost Justification. On the subject of productivity, Lefebvre admits production on the robot-equipped

lines is not exactly eye-popping. The number of parts produced per hour is about 100 fewer than what was turned out when operators were feeding the equipment.

"Bear in mind, however," says Lefebvre, "that a decrease in production of parts per hour does not necessarily mean a decrease in total parts produced over an extended period of time. The Unimates will maintain a more consistent pace than a man, thus accounting for approximately the same eventual output."

The output on the two inner rail lines is determined by a piercing operation at the end of the line, not the speed of the robots. Also a problem, says Lefebvre, is that A. O. Smith and Unimate have been

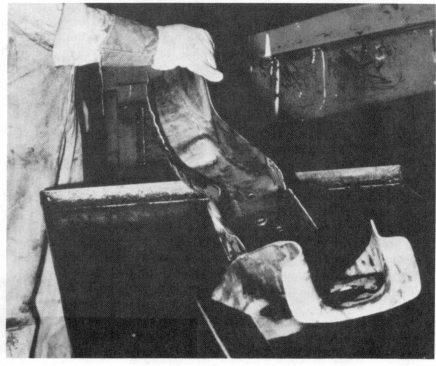

Operator loads "A" frame inner tail blank onto automatic blank feeder which inserts the part into the first of three 800 ton forming presses on the line. To date, A. O. Smith has been unsuccessful in automating this operation

Operator holds inner-rail following its final forming by the last of three presses

unable to get robots to repeatedly and accurately pick up and orient blanks in the first press on the line. The problem is stack orientation which is not totally attributable to automation.

The size of the parts being handled influences productivity, too. The weights and sizes of parts have a direct correlation to speed. The larger and heavier the part, the slower the operation. Obviously,

177

SEQUENCE OF OPERATION WITH TWO UNIMATES IN SERIES

#1 UNIMATE

Step # 1 The Unimate begins by looking at the first press to see if the ram is on top center, 300° to 5°, and whether the press made a 360° rotation indicating to the Unimate that a part was formed.

Step # 2 If conditions in Step #1 are met, the Unimate automatically advances into the press and energizes a circuit which deactivates the press clutch preventing any possible running of the press ram.

Step # 3 Once in position, the robot grasps the part and resets the condition met in Step #1.

Step # 4 Unimate raises the part.

Step # 5 Unimate retracts the part and breaks a photocell beam which signals that it has a part in hand and should proceed to the next step.

Step # 6 Rotate boom 180° and rotate part 180°. While these motions are simultaneously occurring, the Unimate will reset the circuit allowing the press to cycle once again. Also, it scans at the second press to see if the ram is at top center, and if the last part was removed from the die area.

  If all of these conditions are met, the Unimate will advance into the second press.

Step # 7 While moving into the second press, the Unimate will deactivate the clutch circuit.

Step # 8 Once in position, it will set the part down.

Step # 9 Unclamp.

Step #10 Retract from press.

Step #11 Rotate 180° back to the first press, reactivate the clutch circuit for the second press, energize the clutch circuit allowing the ram to come down, resetting all conditions met previously in Step #6.

#2 UNIMATE

Step # 1 Looks at the second press to see if the ram is on top center and whether it made a 360° rotation, indicating to the Unimate that a part was formed.

Step # 2 If prior conditions in Step #1 are met, the Unimate will advance into the press. While moving into the press, the Unimate will energize a circuit which will deactivate the press clutch circuit preventing any possible running of the press.

Step # 3 Once in position, it grasps the part by its jaws and resets the conditions met previously in Step #1.

Step # 4 Unimate raises the part.

Step # 5 Retracts and breaks a photocell beam, telling itself that it has a part in hand and indicating to #1 Unimate that a part has been removed.

Step # 6 Rotate boom 180° and reset the circuit allowing the press to cycle once again. Unimate looks at the third press to see if the ram is on top center and whether the last part was removed from the die area.

  If all of these conditions are met, the Unimate will advance into the third press.

Step # 7 While moving into the third press, the Unimate deactivates the clutch circuit.

Step # 8 Once in position, the Unimate will set the part down.

Step # 9 Unclamp.

Step #10 Retracts from press.

Step #11 Rotates back to the second press, reactivates the clutch circuit for the third press, energizes the clutch allowing the ram to come down, resetting all conditions met previously in Step #6.

the frame members handled by A. O. Smith are quite large.

Maintainability. As for added maintenance fees, Lefebvre says he treats the Unimates no differently than he would treat any other piece of NC equipment. "We have a specially-equipped shop to service the robots and the in-house capability to completely overhaul units. This we do about every 2000 hours."

There is also a continuing preventive maintenance procedure which requires an electrician to sign off on 37 tests; a pipefitter to sign off on nine checks; the lubeman to perform 19 inspections; and the machine repairman to initial 51 items on the check list he follows.

"Attention to maintenance has paid off," says Lefebvre. "Total repairs on Unimates during fiscal 1975 only ran about $3000 per unit. This is a modest figure considering each Unimate had better than 20,000 hours of service at that time; about half the projected life of an industrial robot."

Reduced Labor Costs. Any cost justification exercise pertaining to robots must also address manpower reductions. Each Unimate at A. O. Smith, for example, replaced two operators per shift, or 30 operators per day when the press room is operating on a three shift basis. Lefebvre also emphasizes that the only personnel needed to support the robots is one maintenance man on the first shift and one on the second shift, both of whom are free to perform other tasks if robot maintenance is not required.

Over an extended period of time reduced labor costs can more than offset the $35,000 purchase price of a robot, and the approximate $100,000 costs for interlocking all the robots and presses so that everything stops if there are any cycle interruptions anywhere on the line. "When labor savings are figured into the cost justification equation," says Lefebvre, "the return on investment for robots is not too bad."

Finally, there are the safety and environmental issues to be taken into consideration. Robots do eliminate the threat of an operator putting a hand in the die. Equally important is that robots can take operators out of the noisy environments that are receiving so much attention today.

So to answer the question "Can robots feed presses?", Lefebvre says: "Yes. It's my firm belief that in the long run a robot will make a press room safer and more economical." □

Brian D. Wakefield

1. *The five Auto-Place robots in this press line provide numerous benefits in loading and unloading operations.*

Robots Prove Ideal for Press Loading/ Unloading

The use of robots

for loading and unloading

the presses in this

stamping line

solves several key problems

inherent with

the earlier manual methods

The task of loading and unloading presses is a natural for the robot. For the human operator, it's a job that can be dangerous, and it is both tedious and tiring. These conditions are high on the list of reasons why the roster of companies using robots continues to increase at a fast pace.

The application of robots for loading and unloading stamping presses at R.E. Chapin Manufacturing Works, Inc., Batavia, NY, has paid off significantly in terms of reduced labor costs, reduced scrap, elimination of die damage, and reduced floor space requirements. Chapin manufactures an extensive line of compressed air sprayers for the application of weed killers, pesticides, and general spraying solutions.

Major elements in the stamping line at Chapin include three presses and five robots as shown in *Figure* 1. The robots are standard Series 50 models designed and built by Auto-Place, Inc., Troy, MI.

OSHA safety requirements, while necessary for operator protection, can slow down production and increase maintenance time because the machines are more difficult to reach. The use of robots in this installation simplifies the guarding arrangement compared to that required with human operators. The entire area is blocked off with a protective fence, but the individual presses are open and readily accessible to facilitate maintenance.

Loading/Unloading Operations. Sheet metal blanks slated for forming into funnels are delivered to the press line on a chute. The first robot picks up a part off the chute, *Figure* 2, and places it into the first die where it is drawn and formed. The second robot removes the part and places it in the die on the second press for trimming and piercing. The third robot removes the part, turns it over, and places it on a pedestal between the second and third presses. Robot No. 4 removes the part from the pedestal and places it in a cam die in the third press where it is reformed and curled.

2. *Parts are delivered to the press line on a chute (left) where they are picked up by the first robot and placed in the first press.*

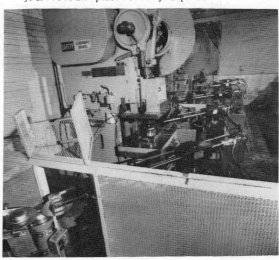

3. *The fourth robot (left) places part in a cam die in the last press. The hand of the fifth robot is shown at right after removing part from the die.*

The last stamping operation presented a critical and time-consuming unloading problem. The curled edges of the part form a vacuum, and the operator had to use a tool to pry the part out of the die. The fifth Auto-Place robot in the line easily overcomes the vacuum, removes the part, and drops it into a chute leading into a bin of finished parts. The loading and unloading operations on the last press by robots No. 4 and No. 5, respectively, are shown in *Figure* 3.

Robot Features. Auto-Place Series 50 industrial robots are pneumatically-powered, limited sequence handling and placement mechanisms for parts weighing up to 30 lb (13.5 kg). A vertical lifting force of 318 lb (143 kg) at 80 psi (552 kPa) proves advantageous for operations such as that handled by robot No. 5 at Chapin. Various types of hands, wrists, arms, and bodies are offered to cover a wide range of application requirements. Modular design allows for parts interchangeability and easy reprogramming.

The Series 50 is programmed with Auto-Place's moving part air logic control system, *Figure* 4. Interfacing with the presses in this application is accomplished with solenoid-operated air valves and electrical pressure switches.

The control system dictates sequential robot motions that are activated by small spool valves arranged in a circuit on a sequencing module. When one motion is completed, the next is initiated. Motions can be individually controlled in speed, independent of time. Since the control proceeds sequentially through a given series of motions to complete a full cycle, the series of actions can be interrupted at any point, or segmented by sensors, limit switches, or other external inputs.

Benefits. Labor savings represent the principal benefit of this robot installation. Although the robots work at a slower rate than their human counterparts, their total output exceeds that of the human press operators. Three incentive piece workers were moved to safer, more interesting jobs, and only one material handler is now required to run the entire line.

Another advantage enjoyed by Chapin is that the robot system eliminates die damage and possible press damage due to the more consistent performance of the robots. Human hand-loading sometimes resulted in uneven placement of parts, resulting in scrapped parts or damaged dies. Since the robots have been installed, Chapin reports that not one piece of tooling has been damaged because of misplaced parts. The scrap rate has also been reduced drastically.

The robot system also makes production planning easier and more reliable. Chapin management can plan each week's production accurately without concern about unexpected employee absences. Once the robots are set up and programmed, they run continuously without distraction. If the part configuration or even the entire production process should change, the robots can be reprogrammed and retooled to perform the next job.

Of the various approaches considered by Chapin, including a progressive die setup, this Auto-Place system proved the most mechanically satisfac-

4. *Auto-Place programmable control incorporates moving part air logic control system. Programming is accomplished by interconnecting small pieces of plastic tubing on terminals on the programming section.*

tory and cost-effective method. The robot system paid for itself within the first year and a half, and Chapin is planning to extend the line.

The system has been well accepted by employees. Chapin informed its people of the problems the company was trying to solve, and there was no negative reaction to the robots. ■

Presented at the Robot V Conference, October 1980

Low Technology Robot Press Loading

By Thomas O. Blunt
General Electric Company

I am the Process Engineering Manager for G.E.'s Range Business which consists of three plants; Louisville, Columbia, Maryland, and Chicago.

Today, I'd like to describe an operation we are automating in our Louisville Range Plant. But first, I want to cover the rationale behind the choice of this operation and the selection of technology.

The Range Business is relatively stable as far as product design is concerned. Our product mix in Louisville has evolved over a number of years and currently consists of a broad range of models and families of ranges. In order to produce these ranges, we cycle over 1,000 sets of dies through over 100 presses in our fabrication shop.

As technology and product mix has evolved, we have acquired a large number of different processes in our fabrication area for the production of sheet metal parts. Since the Park has been in existence for approximately 27 years now, that's a lot of evolution. We have over 300 progressive dies which are utilized mainly for small, high speed parts or parts that have multiple usages over a product line. However, the bulk of our operations in the fabrication area are hand operations. On a simple arithmatic basis, we're faced with an average 7.5 sets of dies per press. With that many die changes required over a given period of time our average press utilization for actual output is less than one half the available hours.

Obviously, our long range goal should be and is to optimize press utilization through redesign of the product, to achieve commonization and combination of parts. As products evolve, we have the opportunity to change some of the tooling concepts and practices which have developed over the years and that's a program to which we are firmly committed. However, we recognize that we will have a large number of hand die operations for some time to come. So, we have developed a system which allows us to realize productivity gains on the high turnover, short run kinds of parts that constitute the obstacle to our success, obviously, is the very low potential savings because of low average press utilization rate and the single shift.

Our first problem was choosing a target operation. From a systems concept development standpoint, we felt we would be better off solving most of the problems up front. So instead of looking for a simple application, we looked for a hard one. It's called the gusset blanking operation. Gussets are reinforcements used in the oven door. They are hand fed flat blanks, not drawn parts, which puts the maximum demand on locating accuracy because there is no shape to the part to help locate it in the die. And, in this particular operation, both ends of the operation are variable. The die sets have three pins on two prime sides that the operator slides to blank against for location in the press and, on the other end, the blanks are fed from a gondola. The material used in the gussets is scrap from the operation that pierces the front hole in oven bodies. With our plant being as old as it is, we have a fairly active press replacement program. It happened that the press on the gusset operation was due for rotation. We had ordered a new press for delivery and installation in 1980. It became obvious that having to shut down, tear out and disrupt the operation to put in a new press would be an ideal time to put in a new piece of technology. A brief description of the gusset operation: it utilizes two sets of dies. Through these dies we process six different kinds of scrap blanks. The press runs an average of eight hours per day and the pay rate on the operation is 310 pieces per hour. The gusset blanking operator used to fish in a tub, build a small stack of blanks in front of his work station, hand feed the parts to the press, trip the press, reach in and grab the scrap and drag the scrap ring and the two parts out of the press, knock the left and right parts loose into two separate gondolas, and then dispose of the scrap. When the decision was made to buy a replacement press, one of the options exercised was to buy a part extractor for the press, even if the blanks were hand loaded, the parts would be automatically extracted with left and right pieces separated out the back of the press into gondolas. The task of automating this operation was made a little bit easier.

Choice of technology on this job becomes critical in that the accuracy requirements, large number of die changeovers on each press, and the resultant probable number of positions for part location, almost demand a high technology programmable device. However, almost no press in the shop, as described earlier, even approaches a full man shift worth of savings. We decided to examine the three gross categories of devices available to us: high technology robots; low or lower technology robots; and the iron hand class of loading devices. High technology machines provide good accuracy and their programmability is excellent and they can handle a larger number of parts locations. However, they are large and we are limited for floor space, and their cost is high. Low technology machines with fixed stops offer outstanding positional accuracy. Their programmability is limited and very slow, but they are physically smaller and considerable lower in cost. The iron hand class of press loader offers very good positional accuracy, virtually no programmability without major set-up, their size is small, and they are the lowest cost. The machine we eventually decided to use is the PRAB 4200 and, in this particular operation, as we will describe it in a minute, we feel that we have achieved the maximum price leverage on the system by combining a less complex machine at a lower price, with smarter tools, to cope with some of the variables in the process and overcome a basic lack of quick programmability. What's becoming clear is the basic philosophy that we are in fact automating the press, not a collection of particular parts. If we run seven or eight parts through a press and six of them fit the automation and one of them doesn't, we'll rebalance that part into some other press. Eventually you'll see a number of automated systems and probably an equal or greater number of hand operations to take care of the parts that just are too difficult or not profitable. One of the basic requirements of this project, from a cost standpoint, is that an absolute minimum of changes be made in the dies themselves. Most of these are die sets that have been in existance for a number of years, and we have a program to up-grade dies as we buy replacements, but we just couldn't afford to go in and arbitrarily change out many sets of dies and still realize an adequate savings.

The layout after installation of the robot has Press 1, which is the gusset blanking operation, serviced by a destack device and the robot. A small slider belt conveyor carries the scrap away from the operation to a gondola mounted outside the work area. The fork truck doesn't have to get too close to the automation in order to remove the scrap. Left and right parts are ejected from the back of the press and go into separate gondolas for eventual transportation to Press 2 which is the secondary operation. It would have been possible to convey the parts from Press 1 to Press 2 on a conveyor but since Press 1 operates at a significantly faster rate than Press 2, we felt it was advisable to decouple the two operations and provide an accumulating float. The classic accumulating float of course being the gondola. The operator of the second press is charged with the responsibility of building a stack of material for eventual use in the Press #1 destack device. Another reason for decoupling the two operations is, with a conveyor between the presses, the second press operator would automatically shut the first press down when he serviced the first press. The system is designed so the first press can continue to run while it is being serviced with new blanks. I know I have spent 10 minutes telling you why and no time telling you how. I feel that the justification of a system, whether it's robotic or hard automation, is probably more important than the technology involved in doing it and I wanted you to understand why some of the things were done the way they were. Let's talk about the actual mechanics of the system and how it operates.

The robot is a PRAB model 4200. It is a four axis machine; three major axes and one minor wrist axis hydraulically powered. It's programmable through a pin drum memory circuit. The machine has an end of arm capacity of 100 pounds in the heavy duty model and, because of its fixed stop design, can achieve accuracies of .008" in positioning parts. Extreme accuracy and relatively low cost of the machine were the two positive driving elements in the choice of this machine. It's disadvantages are the relative length of time of re-programming which is long, especially if the physical position of the machine has to be changed, because you have to physically move stops in the machine; it's speed, which is approximately 2 feet/second in its major axis; and its inability to function in two axes simultaneously. Like everything else, there's good news and bad news.

In our application the machine is mounted on a wheeled cart with a home position plate sunk in the floor. Whenever die changes are necessary in the press, we loosen the master locating device, push the machine out of the way to give access for the die fork truck, push it back in position, bolt it back down. It loses no accuracy in this mode and it allows us quick access to the press without major tear-up of the system.

Three services are required: 460 volts, water for the cooling system if required, and air for the end of arm tooling. The end of arm tooling in this application is of our own design. You see its attachment method is suction cups, which work fine on flat blanks. The arm is articulated, in effect, applying a fifth programmable axis to the machine. A small air cylinder acts as a positive driving force, stroking the hand in and out. The cylinder does two things - it assists the robot in removing things from the destack device and, in the final movements of the arm into the die set, it acts as a resistance or a cushion supplying compliance to the wrist to help locate the blank against the locating pins, very much like the human hand does.

We have subsequently developed a second generation hand which has compliance in two directions at 90 degrees to each other. The entire hand can be mounted in various configurations to allow the robot to push the hand forward, sideways, or pull it back. Wherever a set of dies has its locating pins, the hand can be mounted to function in that direction. The compliance cylinder on the hand is operated through a valve mounted in the robot's body and it is controlled as separate function in the robot's memory.

The destack device presented some very interesting problems. The robot functions in circular coordinates. Visualize a vertical stack of material. As the robot works from the top to the bottom of the stack, it describes an arc. The machine has no positional feedback system; it goes to the extreme of an axis and hits a hard stop. As the robot unloads the stack from top to bottom, its pickup point relative to the vertical plane of the stack changes as much as .8" in a 16" high stack. The .8" would, under normal circumstances, result in mispositioning the piece by .8" in the die. To overcome this problem, we tipped the stack so that it is perpendicular to the tangent point of the arc of the arm in the up and down movement. Then we hit upon an interesting thought: if the robot's control system is not capable of coping with slight programmable movements which a sophisticated control could, and the problem evolves around a circular or curved coordinate of the arm trying to work through a straight up and down stack, then it appeared that the solution might be to curve the stack. We constructed magnetic fanning devices with a stainless steel sheet in front of the magnet. The sheet is curved from top to bottom to match the same arc as the robot hand as it moves from the top to bottom. The top three or four sheets, as they separate magnetically, were drawn to the two fanning devices, thereby curving the stack in two dimensions. Our accuracy of pickup is about .050" from top to bottom through the stack. The robot starts in a fixed position, high level, and moves down through its stroke. There's a sensing probe on the arm. When it contacts the top of the stack, it sends a signal to the control and says, "Stop here and go to the next point in the program." By this time, the suction cups have contacted the sheet and are firmly attached. As the machine starts back up, another signal is sent to the small cylinder retracking the wrist. This action pulls the sheet away from the two magnetic fanning devices so that the blank will not drag up the face of the magnetic fanner and eventually wear it out. You'll notice there are two stack positions on the destack device. One position is in use. The other is a preload and is out of the work zone of the robot. Being a fixed stop machine, it's physically impossible for the robot to overtravel and get outside of its predetermined work envelope. This preload position allows the operator, at his convenience, to service the area. Parts are brought to the area in a gondola. Eventually, we are going to rearrange the operation at its point of production so that parts come over in more or less reasonable stacks. Right now, however, they are in tubs. The operator pulls 8 or 10 pieces out of the tub, throws them into the pre-stack position. He can build up to a 16" stack of material. When the last piece is taken out of the use position a "No Part" sensor will stop the robot and turn on a light, signaling an operator that the use stack is empty. At this point, he swings the stop out of the way, and the stand-by stack rolls into position, and self locates. The operator then hits a "Resume" position on the control. The robot will pick up in cycle and continue to operate. The robot, by the way, operates the press much as an operator would. It initiates the palm button circuit with a programmed signal from its control. This allows us to still maintain dual palm button capability. In case of a catastrophic failure of the loading device, we can run the press with an operator.

If all we had to deal with was one blank and one set of dies, my talk would be over. However, we have multiple sets of dies and multiple sets of blanks in this operation. On this particular press we were able to force all the dies to a common center line. The dies are mounted in the press on a 45 degree angle. This was done for the automatic part extractor on the back of the press because we have a right and a left hand part blanked simultaneously. We want those two parts fed into separate containers. By turning the die on a 45 degree angle and putting a center web or a separator in the middle of the press, we naturally separate the two parts, so we put all the dies in this operation down the same center line. That solves the problem of where to put the blank - we have a common point in the press side to side, no matter what die set and no matter what the blank. We use the diagonal center line of the blank. But we have to change the position of our pick up point to compensate for changes in fore/aft position. In order to do that we came up with our second invention which is the ability to program the destack table. As we mentioned, one of the disadvantages of the PRAB is the necessity of changing fixed stops if you want to physically change the position of the arm. The robot in this operation always stops at the same place in the destack position. Its stop point in the press is always the same place. To program the destacker, we move the blanks relative to the hand, by pulling two master pins in the floor and physically sliding the destack device to a new location which is keyed by number to a new set of dies. In fact, we keep the hand in the same place, and move the stack underneath it. This allows us to change dies fairly rapidly without ever touching the robot. Minor location anomalies are taken up by the hand's ability to compensation for misposition through its basic compliance. Our original plan was to have the robot load a blank to the press, retract its arm to an intermediate position and signal the press to stroke. When the press opened the knock-out ram and the automatic unloader would remove the two parts, the robot would go back into the press, snag the scrap ring with its suction cup on the common center web, remove the scrap from the press, drop it onto a conveyor for removal from the area, and continue the operation with a new blank. It took too much time. We were able to pick up 150 pieces/hour of output by simply adding a programmable air cylinder to the back of the die. Now after the press stroke the ram travels up, the parts are automatically extracted, the air cylinder fires and literally flicks the scrap out of the press onto a removal conveyor. This is an instance where the addition of a couple of hundred dollars of air cylinder resulted in a dramatic increase in output of the operation.

This system will cost approximately one half of the cost of a high technology robot system. We don't have absolute flexibility and, in fact, will have to pick and choose our opportunities very carefully, but then we do anyway, don't we?

A Prab robot and multifunctional support equipment automate low-volume blanking of gussets used in electric ranges.

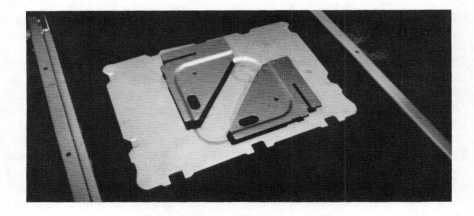

Right and left-hand reinforcing gussets are made from scrap blanks coming from a prior operation. Six different blank configurations are run through the automated system.

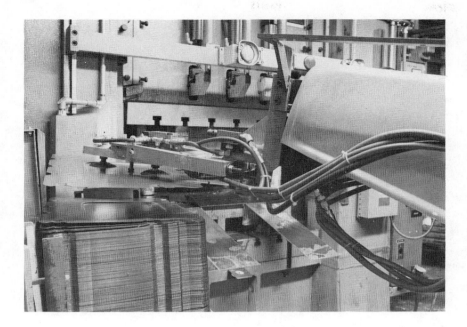

Stacking the blanks in a curve to match the arc travelled by the robot hand assures that each blank is picked up at the correct point for accurate positioning in the die.

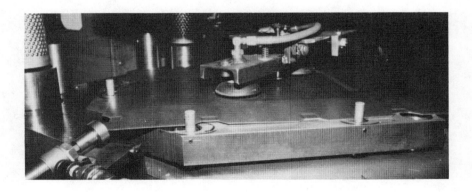

Robot hand with two suction cups places blank in die which is oriented at a 45° angle in the press. The small hydraulic cylinder at left is programmed to eject the scrap ring automatically.

CHAPTER 6

FORGING AND HEAT TREATING

Commentary

Forging and heat treating are tasks where robots can relieve people from especially unpleasant working conditions. Heat, noise, dirt, smoke, heavy loads, fast pace and monotonous routines may all be present in these operations.

Robot applications in forging include die forging, upset forging and roll forging. The robots may load furnaces, forging presses, headers, and trim presses and lubricate dies.

In heat treating operations, robots may load and unload furnaces and quench parts.

Special attention is necessary to protect the robot from heat and shock loads and cooling of the robot's hand tooling may be required.

Robots applied to these questions offer the user significant benefits not only in labor cost reduction, but also in removing people from undesirable environments.

Reprinted from Metal Progress,
July 1980

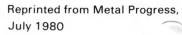

How Industrial Robots Are Automating Forging

By John E. Mattox

WHETHER AND HOW industrial robots might be used in forging shop operations are questions that frequently occur to operating management when considering possibilities for automation to relieve worker turn-over problems, meet safety requirements, and increase productivity.

Such applications have been expanding in variety and number for several years. Enough experience has been accumulated to reach conclusions and make recommendations.

For the benefit of managers evaluating methods, including industrial robots, we suggest they should consider:

1. How industrial robots work and what they can do.

2. What procedures have been developed for evaluating justification.

3. What results have been in forging shop applications.

4. What the general limitations are.

It is the programmable, general purpose industrial robot on which we will concentrate. They have been produced in large enough quantities to have established a measurable reliability track record. An established, over-all track record allows a prediction of probable downtime for unscheduled repairs. We consider an uptime of 98% to be essential for industrial robots.

In contrast with hard automation that is custom built and rarely economical for small or medium volume production runs, general-purpose industrial robots have been shown to be economical for this kind of production in a number of cases. Some suggestions for identifying those cases within the forging industry will be given. But first let's describe the nature of a general purpose industrial robot, using the Unimate robot as an example.

How Industrial Robots Work

Actions of the hydraulically powered arm of an industrial robot are directed and controlled by an electronic memory system. Timing of the actions is synchronized with external equipment — forging press, conveyor, and other equipment — by means of electrical connections. An external limit switch will signal the industrial robot when a press is open and that it is permissible for the arm to move its gripper into the die area. Relay contacts are internal to the robot and a set can start the cycle of equipment such as a press when the robot's gripper has been withdrawn from the die area. In this way, an industrial robot will automate a previously manual operation.

The end of the industrial robot's arm is equipped with a pneumatically operated gripper, or will accept a tool. In forging applications, the gripper (hand) will be most common, and it can be a single gripper or multiple grippers that are independently actuated.

Three articulations are provided for an arm: move up or down, move left or right, and move in or out. The articulations are also referred to as "axes" or "degrees of freedom."

Three additional degrees of freedom can be provided at the robot's wrist to orient the gripper at the end of the arm: twist (roll), bend (pitch), and yaw. Gripper closure is not counted as a degree of freedom. For some furnace-loading, press-loading, and other jobs, three degrees of freedom are sufficient, but most jobs require five. With five, the gripper can be oriented to pick up or deposit a part at almost any point within the work area. The sixth degree of freedom is needed in a few specialized applications.

Typical maximum load capacities, at the robot's wrist, are 300 to 450 lb (135 to 205 kg) when the arm is fully extended. These are the weights of gripper plus work piece gripped.

A job is taught to an industrial robot by switching its controller to a "record" mode, using a "teach" control to move the arm and hand through the desired sequence, and recording the program in the robot's memory. Included at strategic places in the program are commands to issue control signals (for instance, to initiate a press cycle as soon as the robot's hand has withdrawn from the die area) or to wait after a particular step until a signal is received from external equipment (for instance, when a hot billet is in position at the billet pickup point).

Teaching a sequence of steps — a program — to the robot can proceed as quickly or slowly as the setup man finds convenient. The actual program playback will cause the robot to perform the taught sequence smoothly and at full speed, with the teacher's pauses and waste motions eliminated. Programs that will be re-used can be saved by recording them on cassette tape by an accessory Program Storage and Verification Unit, then "playing" them into the robot's memory, as required.

Loss of ac power does not cause amnesia. A program in the robot's memory is retained until a new program is substituted. As few as 100 steps of memory capacity suffice for most jobs, but the memory capacity of the Unimate industrial robot can be increased in modular increments to 1024 steps.

Repeatable positioning accuracy is 0.050 in. (1.3 mm) for standard size industrial robots and 0.080 in. (2.0 mm) for the larger industrial robot. The temperature environment can be within the range of 40 F (5 C) through 120 F (50 C). Grippers that handle hot billets or slugs may be periodically immersed in a tub of water or have continuous flow, water cooling channels built into them.

Workplace Considerations For Industrial Robots

More insight into how industrial robots work is provided by taking a look at workplace considerations.

In a new facility, equipment can be located to permit immediate or future use of industrial robots. In an existing forge shop, equipment teamed with the industrial robot may have to be repositioned. Conveyors will often remedy location problems. For instance, hot billets from a furnace too distant from where a press and robot are located are usually transferred by a conveyor to a pickup point where the robot can reach the billet for loading it into the press.

The robot then places the formed or preformed shape on another conveyor for transfer to the next operation.

Existing conveyors will be acceptable for presenting a workpiece at the robot's pickup point only when they assure consistent workpiece orientation at the pickup point. Mechanical aids such as gates or guide rails can often be devised to permit use of existing conveyors.

A more serious difficulty may be workpiece irregularities which interfere with consistent orientation or with the robot's ability to grip the workpiece. In these cases, prior operations must be controlled more closely than they were when the robot's job was done by a worker. For instance, irregularities in the shape of the sheared ends of a billet may hinder the robot in gripping each billet correctly. This condition would require a change at the shearing operation.

As a general purpose machine, any industrial robot must be flexible enough to handle a variety of tasks. Expandability of the robot's memory permits capacity to be added in the field if a job requiring a much larger number of steps is the next assignment. Grippers ("hand tooling") are readily changed. Models with three degrees of freedom can be retrofitted to provide a total of four, five, or six degrees, as increased dexterity becomes necessary for later jobs.

An "alternate program selection" option should be specified when the robot is ordered if there is a known or anticipated requirement for the robot to switch between multiple programs occupying the memory at the same time. Selection of a particular program is initiated by a signal from an external source.

For instance, if a robot is to load two adjacent presses alternately, this option permits the robot to switch automatically to operating one press only if the other press should go out of service.

Justifying the Investment in Industrial Robots

A starting point is to identify likely jobs with the aid of a tabulation of jobs. Each job is then numerically rated. Engineers at one large company rank each forge-shop job on the basis of the following factors:

1. Part weight and shape and changes in shape occuring during the operation.

2. Floor space available at the job site, location of existing equipment, and space for possible rearrangements such as furnace-conveyor or trimpress positioning to put equipment within the reach of the robot.

3. Number of degrees of freedom that the robot must have — which influences cost.

4. Magnitude of changes or improvements required for existing equipment.

5. Possible labor savings and production gains.

The resulting rough ratings can be refined by examining the most likely jobs in the light of:

1. What functions the robot should perform and what capabilities it will need to perform them.

2. How the part can be gripped and what changes may be required so that a part can be gripped.

3. Interlocking to sychronize equipment, to protect against malfunctions, and to insure safety.

As for economic justification, information gathered from installations of about 2000 robots and more than 8 million operating hours has resulted in two rules of thumb:

First, a complete installation with accessories for two shift operations, five days per week commonly provides total payback on the investment within two years.

Second, the rate of return on investment is seldom less than 25% on two-shift operations.

One of the key questions is: Does the robot provide a gain in productivity over the human worker it replaced? In a forge shop, an industrial robot may be faster than a worker but provide no real gain because the forge press cycle time or furnace cycle time limits production rate.

Another key consideration is workpiece weight. If a part is significantly heavier than the average, the initial investment for a robot will be higher than average because heavy duty, load handling capability is required.

Results in Forging Shops: Five Case Histories

In a pancaking and high energy rate forging (HERF) press operation, a hot billet is picked up from an induction heater by an industrial robot and placed in a HERF press for forming into a round, pancake-shaped preform. The robot's gripper has dual cavity fingers: one cavity to hold the billet, the second cavity for the larger diameter preform.

After cycling the first HERF press, the robot transfers the preform to a second HERF press where a knob is extruded and then cycles that press. Finally, the robot removes the part from the second press, transfers it to a tote bin, and repeats the entire sequence. Weight of the part is approximately 56 lb (25 kg). The robot replaces two workers.

Application No. 2 — In the trimming of impact-forged parts, a robot works with a forging press and trim press. The robot transfers the finished forge platter from the forge to the trim press, cycles the press, reaches into the press for the scrap, and discards the scrap on a conveyor. The part drops through the die. Because the hot part is flexible after removal from the impactor, the gripper is designed to keep it from bending during transfer to the trim press.

For the protection of expensive dies, heat sensors signal the robot to discard the billet for reheating, omitting the usual intervening steps, if temperature is out of limits.

This automated operation is more reliable and part quality is better than when it was performed manually. Production rate is 400 parts per hour.

Application No. 3 — In a busting and bending operation, a chute delivers a heated billet to a pickup point from which the robot transfers the part to a press. The robot continues to grip the billet to insure correct positioning during the busting operations. Next, the robot shifts the part to a bending press. During the press upstroke, a pusher moves the part on to the next operation. After cooling its fingers in a tank of water, the robot repeats the cycle.

Application No. 4 — In a busting, drawing, and piercing operation, 80 pound (36 kg) billets are forged into cylinders. After being busted to the proper size for a combination deep draw and punch operation, the billet is put into a drawing die by the industrial robot. The forge press operator drops a collar onto the billet and initiates the press cycle, during which the part is lengthened and a hole is formed. The robot then turns the part over and places it on a piercing die where the end plug is punched out. The finished part is removed by the forge operator while the robot turns toward the next busted billet to repeat the cycle. Billet temperature is 2200 F (1205 C).

Application No. 5 — This is a three pass upsetting

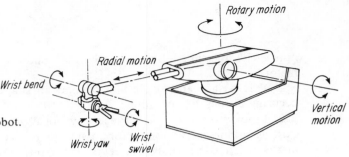

Six degrees of freedom of a Unimate industrial robot.

operation for forging auto tie rods with the aid of an industrial robot. The robot grasps the bar located at the pickup point outside an induction heater, transfers it to the first horizontal die, and signals the upsetter to cycle. At the completion of this first pass, the robot turns the bar 90° and positions it in the second horizontal die. After the second pass, the robot places the bar into a vertical die, signals the upsetter to cycle, and finally places the formed part in a floor tote.

The robot's gripper is offset to permit it to maneuver around the pickup point end stop and the upsetter back stops. A wrist-swivel action by the robot allows the part to be twisted out of the upsetter dies.

A slide mechanism feeds bars to the induction heater. Heated bars emerge and are accurately positioned for pickup by an end stop. An infrared detector focused on the bar at the pickup point provides a go/no-go signal that establishes whether the bar has been heated properly and may be formed. The robot is programmed to reject improperly heated bars. Production rate is 140 parts per hour.

Limitations in Automation of Forging Operations

Not every forging installation can or should be automated. In three general cases, it may not yet be feasible:

1. When a forging hammer is to be automated, but it is difficult to prevent the part from adhering to the upper half of the dies.

2. When an existing forging installation could be automated but cost of relocating equipment, to put it within the reach of the robot, is greater than the benefits expected.

3. When weight of the billet is beyond the load capacity of persent industrial robots.

Until recently, weight was the chief limitation. Today maximum weights that can be lifted at full extension of the robot's arm are in the range of 300 to 450 lb (135 to 205 kg). The weight limit usually includes the weight of the robot's hand tooling.

Remaining difficulties are commonly confined to arranging the marriage between modern automation technology and the traditional technology of forging. Process irregularities are prevalent in forging, making it natural to rely on a worker. He can compensate for part temperature variations, for die design shortcuts, for poor die condition, and for variations in billet

shape after cutting or shearing, for instance.

Successful automation of forging requires better control of irregularities. Tightening controls on conditions in the forging operation and on previous operations becomes part of automation. In other words, the whole system must be upgraded. Key considerations include the following:

1. Revising the equipment layout, if the installation to be automated is an existing one.

2. Removing the heated part from the furnace and orienting it for pickup by the industrial robot.

3. Checking the temperature of each part.

4. Modifying the press control so it can "talk" with the robot.

5. Giving special attention to the die.

Approaches for dealing with these elements are relatively routine but the die warrants more comment. Cost saving shortcuts in die design can be compensated for by a man but not by an industrial robot. Dies for automated forging are likely to be more expensive because they require more design time.

If the robot needs assistance in locating the billet in the nest, die modifications can provide it. For instance, a chamfered lead-in or a pneumatically operated kick-in pin will give an assist.

Die care is also essential. Difficulties are frequently traced to unsatisfactory die lubrication or to worn dies. A worker can get by with a worn die for a time and still produce acceptable parts. The industrial robot is not so forgiving.

Production rate in a forging operation is seldom a limitation when choosing to automate with an industrial robot. A robot is generally as fast as or faster than a man. However, if the part is small, a man may be faster.

A short term rate of one part every 75 s is common in a manual operation, with the furnace or billet heater pacing the man. Fatigue and distractions cause longer term production rate to be less, but introduction of automation remedies this problem, leading to more accurate time estimates, better production planning, and greater over-all production rates.

These are some of the reasons why the automation of operations such as roll forging, upset forging, press forging, controlled hammer forging, and coining are proving to be well worth the effort, and the investment in industrial robots. ✪

For More Information: Mr. Mattox is applications engineering manager for Unimation Inc., a subsidiary of Condec Corp., Shelter Rock Lane, Danbury, Conn. 06810. Telephone: 203/744-1800.

Presented at the Robot II Conference, October/November 1977

Robot Application In Aluminum Forging
By Craig A. Hoskins
Aluminum Company of America

Loading, unloading, and lubricating aluminum closed die forge presses can be a dirty, fatiguing, monotonous job. High die and stock temperatures, flaming lubricant, stock weights over 40 pounds, and high noise levels all contribute to worker fatigue, slow production, and increasing scrap rates. When volumes are high enough to absorb setup costs, hydraulic powered medium technology robots can be cost effective. Their capability is well matched to the loading or unloading task, their reliability is high, and payoff is more than acceptable. They can be integrated with the press, conveyors, and die lube systems to provide a complete automatic forging complex.

HISTORY

Today the Alcoa Cleveland Forge Plant uses a mixture of old and new technologies to produce a wide variety of aluminum forgings. The work is of two basic types, aerospace and commercial, with commercial work the growth segment today. Cost reduction programs are a continuing effort to hold prices while maintaining margins. Of the many potential cost savings associated with particular product, material handling improvements are generally the most lucrative. Material handling improvements fall into two basic categories; hardware which assists the worker and makes him more effective, and hardware which replaces him. Within the framework of replacing the man, both hard automation and programmable automation must be considered. Hard automation has been attempted without success, mostly because of the jobbing shop nature of our business. Many attempts have been made over the last ten years to install programmable robots into the production system. Only recently have we been successful at operating robots in production on a month-after-month basis. Failures in the past were mostly attributed to 1) short production runs, 2) improper synchronization of automatic loading and manual die lubrication, 3) assigning the project and responsibility to someone outside the normal production ranks, 4) strong hourly worker resistance, and 5) insufficient technical skill for maintaining the robots. The robots were of a programmable hydraulic servo system configuration. These experiences created a resistance which had to be overcome before significant automation could be installed.

FIRST APPLICATION

In the past several years the following ingredients to make a successful robot installation became available.

1) Single product production runs of sufficient size to justify setup and automation costs.

2) Increasing business levels.

3) A reduction in the hourly worker's apprehension of automation.

4) The addition of a variety of sophisticated equipment through the plant with the associated maintenance skill.

5) The development of automatic die lubrication to supplement any automated handling.

6) The desire to improve working conditions.

7) Ever increasing wage rates and fringe benefits.

8) Need to increase production rates and avoid major capital expenditures by better utilization of existing equipment.

Our first attempt involved automatic loading and unloading a mechanical forming press with a transfer device we designed and built. This unit was assisted by an automatic device which provided the necessary die lubrication. Part weights ranged from 35 to 45 pounds and temperatures from 600 to 800 degrees F. The loading device was air powered and operated on a three-shift basis. The transfer device eventually was removed from production, but only after a considerable effort by many individuals to solve a variety of technical problems. A process revision contributed also to the decision to retire the handling device, but we learned some valuable lessons.

1) After successfully automating an operation, equipment reliability is the most important consideration.

2) Anticipation of real shop operating conditions is necessary during design.

3) Systems must be selected and designed within the abilities of general electricians and mechanics to trouble-shoot and maintain them.

4) Purchase of proven equipment with a good track record should be considered prior to designing a one-of-a-kind.

This transfer device did process a large volume of product and proved to many in the plant that automation had many advantages and that our efforts should be extended.

Our second effort consisted of installing two Prab robots and two Transcon conveyors. One robot transferred the part from the conveyor to the die and the other transferred the formed part from the die to a second conveyor on the back side of the press. The die design provides the necessary part locating and the press and lubrication systems operate automatically. The grippers were provided by the manufacturer and required only minor modification. After initial startup and shakedown, maintenance problems were few and easily corrected. Since September 1976 maintenance problems have been manageable. The most recent problems involve fatigue life of robot electrical components. The solution seems to be periodic replacement of some components prior to anticipated failure.

As one might expect, manual peak production rates exceeded the best robot rates. However, the bottom line is the eight-hour-shift or three-shift production. With these criteria, robot installation generally exceeds manual operation. These two robots replaced two men in the press crew. The number of human problems which interfered and reduced production were reduced. Greater production resulted as well as fewer personnel problems for the foreman. These two jobs were the worst jobs in the entire crew, their elimination helped generate hourly worker support. Prior to installation, representatives of the press crew were taken to the robot manufacturer to familiarize them with the hardware and to answer any questions they had. We feel their support for a project of this nature is as important to its success as any technical aspect.

The robot press operation was the second forging operation performed in line. The first was a 3000-ton press performing the forging operation with manual loading, unloading, lubricating, and press operating. This crew, facing a challenge, ran at high enough rates to back up the robots at the second press. We soon learned that many improvements could be made to the robot programs to improve their cycle time. After a number of program revisions, a 30% improvement in cycle time was achieved over the original installation.

SECOND APPLICATION

The second major robot installation was at the 3000-ton forge press in this same line. The plan was to load, unload, and lubricate with one robot. A special gripper and lube gun were designed in house. Several modifications were made after testing to insure reliable gripping of the part and proper die lubrication. We also added provision for gripper cooling because of the high temperatures encountered in the operation. After a number of dry running hours, the installation was made smoothly and took only two days. A large amount of preliminary work had been completed weeks before during a planned shutdown. Problems were dealt with as they occurred over the first weeks and months. The application turned out to be far more difficult than the first installation we attempted. The combined gripper and part weights were at the manufacturer's recommended maximum. Also some motions were run at their maximum. These factors together with the high temperature conditions worked to press the equipment to, and sometimes past, its limit. The weak areas which the robot manufacturer felt could compromise other applications have been dealt with and those which were related only to our application have been corrected.

By far the greatest problem is the integration of an automatic robot with a manually operated press. Even with all our efforts, the press was still capable of damaging the gripper. The robot was interlocked with the press to power the arm out if the press came off its top limit. Unfortunately the press cannot be easily inhibited until the robot is in a safe location. Revisions are planned to the press control system for an automatic cycle where the press will also be interlocked with the robot to prevent lowering the ram when it is in the die space. An automatic cycle will provide the press operator more time for watching and adjusting the equipment. This should result in higher productivity and forging quality.

One occasion when the gripper was forged and totally destroyed, the robot was removed from production. The press returned to manual operation until a new gripper could be installed. Production had become accustomed to operating with two less men in the crew. It was difficult to find six men over three shifts to reman the press until the robot was ready. The result was a great deal of pressure on engineering and maintenance to return the robot to production. Needless to say, we maintain a spare robot and gripper now to minimize any downtime.

Beyond the cost savings of crew reduction of six men over three shifts, we noted a substantial improvement in product quality. It seems the lubrication pattern the robot generates is more uniform and consistent from cycle to cycle. The resulting reduction in scrap rate is sufficient to justify the robot installation.

This robot installation is successful because of the extra effort by many people to solve the technical problems during a time of high production requirements and a commitment to the project's success by all of management from the Product Manager down.

FUTURE

We have active projects to apply medium technology robots like the Prab robot where their capability meets the task requirements. Also where the tasks are more difficult, we plan to apply more sophisticated types programmable in a number of axes. In our applications so far where a hot, approximately 40 pound part, requires transfer from point "A" to point "B" the Prab robot has proven adequate. In addition, its simplicity is advantageous from a maintenance standpoint. Problems are easily diagnosed and corrected. Abuse and external damage are easy to repair.

Work tasks which may require more complex motions such as tracking a conveyor, palletizing, program branching, or just a large number of movements in one cycle require multi-axis, servo controlled robots. With the addition of sophisticated equipment to the plant in recent years, we feel our ability to maintain these more complex robots is improved to an acceptable level.

OUR CHALLENGE

With the addition of more complex automated equipment to the plant, an accompanying change in procedures, attitudes, and basic philosophy needs to be considered. The equipment cannot be left to run until failure. A preventive maintenance program should be established with specific items planned for daily, weekly, monthly, or yearly maintenance. Problems need to be dealt with during planned shutdowns. Sufficient spare parts and/or complete robots need to be stocked. Personnel must be specifically trained to operate and maintain the equipment. Setup, operating, maintenance, and engineering responsibilities need to be clearly defined. A continuing program of design improvements needs to be established to gain higher production rates and reliability. Finally, management should consider retraining displaced workers to maintain the new equipment. Maintenance expense will always rise with the addition of automated equipment, but it should be more than offset by increased production and fewer production workers.

MANUFACTURER'S CHALLENGE

There is no question we are proceeding down the road to flexible, programmable automation. The only issue is the rate at which to proceed. Material shortages, lack of skilled labor, employee health and safety, and the threat of foreign competition are encouragements to productivity improvements through industrial robots.

The manufacturer's challenge is four-fold 1) to improve the capability and reliability of robots as field experience grows, 2) to upgrade customer training programs, 3) to provide complete operating and maintenance information to customer, and 4) to assure that all applications are successful.

Presented at the Robot III Conference, November 1978

Robots In Hot Forging Operations

By Ronald G. Powell
Norris Industries

Robots indefatigably perform human like tasks such as grasping, carrying and manipulating either hot or cold objects with precision and consistency. They can lift and manipulate objects in ways and in environments that human operators would be unable to endure. Robots can be "taught" new tasks, in considerably less time that it takes to re-train a human operator.

There are many things to consider in selecting and installing the robot. Design criteria for tooling the robot, layout and interface with existing work stations are discussed, and line drawings submitted.

WHY A ROBOT WAS CONSIDERED FOR A HOT FORGING OPERATION

There was a need to increase production and reduce operating costs on a particular product being produced in the hot forging shop. The forging was being made on a 2500 Ton Clearing Mechanical Press, employing two men with a production rate of 150 parts per hour. Industrial Engineering conducted time studies and evaluated modified tooling systems with a view toward increasing production and reducing operating costs. All of the proposals eliminated one operator, but to meet the minimum new production rate of 185 parts per hour, some form of automated material handling was required.

Several designs of "hard automation" were studied and rejected for a variety of reasons, some of these being cost, size, flexibility, and also a major tear down was required to facilitate a die change. The need for a form of programable automated material handling was apparent and the decision to purchase a robot was made.

SELECTING THE ROBOT

Having determined the need for a robot, careful consideration was given in selecting the robot. These considerations are listed -- not necessarily in order of preference:

a) <u>Cost</u>: To include purchase, site preparation, installation, special tooling, training and start-up.

b) <u>Load Capacity</u>: The load that the robot was required to manipulate in all axes of freedom.

c) <u>Axes of Freedom</u>: Those motions necessary to pick-up the hot forging transfer between dies and unload on exit conveyor.

d) <u>Power System</u>: Actuating the various degrees of freedom of the robot, i.e., pneumatic, electric, hydraulic.

e) <u>Logic</u>: Control System for the robot.

f) <u>Memory</u>: Capacity to store information.

g) <u>Programming</u>: How, and how much time to program and reprogram the robot.

h) <u>Flexibility</u>: The ability of the robot to be relocated, and reprogrammed on different equipment.

i) Any special maintenance and/or programming skills required.

j) <u>Environment</u>: The ability of the robot to perform in dust and dirt laden atmospheres, such as those found in our hot forging shop.

k) <u>Physical Size and Weight</u>: The space which was required to install the robot, and did the weight of the robot require a specially prepared foundation.

l) <u>Performance</u>: Robots "track record" and the availability of witnessing a demonstration of the robot in a comparable environment to one in which we contemplated using it.

m) <u>Cycle Rate</u>: When programmed, the total cycle time of the robot to complete our sequence of operations.

n) Availability of service and spares.

<u>INSTALLING THE ROBOT</u>

With the above considerations satisfied and a purchase order placed with the vendor, preplanning for the installation began.

Successful installation must include defining the areas of responsibility and communication between all parties concerned. In order for the vendor to design the "hands", a plan and elevation of the press, die stack, stripper, knockout, incoming and exit conveyor was drawn up and submitted (see sketch). In addition, a graphic program was prepared and submitted (see sketches) to the vendor.

Meetings were held with the Electrical Engineer and the Mechanical and Electrical Maintenance Foreman (in attendance). Among the items discussed at these meetings were:

a) Who will attend the vendor's training school?

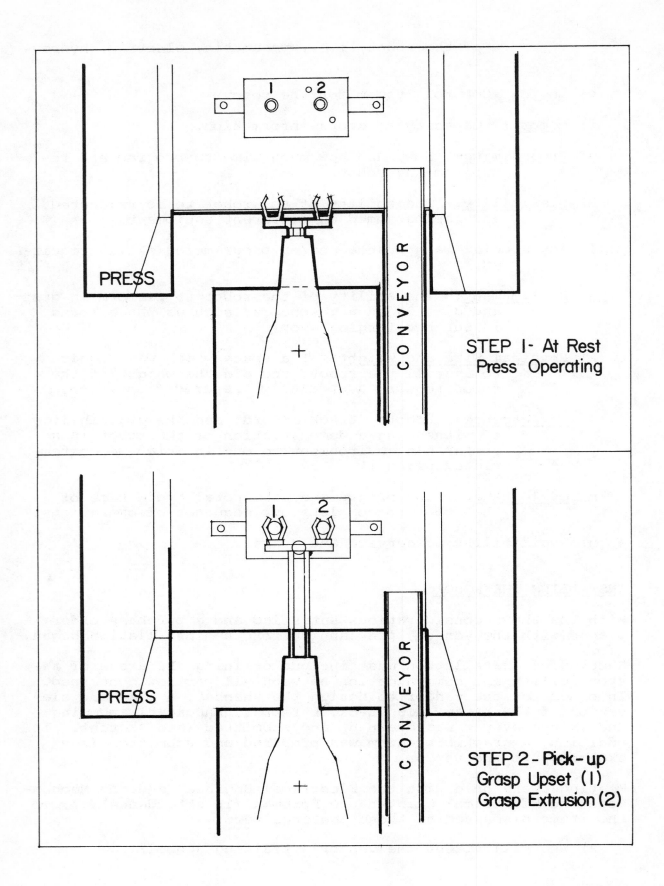

PRESS

CONVEYOR

STEP I- At Rest
Press Operating

PRESS

CONVEYOR

STEP 2 - Pick-up
Grasp Upset (I)
Grasp Extrusion (2)

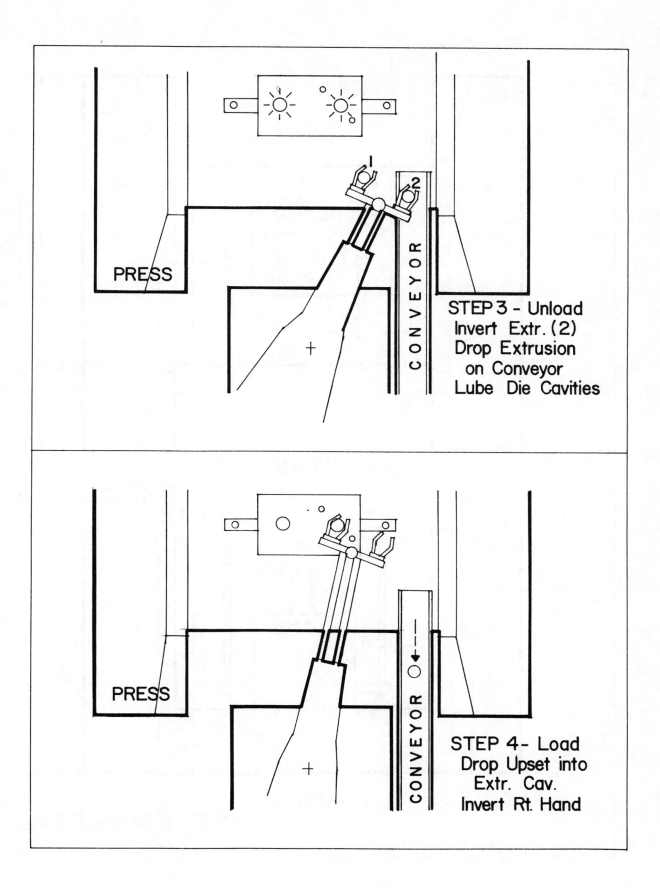

PRESS

CONVEYOR

STEP 3 - Unload
Invert Extr. (2)
Drop Extrusion
on Conveyor
Lube Die Cavities

PRESS

CONVEYOR

STEP 4- Load
Drop Upset into
Extr. Cav.
Invert Rt. Hand

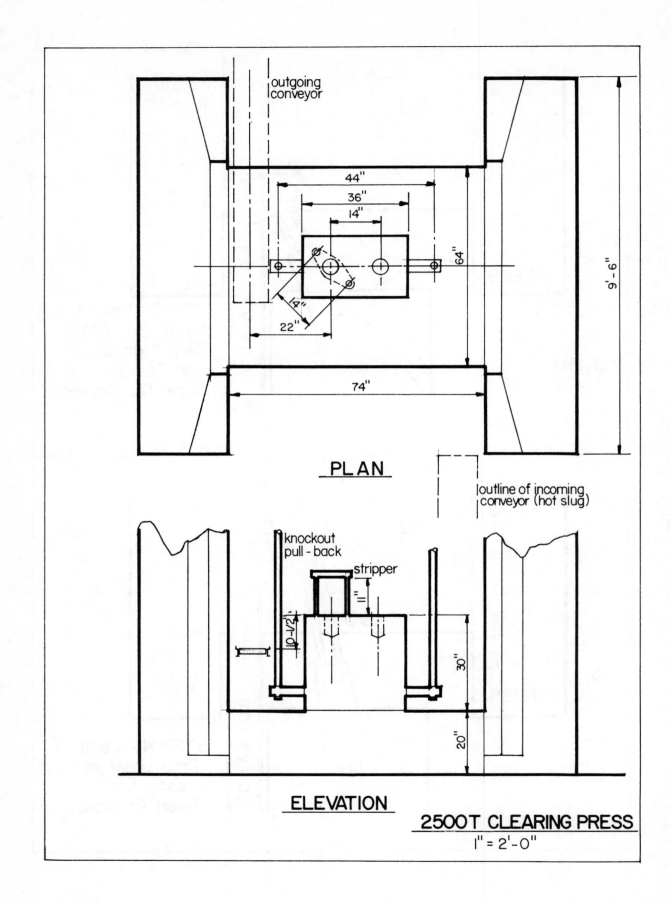

PLAN

outgoing conveyor

44"

36"

14"

64"

14"

22"

74"

9'-6"

outline of incoming conveyor (hot slug)

knockout pull-back

stripper

1'0-1/2"

30"

20"

ELEVATION

2500T CLEARING PRESS

1" = 2'-0"

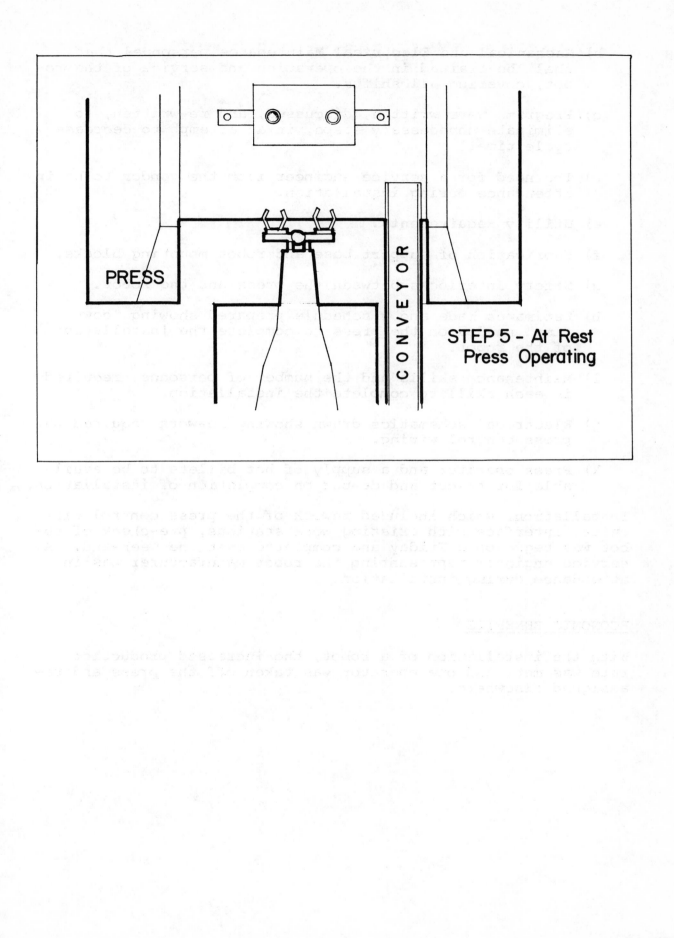

PRESS

CONVEYOR

STEP 5 - At Rest
Press Operating

b) Mechanical and Electrical Maintenance personnel that shall be trained in the operation and service of the robot, covering all shifts.

c) Programs were written, discussed, and re-written, to eliminate unnecessary steps, in an attempt to decrease cycle time.

d) The need for a service engineer from the vendor to be in attendance during installation.

e) Utility Requirements.

f) Fabrication of support base and robot mounting blocks.

g) Safety interlocks between the press and the robot.

h) Estimates made and a schedule prepared showing "down time" needed on the press to complete the installation of the robot.

i) Maintenance skills and the number of personnel required in each skill to complete the installation.

j) Electrical schematics drawn showing re-work required on press control wiring.

k) Press operator and a supply of hot billets to be available for tryout and de-bug on completion of installation.

Installation, which included rework of the press control circuits, interface with existing work stations, pre-check of robot was begun on a Friday and completed over the week-end. A service engineer representing the robot manufacturer was in attendance during installation.

ECONOMIC BENEFITS

With the installation of a robot, the increased production rate was met, and one operator was taken off the press and re-assigned elsewhere.

PHOTO 1

PHOTO 2

PHOTO 3

PHOTO 4

PHOTO 5

PHOTO 6

Presented at the Robot IV Conference, October/November 1979

Thoughts and Observations on the Application of Industrial Robots to the Production of Hot P/M Forgings

By A.L. Alves
Engineered Sinterings & Plastics, Inc.

When the thought of P/M hot forging occurred to the writer some 15 years ago robot engineering had no part whatsoever in the beginning of this research project. Our company, Engineered Sinterings and Plastics, Inc. - ESP - was well established in the fabrication field of powdered metal parts. We were well experienced in the behavior of sintered ferrous and non-ferrous materials. We understood the flow characteristics of the metal powders, both before and after sintering. We knew the problem of handling individual parts after sintering and the next coining or repressing to obtain higher densities. But at the time no particular attention was given or seemed necessary on how the hot preform blanks could be handled - first placing the cold blanks in the furnace and, second, removing the hot blanks from the furnace with the least delay and locate in the forging die. But when the mind has the fervent ambition to accomplish something it often finds it easier to overlook the multiple details and get started at some point or another, even when one does not know whether the right starting point has been chosen.

The following eight years could be called the period of cut and try. This included deciding on a suitable furnace to heat the blanks. The sintering furnaces were not adequate for the concept the author had in mind and for a while we decided to investigate induction heating. First, we leased a small unit; next we bought a larger one and after a period of time it was necessary to change direction completely. In sintered metals one must always bear in mind anything that is pressed and sintered has porosity. It is actually a kind of sponge and the heating of the blanks for forging is most important because without a protective atmosphere oxidation to some degree, depending on the time and temperature, will occur. Another difficulty in promoting uniform heating rate was that some of our parts had several levels and in some cases we might have a section 1" thick and the same piece would have a section 1/4" thick. The required adjustment of the heating coils was not always easy to accomplish and the thinner section would burn before the thicker section was of the right temperature. Our next step had to be investigating the furnace problem.

Standard forging drop hammer shops were observed but this observation was not of a great deal of value because all forging observed at the time involved solid metals, and oxidation was not a problem. While this thinking activity was going on we were investigating other fields to learn more about the flow of porous hot blanks using whatever crude method seemed appropriate and actually produced a small key ring screw driver and displayed it in Chicago at the ASME Design Engineering Show in 1969. The tools for forming the blank were changed two or three times but we gained the knowledge that was subsequently useful later on how the forging blank had to be produced to result in the best hot P/M forged piece with variable thicknesses. Just as a matter of general interest to the engineer, both the blank and the finished piece are shown on figure 1.

At the period just reviewed above the concept of utilizing a robot had not occurred to the writer. However, a final decision had been made on the type of furnace that would best serve ESP's needs as a job shop of small and medium quantities of hot P/M forged parts. A furnace with a rotating hearth was decided on, equipped with all the facilities for using the necessary protective atmospheres with an accurate indexing system for positioning the forging blanks and with a maximum temperature of 2100°F. Various experiments were made on the time required to heat blanks of various thicknesses and weights so the rotary hearth could be run at the right speed and thus determine how much weight of parts could be placed in the furnace when first charged. The design conclusion was that we needed three shelves so at certain speeds each part would be given enough time at the chosen forging temperature. It was determined at that time that the installation should be as shown on figure 2. The rotary furnace was designed with two doors so the 150 ton No. 2 press and the 300 ton No. 5 press could be used with one door only and if large parts had to be forged with the 500 ton press No. 4 the second door gave access to the heating work located on any of the three shelves in the furnace.

Now we had the furnace, we had the presses and we had production orders. The parts had to be forged and handled, both cold and hot, and for the moment the old blacksmith method of handling the work was given a trial. Equipment item 6, figure 2, was not in place; it was not even thought of at the time. That space was reserved for the operator. The reader will notice that figure No. 3 shows

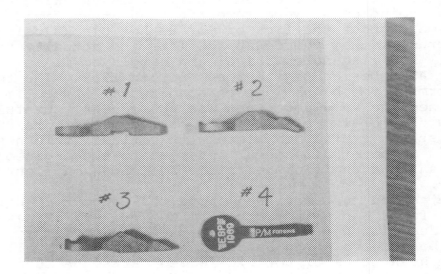

Fig. 1

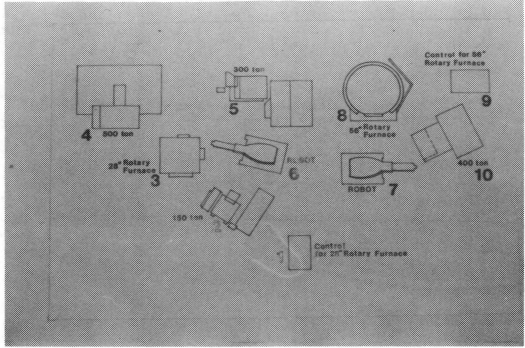

Fig. 2

the operator placing the hot piece in the die after he removed it from the furnace as shown by figure No. 4.

Actually, using this blacksmith approach at the beginning involved a press operator, the furnace operator with the tongs, and another operator to place the cold piece, properly oriented, so the furnace operator could pick it up, place it in the furnace and subsequently remove it so it could be placed properly in the forging die. This approach was not only costly but the operator could not stand the heat very long because every time the furnace door was opened a strong fire curtain had to be created to minimize the introduction of air into the furnace, which showed later as oxides trapped in the piece after it was forged. Now the thought of using a robot was very strong. We were just up against a stone wall and a different solution had to be found.

Robot manufacturers were interviewed and the difficulties of the problem were presented. The engineering needs were explained and we received our first lesson on the capabilities of robots available. Up to this time it was impossible for us to develop reliable costs because of the irregular rates of production. The human element concerning exposure to heat and resulting fatigue were reviewed carefully with the robot people and it was necessary at this time to make a final decision - either try a robot or give up hot P/M forging completely. The development of the technique of hot P/M forging was reasonably mastered but the economics could never be adjusted to the irregular rates of production and the extreme resulting high costs.

In March, 1975, a purchase order was placed for our first robot. This was done after a complete analysis of the problem was understood by the manufacturer of the robot. Our organization began the study of robot potential in our work application. This new technique had to be mastered to solve the problems we had wrestled with for so long. It was decided to send two of our key employees to the robot school to learn to program the new machine under various cycle conditions. The robot equipment piece No. 6, figure No. 2, was placed so it could easily serve both the 150 ton press and the 300 ton press. A new period of additional learning had to be devoted to the time and motions involved in the process. Our first experience with the new robot eliminated one of three operators and we learned a great deal about the additional tools needed to handle the various shapes, both hot and cold. The human drudgery

Fig. 3

Fig. 4

of working under very difficult conditions was eliminated.
Production with more regularity was now a proven success
and the first robot stimulated our people to find ways
to improve production rates. Further automation was
promptly started.

A proper hand with two fingers was built for the robot and
certain alterations were made on the fingers to suit the
different shaped parts to be handled. Figure No. 5 shows
the two fingers holding a forging blank. Production standards were developed with this set-up. Ordinarily, for
each motion the robot makes up and down, or in any horizontal position, it consumes a theoretical 0.8 seconds.
After several weeks of production we learned to program
the robot very efficiently and the cycle time was reduced.
While production was going along smoothly the unexpected
happened: the trigger being forged stuck to the top punch
without the press operator noticing it. The robot continued on its programmed routine resulting in double loading the die by press operator and the tools were damaged.
This showed us a condition which could happen with other
parts in hot P/M forging because the process facilitates
producing finished shapes that are not possible in regular
forging and one does not always know whether the greatest
friction is going to be in the bottom punch or the top
punch. A solution to avoid this same thing happening in
the future had to be found, and it was.

Let's now refer to figure No. 6. You are looking here
at the front of the press. There is a funnel and a steel
tube and as the pieces are ejected from the die the operator places the finished forged part in the funnel. It
slides down the tube giving an appropriate signal that
the priorities are met and the robot and everything else
keeps going on its repetitive programmed cycle.

It is appropriate at this time to bring to the reader's
attention that full advantage of the robot's electronic
memory is being used. The press, the furnace door, the
furnace index mechanism for the rotating hearth, the automatic lubricating of the tools, press ram lockout, and
operator protection, are all electrically interlocked, including the already mentioned discharge of the finished,
forged piece. If any one of the required operations of
any of the equipment stops or breaks down the robot will
cease working. In other words, the robot will not enter
the furnace to deposit another cold forging blank unless

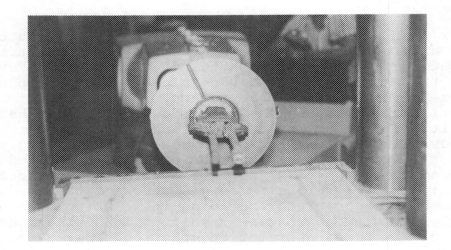

Fig. 5

Fig. 6

the door opens fully at the exact time. This circuitry
includes the robot moving from one shelf to another inside the furnace. If the last hot piece on a shelf has
just been picked up and placed in the die for forging
the new cold piece has to go on another shelf according to the next program selection. The versatility of
the robot sometimes appears limitless, depending strictly
on the ability of the programmer to adapt the robot to
whatever new condition develops to forge a new part.

When the cycle runs smoothly from beginning to end the
program is taped and placed on file so the next time the
same job is run the forging program can be put back into
the robot memory. Sometimes minor adjustments are needed.
At this time engineering thoughts have to be given in
other directions to improve production, which means cutting the cycle. This will be discussed next:

It is hoped that the description given of the use of the
robot is understood to appreciate this new method of
handling and placing parts within the accuracy required.
Our experience prompted us to buy another robot. You
will notice figure 7 which shows a larger press and a
larger furnace and the same type of robot. Certain
changes have been made following our experience with the
first attempt to work out a feasible, economical method
of producing high density sintered metal hot forgings.
The adition of this new forging unit is the result of
increased demand and to facilitate producing more than
one part at a time. The new set-up will not produce any
faster than the first one. We have to find different
avenues to speed up production as already mentioned.

The first robot hand developed has two fingers. Our
Chief Engineer felt he could design a hand with three
fingers as shown in figure 8. The advantage here is that
the robot can drop into the right place in the furnace a
cold blank and pick up a hot blank simultaneously. When
it reaches the die another cold piece is already in place
and as one finger opens to drop the hot piece into the die
it picks up, simultaneaously, another cold piece. In other
words, the action is when one finger opens, the other
closes. The cycle using this new method is exactly the
same but it has been reduced by a third in time. This improvement requires better and more control. One can appreciate that as the hearth rotates it has to locate the hot
piece in exactly the same point in space for which the
robot has been programmed. All the equipment has to be

Fig. 7

Fig. 8

maintained in first class condition because the robot has been programmed to pick up and deliver the pieces on exactly the same location. A good look at figure No. 9 will reveal to the observer how smoothly the new fingers drop the hot and pick up the cold pieces simultaneously.

In our early development with the first robot with the two fingered hand an operator was used to place the cold piece in exact location for the robot to pick up and go on into the furnace. That was followed by the development of a feeding device which delivered the piece where the robot was programmed to pick up. This eliminated an operator but it was not the final solution. A tremendous amount of research by our engineers had to be done to utilize the capabilities of the robot.

The writer does not wish to transmit the idea that this is the ultimate feeding system in our industrial system. There are disadvantages which really are no fault of the robot. A rotary hearth furnace used at minimum temperatures of 1800°F will gradually develop some changes, such as warpage, and there will be times when the robot is programmed to place a piece oriented just right and when it doesn't happen it might miss the piece next time around. With the blacksmith method, as illustrated at the beginning of this paper, you can live with bad conditions and keep production going. In other words, perfection is not attained but the improvement over the old methods is so vast that one can easily tolerate the shortcomings. Even the robot has its periods of malfunction, but when the organization is developed to handle the new manufacturing technique the advantages of the new equipment outweigh the occasional disadvantages.

There are times when two or more cavities to forge two or more pieces at a time can be developed if the volume involved warrants the extra research and tooling cost.

Another factor that should be brought to the attention of the reader is that it may be more convenient or more economical to use the two fingered robot hand. This happens if the quantities are small. The main reason here is that it is easier and more economical to build a two fingered hand than a three fingered one.

Anyone deciding to make use of the modern robots in the hot P/M forging must be aware that tool design is somewhat different. The orientation of the parts with reference to the tools is critical, especially with double

Fig. 9

gears, or both external and internal gears. An important thing for the engineer to think about is that the
memory of the robot is available; the trick is to learn
how to use it. It is by no means an easy task but with
persistence we have developed now two hot P/M forging systems, using only one press operator with the only duty
being to place a cold part for delivery to pick-up area
and to see that the piece drops into the die properly and
when the part is forged, to remove it and put it into the
funnel as mentioned before so the cycle keeps moving as
programmed.

Presented at the Robot V Conference, October 1980

Upset Forging with Industrial Robots

By John Saladino
General Electric Company

Introducing robotics into manufacturing is quite a challenge for the project engineer. First applications of this technology often involve substantial learning by trial and error. Forging applications present interesting and unique technical challenges. This paper shows one approach taken to introduce robotics into upset forging of jet engine blades.

BACKGROUND

In the spring of 1978, a department within General Electric's Aircraft Engine Business Group decided to investigate the use of robotics in their operations. A project engineer was assigned and the company's corporate consulting group on robotics was contacted for assistance. A survey was then conducted for potential applications. Each application observed was ranked according to two criteria: <u>technical risk</u> and <u>potential payback</u>. It was felt the first application of robotics must be successful (prototype failures are long remembered!), so the criteria of technical risk was weighted more heavily.

The forging operations observed appeared to offer the best opportunities for introducing robotics as a form of flexible automation: sufficient <u>volume of parts</u> to offer good equipment utilization (millions of airfoils per year), highly <u>repetitive processes</u>, and medium to small <u>batch runs</u>.

Figure #1 shows the sequence from raw material to final product (materials are primarily titanium alloys).

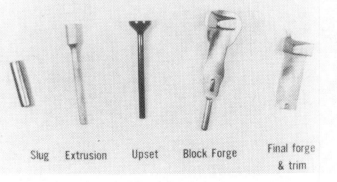

AIRFOIL FORGING SEQUENCE

Slug   Extrusion   Upset   Block Forge   Final forge & trim

Figure #1

For each operation, the part is heated to approximately 1750°F then placed in a forge die. Trying to maintain orientation between operations was not considered practical since parts needed to be cleaned and descaled after each forging.

Figure #2 shows a typical manual workstation. An operator takes a part from a bin with a set of long tongs, places it in a rotating hearth furnace, removes a heated part, places it in the die, cycles the press, ejects the part, and places it in another bin.

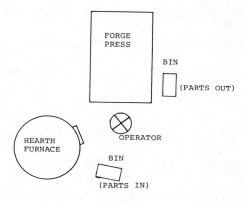

Figure #2

While the process is similar for all forging operations, there are several subtle and significant variations which made the Upset operation a first choice:

- A minimal amount of work was done to the part at this operation. The amount of part-to-die contact was small (the lower die held the extrusion while the upper die "flattened" the head). This meant fairly long die life and the need to visually inspect only a few pieces per hour.

- All other operations except Upset involved coating the parts with a lubricant to promote die life. The lubricant was soft and easily scratched. Handling of coated parts by automation devices posed real concerns.

- All other operations except Upset involved lubrication of dies between press cycles. The lubricant was graphite and water. Automating die lubrication posed concerns also for material flow is affected by the amount and location of graphite on a die and operators would often alter their spray technique to compensate for die variations. Prior attempts at automating die lubrication were not very successful.

ROBOT SYSTEM

To convert from manual to robotic operation involved a total systems approach. The integration of hardware, use of sensors, and overall system control were the key elements in the system. The goal was to have a workstation that needed the minimal amount of human intervention. Ideally, the system would run unattended.

HARDWARE

There were to be six elements of the robot system as shown in Figure #3.

- Parts Feeder/Orientor: a vibrating orientor which positions parts in a track for pick-up.

- Rotating Hearth Furnace: the existing furnace equipped with a carousel rack for part loading and a stepping motor for indexing.

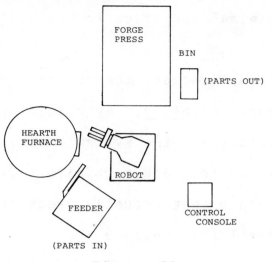

Figure #3

- <u>Press</u>: the existing 75 ton mechanical press.

- <u>Robot</u>: a 5-axis Unimate 2000 with standard hand and stainless steel gripping fingers.

- <u>Controller</u>: a programmable controller which monitored all sensory information and the interaction of the above components.

- <u>Sensors</u>: included optical and mechanical types, they provided information to the controller on the status of the system.

There was a definite philosophical approach on how to implement the hardware. Where possible, all hardware was to be standard off-the-shelf equipment. This was done to minimize the risk. Start-up failures tend to fall into two categories: <u>component failures</u> and <u>system failures</u>. Component failures are resolved by debugging devices or tooling while system failures are resolved by debugging the interaction between devices (interface issues). By purchasing standard equipment, the likelihood of start-up failures would thus be minimized. This was done with a minimal sacrifice to overall system performance.

SENSORS

In any robot application, the project engineer needs to go through "contingency" planning. By examining all of the possible failure modes for the system, the system depends on sensors for providing feedback for failure detection. It is also crucial to relate what actually needs to be measured with a sensor that can reliably measure it.

The following is an explanation of the sequence showing where, why and what kind of sensors were needed:

- The process begins when the operator dumps parts into a 15 cu. ft. bin. Parts are then elevated by a conveyor out of the bin and dumped onto a vibrating orientor (a pair of parallel adjustable tracks) to allow the parts to hang by their heads. The parts then slide down a gravity feed track to an end position. A limit switch sensor is used to keep the track full of parts and signals the feeder to run when the queue gets too small. At this point, the parts are available at a known location and known orientation for the robot fingers to grasp (Figure #4). But the robot does not know for certain if there is a part there. What if there were a jam-up or the feeder ran out of parts? To answer this question an infrared L.E.D. (light emiting diode) sensor was used to scan the end position for an available part. This sensor was selected for three

Figure #4

reasons:

(1)   Unaffected by changing ambient light conditions;

(2)   Non-contacting, hence longer life;

(3)   Unaffected by dust in the air.

● The robot then grabs the part by the head and
transports it to the furnace (Figure #5).

Figure #5

Before entering the furnace several questions needed answers:

(1)   Does the robot have the part between its fingers?

(2)   Is the furnace door fully opened?

(3)   Is the furnace table in the proper position to accept another part?

(4)   Is the furnace at an acceptable operating temperature?

The question of part in gripper was answered by passing the gripper over another LED sensor. The profile of the hand with a part was wider than the gripper without a part and was easily detected by the LED. This same sensor was used again to verify the robot came out of the furnace with a hot part. The opening of the furnace door was sensed by a simple lever-type limit switch. The question of table position was answered by integrating the stepper motor drive control. The indexing was done as a step function and the controller provided output indicating the motor did energize (a change in state) and it did step through the prescribed number of counts for a proper index. The furnace temperature was monitored by thermocouple sensors which looked for an overmax. or under min. condition.

• After a hot piece is removed, the furnace door closes and the table indexes. The door is opened and closed by energizing a solenoid valve which activated an air cylinder on the door. A limit switch built into the robot arm verified the arm was fully retracted and clear before the door was closed.

- The robot proceeds to the press and needs to know if the press is open before inserting a hot part. This was accomplished by using a limit switch mounted on top of the press ram to verify it is in the fully up position.

- After proper nesting of the part in the die, the robot backs off and cycles the press (Figure #6).

Figure #6

Once again the limit switch in the robot arm verifies the fingers are clear before cycling the press. But how does the robot know the part is properly nested or even in the die? This was sensed by another LED mounted in the die holder looking for the tail of the part hanging through the bottom of the die. This insured a part was in the die and in far enough to be properly nested. Furthermore, the sensor could be adjusted to look for infrared light within a certain range. By setting the range properly the sensor would only "see" parts that were at a minimum acceptable temperature. This also assures the press would not cycle if the robot were to successfully nest a "cold" part.

- After cycling the press, the robot activates an air eject system to get the part out of the die and into a basket. To sense a part was out of the die posed an interesting problem because, after being hit, the part may or may not be hot enough to be "seen" with an infrared sensor. Here is how the problem was resolved. Two infrared LED sensors were mounted perpendicular to each other in the horizontal plane to scan across the top of the die. One sensor had a light source aimed at it and the other did not (Figure #7).

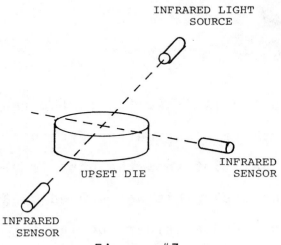

Figure #7

If a part was properly ejected, one sensor would see light and one would not (a yes-no logic condition). If a part

remains in the die and is hot enough to be seen, then
both sensors "see" it (a yes-yes condition). If a part
is in the die and not hot enough to be seen, then it
blocks the light source of the one sensor and neither
sensor "sees" light (a no-no condition). Therefore,
only the yes-no logic condition guarantees the part is
out of the die, regardless of whether it is "hot" or
"cold."

CONTROL SYSTEM

The control requirements of the robot workstation went
beyond the capabilities of the robot controller. Instead a
small PC (programmable controller) was used. All devices
including the robot and sensors came under control of the PC.
The console pictured in Figure #8 was the primary interface
between the operator and the robot system. To activate the

Figure #8

system, the operator turned on the main power switch, selected
the mode of operation (load the empty furnace, run with a full
furnace load, unload the full furnace), and pushed the start

button. The PC did the rest. If for some reason, a sensor detected an error condition, the entire system would "freeze" and a buzzer and alarm light would alert the operator as to the cause and location of the malfunction. After fixing the problem, the operator would push the reset button to clear the error and then the continue button. This allowed the system to recover from the point at which it "froze."

For each new set-up the operator would also enter an inspection count and the PC would "count" press cycles and alert the operator when an inspection was due.

To insure the control system could be easily understood by maintenance people without the need for exotic training, all PC programming was done via ladder diagram relay logic and all I/O was handled via 120 volt AC modules and relays. The total memory requirement for the PC was approximately 1000 words.

CONCLUSION

When the system was installed and fully debugged, there was a significant gain in productivity observed. In general, the robot produced about 30% more parts per unit time than its human counterpart. The workstation was designed to run unattended with a minimal amount of human intervention required with the hope one operator could monitor and service three similar workstations. When this is accomplished the full impact of the cost savings of the robot systems will be realized.

The approach taken to introduce this prototype robot system was a very conservative one. Novel approaches that may have enhanced the gains of the system were not pursued if they involved a real question of risk. The success of the prototype was seen as the vehicle for de-mystifying the technology and paving the way for other, more exotic applications.

In Britain
Robots Prove Their Worth In Forging and Deep Drawing

A British company is successfully incorporating robots in forging and deep drawing operations, freeing human workers from a tough job in a hot and dirty environment

This Unimate 4000 robot feeds an Ajax upsetter (foreground) from an 18-billet rotary hearth furnace. After forging, the robot places the billet onto a conveyor and repeats its full cycle.

Although British industry has been slow to adopt robots for forging applications, one multidivisional company, Tube Investments Ltd., is pioneering the use of robots in forging and deep drawing operations.

TI Ltd. first looked at robot possibilities in its Birmingham plant where an 18-billet rotary hearth furnace, designed to feed an Ajax upsetter, proved to be manually impractical because the three-man team assigned to the project could not keep pace with the capacity of the furnace. In 1978, following a successful trial period, the company installed a heavy-duty Unimate 4000 robot which quickly proved its ability to keep up with both the furnace and upsetter.

With one man to oversee it, the robot now handles 400 forgings for automotive axle cases and hub ends per shift. The furnace is still loaded manually, but the next step, according to Maurice Cooksey, engineering executive, is an automatic loading system to reduce physical strain even further.

When the first of the 18 billets in the furnace reaches a temperature of 2012° F (1100° C), the robot places it in a descaler. At the same time, it repositions the tube in its gripper, since the tube moves slightly as a result of the heat.

The robot then places the billet in the upsetter which it triggers sequentially through up to four forging operations. The operations are interlocked to prevent malfunction. After forging, the billet is placed on a conveyor and the robot swings around to repeat the cycle.

Cooksey says that TI Ltd. is now investigating replacing another manual tube forging operation in the plant with a Unimate robot working in conjunction with an induction heating furnace. Induction heating would give better heat cutoff for more precise length of heat, and it would reduce irregularities in the shape of the forging.

Following the company's success in its Birmingham plant, TI's development services division investigated a robot application for the small cylinder division at the company's plant in Chesterfield. In this plant, high-pressure cylinders, which work in excess of 2000 lb/in² (14 MPa), are produced to meet a world demand for fire extinguishers and brewery and medical equipment.

After a trial period, a Unimate robot was installed to perform hammernecking on the cylinder. This application has proven so successful, with less that 2% downtime on the robot, that similar installations are now being considered.

In the sequence, the robot takes blanks weighing up to 40 lb (18 kg) from pallet to furnace, feeds the cylinder blank—whose open end is now at forging temperature—to the hammer, and places each necked cylinder onto a conveyor.

Because the robot is able to meet the demands of both hammer and furnace, throughput of forgings is increased to about 60 per hour. "This brings a significant saving in gas," says Bruce Gilson, senior production engineer. "But even more important are the environmental advantages."

The latter are demonstrated in the manual operation alongside the robot system. Because of the heavy nature of the work, furnacemen are plagued with back problems, which lead to enforced absenteeism. This is one reason why TI/Chesterfield is looking hard at other potential robot installations.

One likely robot application, according to Gilson, is cylinder prooftesting. Throughout the test a Unimate robot would handle the largest cylinders, filled with water and pressurized for up to two minutes to nearly twice working pressure. After being mirror-examined for base end leakage, each cylinder would be passed to a conveyor by the robot. If a cylinder was faulty, the robot would reject it automatically.

Tube Investments, viewing the humanitarian aspect of robotics with deep interest, is delighted at the workforce's positive attitude. Thanks to the relief from heavy labor offered by robots and to TI's foresight in explaining their advantages from the outset, the robots have been welcomed by employees.

"We have done this with teach-ins and constant consultation on the shop floor," explains Gilson. "Now, having seen the robots in action at Birmingham and Chesterfield, the operators want to know when they will get more of them." ■

Reprinted from Robotics Today, Summer 1980

Two-arm Robot Uses One Control System and Power Unit To Cut Cost and Simplify Operation

A two-arm Versatran robot operating off one hydraulic power unit and with moves

coordinated by a single electronic control provides versatile loading/unloading system

ROBERT N. STAUFFER
Editor

A dual savings—one in terms of equipment, the other in terms of floor space—is one of several major advantages of a "two-arm" robot recently built and shipped by the Robot Div. of Prab Conveyors, Inc.,

Kalamazoo, MI. The "two-arm" designation refers to the fact that the system incorporates two separate Versatran Model FB robots, both operating under the command of a single control and both powered by a single hydraulic power supply. The movements of the two robots are coordinated so that they function essentially as one robot with two arms.

The robots are now installed in a customer plant where they are positioned adjacent to a forging press for loading and unloading heavy duty 400-lb (180-kg) aluminum truck wheels. Wheel blanks delivered on a conveyor to an input station next to the press are picked up by the first robot and placed in the forging die. Following the forging operation, the second robot picks up the wheel and deposits it on an outgoing conveyor.

Simulated production installation shows arrangement of two three-axis robots with their common electronic control and hydraulic power unit.

Placement of the robots on either side of the press leaves the front area clear, permitting easy access for press and die maintenance or repair.

Prab points to lower initial customer cost and an inherently simple system to program and maintain as two key benefits. A single microcomputer control system monitors the operation of both robots, avoids any chance for interference, and eliminates the need for special interfacing and programming between multiple robots.

Movements of the two robots are coordinated in such a way that neither of the robots must wait while the other completes its programmed moves. This is an important factor in keeping overall cycle time to just 11 seconds.

The programmed moves of the loading robot start from a position 5 ft (1.5 m) out from the center of the press. From there the Versatran arm swings over to acquire a wheel from the input station, swings back, and advances to a "home" position 3½ ft (1.1 m) from the press. As the preceding part is removed by the second robot, the arm advances into the press, lowers the next part 6" (152.4 mm) into the forging pocket, releases the part, raises 6", and retracts 5 ft to clear the press. The robot's travel from the "home" position, into the press, and back out to the fully retracted position takes only 3 seconds.

The Model FB robots used in this installation are three-axis units incorporating the Versatran straight-line cylindrical coordinate system. Axes of motion and their standard strokes include: horizontal—60" (1524 mm) at 50 ips (1270 mm/sec); vertical—60" at 50 ips; and swing—270° at 60 ips (1524 mm/sec). Positioning accuracy is ±0.050" (±1.27 mm). Maximum payload is 600 lb (270 kg).

Prab's microprocessor-based, point-to-point control is readily adapted to the coordinated arm concept of robot operation. It has a memory capacity in excess of 4000 points, and can store up to 64 independent programs. The 600 Series control is called one of the most advanced computer controls avail-

Versatran robot at left moves 400-lb aluminum wheel and tooling into position as the gripper of a second robot removes the preceding wheel.

able and has the ability to control up to seven axes of simultaneous motion.

For this application, the control is equipped with modules for controlling just the three axes involved in the press loading and unloading operations. All six axes (three plus three) are programmed for coordinated operation. That is, on the same step in the program, both robot moves must be accomplished before going to the next step in the program. If any one axis in the system stops for any reason, the entire system stops and the control's diagnostics give a status report to the area foreman.

The 600 Series can be programmed to run any one of the 64 programs repetitively (as in this case), or to run continuously from one program to another in any desired sequence. Pro-

gramming can be done from the control console or from a hand-held teaching unit.

Prab engineers note that this development is a first in the industry and results in a much simplified robot system. The two arms are synchronized as in walking beam automation and result in significant increases in speed over the conventional one-arm robot approach.

The forging press requires a dedicated system in which both robots must be operative to keep the system running. Since the wheels are too heavy and hot for a person to handle in case one of the robots goes down, the use of separate controls and power units would not be a factor in keeping the system up and running in the event of trouble. ∎

Reprinted from Foundry Management & Technology, July 1976

Operations Roundup

Robots Thrive In Tough Environments

AUTOMATED and mechanized equipment has entered both foundries and forge shops to relieve men from tedious, hazardous jobs in tough environments. Taylor-Wharton Co., division of Harsco Corp., Easton, Pa., which forges high-pressure steel cylinders for compressed gases on a high-production basis, has installed a general-purpose industrial robot to handle hot, heavy chrome-molybdenum steel alloy billets. It allows two workers to be reassigned to more pleasant jobs and solves a serious labor turnover problem.

The robot selected was a Unimate Series 4000 made by Unimation Inc., subsidiary of Condec Corp., Danbury, Conn.—with load capacity at the wrist of 350 lb. Taylor-Wharton expects the robot to provide a payback in two years on labor cost savings alone. It went into service in August, 1975.

The jobs of the transfer man and his relief man at Taylor-Wharton were to use an air-lift hoist to move hot preforms from a 1500-ton forge press to a 260 x 100-ton draw bench. At the draw bench, each 60 to

at the forge press, ready to grasp the next preform.

The cylinder preform weighs 143 lb. This particular run of preforms is drawn into compressed-gas cylinders with .222-in. thick walls. Forge press cycle time is approximately 45 sec, and robot cycle time is approximately 35 sec, resulting in a 10-sec wait.

The point-to-point industrial robot has an electronic memory with a capacity from 128 to 1,024 steps, depending on job requirements. A "teach" control on an umbilical cord permits an operator to teach the robot a new job by leading its hand through the steps needed to perform the job. Once taught the robot will repeat the cycle endlessly.

It can be reprogrammed for a new job in a matter of minutes. The cycle usually includes issuing electrical command signals to the equipment with which the robot works. The robot also receives limit-switch or similar signals from the other equipment to ensure safety and proper synchronization of the entire, automated operation. The Series 4000 can grasp and place workpieces anywhere within a diameter of slightly more than 19 ft.

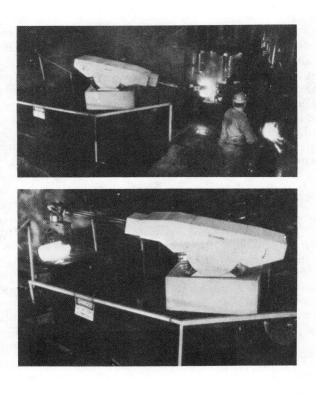

300-lb preform is drawn into a cylinder of the correct size. Production rate for the forging and drawing operation ranges from 500 to 700 cylinders for each of two 9-hr shifts, depending on size of the compressed-gas cylinder being made.

The operating cycle of the robot begins when the forge-press cycle is complete and a piston has pushed the cup-shaped preform up in the forge pot to a position that is accessible to the Unimate. The robot grasps the forged preform, lifts it from the pot, swivels approximately 180 deg to a tilt-table at the draw bench, deposits the preform on its side on the tilt-table, withdraws, swivels back, immerses its "hand" in a tank of cooling water, then stops poised

Reprinted by the Society of Manufacturing Engineers from IRON AGE, February 7, 1977; Chilton Company

Robots lend a hand in a new aus-forming line

They're not producing harrow discs at International Harvester Canada the way they used to. Now, robots are in on the act.

Improved quality, increased production volume and better production control are among the benefits realized at International Harvester Canada as a result of automating a heat-treating and forming line. And, judicious use of industrial robots played an important role in the new setup.

As manufacturing engineering manager Frank Randles explains, "A key problem was how to keep people on the jobs of loading and unloading harrow discs for our heat-treating and forming line."

The opportunity to install a new heat treating method at International Harvester Canada surfaced in 1975. The method in question was aus-tempering, a process of toughening carbon steel by heating the material to the temperature range at which austenite forms. The result is improved toughness and fracture resistance.

"We sat down with our union's bargaining unit," Mr. Randles recalls, "and discussed the advantages of automating the new line. We all agreed that the hard-to-fill jobs were among the toughest—both hot and heavy. The decision was to eliminate three jobs per shift, resulting in reassignment of six to nine workers, depending on whether the heat treating and forming line ran two or three shifts."

The new line went into operation about a year ago.

"Aus-tempering provides a stronger, tougher disc that better resists breakage when the disc hits a rock, compared with the former, conventional heat treatment. In addition," states Mr. Randles, "it puts control of the process in the hands of the staff.

"Instead of being man-paced by the workers handling the hot discs with tongs, the process is machine-paced, resulting in improved planning because volume is more predictable."

Blanked discs of AISI 1085 steel with edge-turned bevels are loaded one-at-a-time onto the entry conveyor of a 1650°F Surface Combustion furnace by a Unimate industrial robot. At the exit end of the furnace, the discs pass through a stamping station where they receive an ID number.

Discs then enter a 550°F salt-bath quench. Upon emerging from the quench, each disc is presented to a second Unimate by a pop-up station. This robot transfers the hot disc into heated dies of a 400-ton Warco press. The press dishes the disc, and the disc is removed from the die area by a third Unimate robot.

The formed discs are then transferred to the entry conveyor of an Ajax washer and dryer by this same robot. Dry discs then proceed by conveyor to the next operation, priming and painting.

The robots take the place of three men per shift.

The production rate for the new operation is close to 300 discs per hour. Five standard sizes of discs are produced. A run of any one disc lasts an average of eight shifts, but a run can last as long as one week. Disc diameters range from 16 to 24 in. in 2-in. increments. Thicknesses range from 11 to 6 gage.

"We make our changeovers," says Mr. Randles, "from one size to the next during the day shift when we have the most people here for changing the die and reprogramming the Unimate.

"When the Unimate has been taught to handle one size, the program is extracted from the robot's memory and stored for future use. Reading a different program from a cassette into a robot's memory is as rapid as the extraction process, usually well under three minutes."

Unimate No. 1, at the entry conveyor for the tempering furnace, has a vacuum-cup pick-up hand which comes down vertically from the end of the robot's arm. The hand then lifts the top disc from the palletized stack of about 50 to 60 discs. The robot then transfers the disc and deposits it on the entry conveyor for the 1650°F furnace.

When no discs remain in the stack, the three vacuum cups come to rest on the open framework of the pallet and a pneumatic pressure sensor detects absence of the usual negative pressure buildup. The sensor signal terminates the program and initiates advance of a new stack of discs into position. When a new stack is in position, a signal restarts the program.

The second and third robots on the line each have one two-fingered hand to grip the disc on the OD. The basic hand tooling remains the same for all disc sizes, and only fingers are changed for gripping different diameters.

All three robots are Series 2000 three-axis Unimates that omit the usual bend (pitch) and yaw gear train found in five-axis robots from the manufacturer, Unimation, Inc., Danbury, Conn. The extra articulations are not needed in this application, thus resulting in a cost saving. Base price of each of these robots was $22,780.

General-purpose industrial robots are programmable, parts-handling or tool-handling machines that will also control and synchronize the equipment with which they work. The robots at International Harvester Canada each have an electronic memory and control system to direct the actions of the hydraulically powered arm.

The memory also sends electrical signals to control the equipment that the robot automates, and receives signals—typically from limit switches or other sensors—from the other equipment to synchronize operations.

Unimates are available with from two to six programmable arm and wrist movements, although the robots at the International Harvester Canada plant have only three degrees of freedom. Memory size is expandable in modular increments from 128 through 1024 steps. Most jobs require less than 100 steps.

Alternate actions and repetition of sub-routine steps, as well as a straight sequential run-through and repeat of a program, are among the capabilities of a robot.

Teaching a new job to a Unimate robot is a matter of switching the control system to a teaching mode and moving the hand of the robot—with the aid of a teach control on an umbilical cord—through the motions. Once recorded, the motions will be repeated precisely, time after time.

Reach of a Unimate can span as much as 20 ft with loads as great as 500 lb.

As each disc emerges from the quench, the powered conveyor carries it to the pop-up station which positions and presents the disc for pickup. Subsequent discs are held by a gate until the robot has picked up the first disc and the pop-up station has dropped.

The robot swings about 160° with the disc and prepares to load it into the die of the Warco press. But, three key conditions must be met before the robot's hand moves into the die area.

They are: (1) That a ram limit switch indicates the die is open; (2) that the Unimate No. 3 has signalled removal of the previous dished disc; and (3) that an Ircon infrared scanner system has signalled that the disc is within the acceptable forming temperature range of 450° to 600°F.

A signal that a disc is too cool causes Unimate No. 2 to switch to a "reject disc" program. If the temperature is too high, that's an in-

Robots load harrow discs from pallets onto a hearth of a heating furnace.

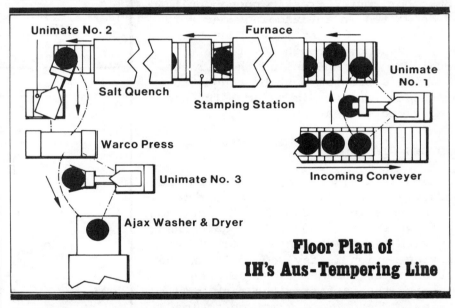

Floor Plan of IH's Aus-Tempering Line

dication that two discs have stuck together. This, too, will initiate a "reject disc" program.

After unloading a formed disc from the Warco press, Unimate No. 3 swings its arm about 180° and places the disc on the entry conveyor of the Ajax washer and dryer. Here the remaining salt is washed from the disc.

Salt buildup both on the press dies and on the robot's fingers is minimized by jets of furnace-heated compressed air.

"Our hourly production rate," says Ralph Pawson, the plant's maintenance superintendent, "is no higher than it was before, but under the previous setup there were days

when we could only run 5 hours a day. Now, we can run around the clock because the new system is much more reliable."

As far as the new aus-tempering and forming operation is concerned, Frank Randles reports that it has definitely met expectations. It has resulted in improved quality, increased volume and savings. The product cost is down due to method changes and labor savings from the transfer of six to nine people. Maintenance labor costs, he adds, are higher than before. □

Reprinted from "Unimate In Action", November 1974, the Application Notes of Unimation, Inc.

Unimation Application Notes*

Furnace Heat Treating

Series 2000 and 4000 *Unimate* industrial robots have been put to work in this manufacturing facility, loading and unloading heat treat furnaces and performing the subsequent quench operation.

APPLICATION DESCRIPTION

In an example of this type operation, a Series 4000, 5 axis *Unimate* is installed as shown in the installation sketch. A rotary hearth furnace is fitted with internal racks to support cylindrical shafts. The *Unimate* unloads and loads three parts for each index of the hearth. The parts are loaded into and unloaded from a horizontal drastic quench unit and deposited on an outgoing conveyor. A standard *Unimate* boom, without boom covers, reaches through the flame curtain to load and unload the furnace. The *Unimate* program must be taught to compensate for the dimensional furnace changes due to heat up.

HAND DESCRIPTION

The hand is designed to minimize thermal conductivity and allow the fastest possible entrance into and out of the furnace. The fingers are designed to cradle the part with six point contact. The finger clamping motion is devised to lower the part onto the furnace racks on opening, and raise the part from the rack upon closing. The *Unimate's* motion to load and unload is only straight in and straight out.

JUSTIFICATION

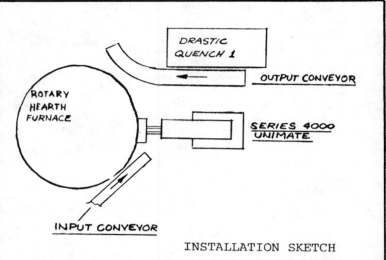

INSTALLATION SKETCH

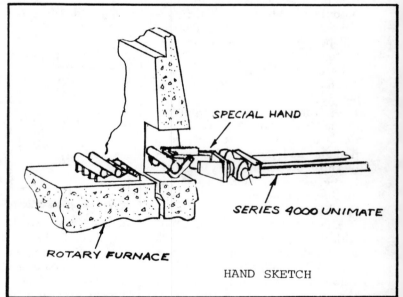

HAND SKETCH

While the replacement of one man per shift on a three shift, seven day a week operation is more than adequate justification based on a quick payback, the underlying justification was the elimination of a very undersirable job - the cause of numerous grievances by the Union.

CHAPTER 7

FOUNDRY

Commentary

In foundry operations, robots are used in a wide variety of applications.

They unload core-making machines, handle cores through washing operations and set cores in molds. They dry and vent molds. They clamp and unclamp molds on pouring lines.

Robots transfer hot castings from molding lines to shakeout machines and remove gates, risers, sprues and flash from castings.

An industrial robot can move people from an environment characterized by heat, dust, smoke, noise and, in some cases, potentially toxic materials.

Presented at the Eighth International Symposium on Industrial Robots, May/June 1978. Reprinted courtesy of International Fluidics Services, Kempston, Bedford, England

Foundaries, Robots And Productivity

By George E. Munson
Unimation, Inc.

The use of Unimate industrial robots in foundries is extensive and has resulted in an impressive list of user benefits.

Over 300 Unimate robots are employed in die casting foundries unloading castings from the die cast machine, quenching the part, performing die lubrication and cleaning, loading inserts, ladling in cold chamber machines and performing secondary operations such as casting trimming and machining. The flexibility of the robot provides efficient use in both large shops with long production runs and small independent foundries where lot sizes of 1000 castings or less are common. Shot sizes typically range from a few kilograms to over 50 kg. Fig. 1 shows a Unimate robot installation in die casting.

While Unimate robots have been in die casting since 1962, their use in investment casting foundries is more recent. The specific task assigned to the robot is to successively dip delicate wax molds (trees) into a series of slurries and fine sands to build up a shell which eventually becomes a casting mold. (See Fig. 2). Finished trees can range in weight from 10 kg. to as much as 175 kg.

This operation is the most critical in this type of foundry. It has been one of the most unpredictable operations in the process because it requires precise manipulation of the trees to achieve uniform distribution of the coatings, free from voids and air inclusions. Human manipulation is, at best, inconsistent. Unfortunately, quality of the mold and hence the finished part is not known until the casting is made. By this time a large investment in time and material has been made in producing what is all too often scrap.

The introduction of robots has removed the art from this process and reduced it to a predictable process of high productivity. For example, scrap rates have been reduced from 85% to under 5%.

In some of the more advanced foundries the entire process is under direct computer control. Process parameters such as slurry temperatures and viscosities, drying temperatures etc. are monitored and adjusted. In addition, robot programs are stored in a disc file, ready to be dumped into the robot's local memory as soon as the wax tree identification (and hence correct robot program) is made.

To a much lesser extent, robots have been applied in steel and iron foundries. To date, their jobs have been limited to removing castings from shake out conveyors, rough trimming, cope and drag preparation and some material handling functions. However, there are a variety of other potential applications which are actively receiving the attention of foundrymen. The remainder of this paper will concentrate primarily on steel foundry operations.

Ferrous Metal Foundries

Ferrous metal foundry operations produce a variety of occupational hazards which virtually mandate the introduction of more mechanization and automation than ever before considered. Air pollution, noise and hazardous tools expose the human worker to an inhuman environment. Also, poor operating efficiency subjects the foundry to very low productivity. In order to discuss the potential for robot use in the foundry it is necessary to have an understanding of its operation.

Typical operations are listed in Table 1 starting with pattern preparation and finishing with machining. Table 1 also contains a breakdown of casting cleaning and finishing operations. These "cleaning room" operations are required to remove sand from the casting, to remove the rigging (gates, risers, runners, pads), and to remove and repair randomly occurring flash and holes. The cleaning room involves the most health endangering and hazardous operations to be found in the foundry. In addition, the cleaning room is the bottle neck in the whole operation and involves high cost and currently, low productivity. This is especially true of steel foundries since typically, the steel foundry is a small lot batch processing operation with lot sizes ranging from 1 to approximately 100 castings.

In studies conducted in both the United States and the United Kingdom the average of cleaning costs of castings as a percent of total production cost was found to be 26.2% (30% in the U.K.). Of the total cleaning costs, 28.6% was attributed to metal removal (cut-off, grinding etc.). It was also found that these figures were greater for small castings compared to large ones. Currently in the U.S., approximately 1,800,000 tonnes (t) castings are produced annually. Further, it has been estimated that 17.5% of metal removal costs could be saved by automation. If finished castings are valued at $2.2 per kg., based on the above figures an industry wide savings of $51,000,000 could be realized in this area of the foundry along. Fig. 3 graphically illustrates general cleaning costs and metal removal costs as functions of casting size (weight).

It is for these reasons of working environment and cost that special emphasis has been placed on robot applications in the steel foundry cleaning rooms.

Cleaning Room Operations

A ferrous metal casting is commonly referred to as a "gate". The gate includes the casting, runners, risers and sprues. The gating or "rigging" includes all excess metal such as the runners, risers etc.

Rigging removal is accomplished by several different methods. Table 2 summarizes these methods, their applicability and advantages and disadvantages. While robots can manipulate abrasive wheels, lasers and circular saws, work at Unimation Inc. has been concentrated on cut-off by oxygen fuel cutting systems. The choice was based on the ability to 1) cut contours, 2) cut fairly thick sections, and 3) cut at fairly high rates of speed.

The Unimate 2005F in the Cleaning Room

A Unimate 2005F robot with continuous path and velocity control has been installed at Fort Pitt Steel Casting, Div. of Conval-Penn Inc., McKeesport, Pennsylvania, U.S.A. to remove rigging from a large variety of gates. (See Appendix 1 for a description of the model 2005F Unimate robot).

The Fort Pitt foundry produces about 4500t of carbon and alloy steel castings annually in average lot sizes of about 30 pieces. About three quarters of this output is made up of various types of valve bodies, butterflies and bonnets supplied to Conval-Ohio and Lunkenheimer, both divisions of Conval Corporation. The remainder of output goes to a variety of low volume consumers. The majority of castings fall into a weight range of 4 to 50 kg. A typical large size gate is shown in Fig. 4. Currently, rigging removal is entirely manual using flame cutting with natural gas.

The Fort Pitt project has been approached on an experimental basis to determine methods and discover pitfalls of this robot application and to establish generic parameters. The total project objectives are to:
1. Categorize product by size, weight, lot size, order rate;
2. Establish economic lot sizes for automation;
3. Study material handling methods in and out of the cleaning room;
4. Devise universal holding fixtures for gates;
5. Evaluate the need and devise sensory feedback where needed;
6. Devise means for automatic control of oxy-fuel system;
7. Evaluate cutting parameters, robot programming, accuracy and cost effectiveness;
8. Recommend process and gate design changes as required.

Because of their fundamental impact, objectives 4-7 have been attacked first. Each will be discussed.

Fixturing of Gates

A holding fixture is required to position the gate in a known position relative to the robot. Because of the large number of gate configurations it will be necessary to have a family of fixture designs based on the information derived from objective #1. The accuracy and repeatability of gate positioning will impact the total cost effectiveness of the cut-off process. This is because inaccurate cuts or cuts that leave large pads (see Fig. 5) will require subsequent operations such as air-arc wash or grinding to produce the finished casting.

Unimation Inc. engineers decided to evaluate a fixture design which was conceived as part of the Steel Founders Society of America Project 100 results (Ref. 1). This fixture design provides X, Z positioning by clamping on the runners of the gate which have been altered in cross section to fit into V blocks. Y positioning is achieved by positioning the gate sprue in a locating cup. While this would seem to be a viable approach that assumes families of gates having different castings could have common rigging, it also assumes that the geometry of the gate is always the same. Unfortunately this is not true. Distortion and variations due to gate dimensions and geometry, temperature of the gate at shake out and rough handling all contribute to errors in position, particularly for gates with heavy castings. The desired positioning repeatability is at least $\pm$ 1.0mm. Additional work in this area will be required. Unless methods are changed prior to arrival of the gates in the cleaning room it will be necessary to sense out-of-position gates and introduce correction automatically.

Sensory Feedback

The amount of undesirable metal left after cut-off (called pads) will determine the need and cost of subsequent finishing operations. Hence it is highly desirable that as close a cut as possible is made. Since there will be dimensional gate variations and fixturing tolerances the spatial relationship between the robot and fixtured gate will want to be fine tuned. This can be accomplished by either modifying the robots (pre-recorded) program or by adjusting the position of the holding fixture through some kind of sensory feedback.

While an analysis has yet to be made, it will be probably be simpler and more cost effective in this application to reposition the gate fixture rather than adapt the robot's program. Through the use of a video system (Ref. 1, 2, 3) key features of the gate would be scanned to servo the fixture into juxtaposition with a computer stored datum.

An even simpler scheme being considered is to cast reference points strategically into the gate that can be proximity or force sensed for fine tuning the gate position.

Automatic Oxygen-Fuel System

To automatically control the oxygen-fuel system to the Unimate robot mounted cutting torch a special control board was constructed, Fig. 6. The control provides for pressure regulation and flow control of both oxygen and fuel gas in a manual mode as well as automatically as a function of programmed Unimate robot I/O functions. The control also provides a dual mode of oxygen regulation, high pressure for high pre-heat temperatures and low pressure for cutting temperatures. This is an energy conserving measure.

Also interlocked with the robot program is an electric ignitor (Fig. 7) mounted adjacent to the cutting torch nozzle. Thus, ignition is automatic and further fuel economics can be realized by turning the flame on only when the robot is about to make a cut.

Unimate Robot Cutting Performance

A single casting gate as shown in Fig. 4 was used in the initial evaluations. This is 10cm gate valve body. The complete as poured casting weighs 68 kg. The finished casting weighs 23 kg. Material is low alloy carbon steel. Gate location in the fixture was established with reference to "as cast" surfaces of the casting itself.

Fig. 8 shows the gate being flame cut by the Unimate robot system. Four cuts were made. Referring to Fig. 5, cut "A" removes a large riser. The remaining pad is a cosmetic surface only. Cuts "B" and "C" are at a bolt flange which must be later machine finished. Cuts "D" and "E" are actually made first and simply reduce the rigging to a manageable size for remelting. Making these cuts while the gate is fixtured reduces subsequent material handling costs.

Teaching (programming) the continuous path robot for the four cuts was done in less than 10 minutes once the correct torch attitude and flame settings were determined. This involved considerable experimentation. It become quite obvious that there is an art in the process which a human operator develops with the aid of his sensory perceptiveness. Attitude of the torch axis with respect to surface contour, distance from the workpiece, velocity of tip trowl and oxy-fuel settings are critically interrelated to achieve satisfactory cuts.

Results of cutting performance can be summarized as follows:
1. Cuts made by the robot were smooth and repeatable in areas of clean metal. The quality of the cut was superior to those made manually _u- ..
2. Cutting rates were in line with manufacturer's cutting guides (e.g. 500mm/sec for 25mm thickness; 380mm/sec for 50mm thickness). However, for cleaner smoother cuts using smaller tips, rates were often 50 to 75% of the listed rates.
3. Positional repeatability of the cut (gate to gate) was better than 1.5mm and quality and cutting times were consistent except when material variations such as sand inclusions were present. In such cases the flame was diverted and sometimes the cut was interrupted.
4. Pre-heat times (dwell time required to raise metal to cutting temperature) were totally unpredictable and upon the cross sectional area of the pre-heat zone. For thin edges, pre-heat time varied from 20 to 40 seconds. For thick sections pre-heat time varied from 60 to 80 seconds and often was never achieved. These conditions prevailed even with the dual mode energy level control.

These results suggested several immediate courses of action:
1. Further evaluation of torch tip sizes to optimize cutting speeds, quality of cut and energy consumption.
2. Evaluation of offset tips to facilitate programming in tight quarters of the gate.
3. Determine how to speed up and increase predictability of pre-heat times.
4. Investigate means for sensing improper cutting due to material defects so that the cut can be automatically terminated. This is desirable, particularly on large castings, since the bad cut may require expensive repair welding or result in scrapping the casting.

As of this writing some results of pre-heat investigations are available. In summary it can be stated that pre-heating at thin sections takes place rapidly and predictably. Furthermore, thin section pre-heating results in cutting temperatures that are sustained even when immediately followed by thick sections. As part of this investigation various cross sectional areas were evaluated and the results are shown in Fig. 10

Unimation Inc. engineers have been advised by foundrymen that the casting in of "pre-heat tabs" is feasible. Experience to date suggests this will be the solution to pre-heat predictability and efficiency.

Summary of Automated Flame Cutting

Investigations into other cut-off anomalies are continuing. It can be concluded that metal removal by programmable, continuous path industrial robots is feasible and will produce superior, cost effective results compared to manual means. However, the application is complex and will require not only the dedicated efforts of roboticists but the direct cooperation and support of foundrymen whose business it is to produce a huge variety of product in small batches.

Conclusions

Metal removal by flame cutting is but one of the foundry cleaning room operations which can be done by the industrial robot. Air-arc wash, grinding and plasma-arc finishing processes will also come under consideration.

In addition other foundry operations such as mold venting, core setting shake out removal and even, perhaps, repair welding are potential robot applications. Already, thirteen Unimate robots are being installed in a major U.S. foundry to spray and dry cope and drag molds (see Appendix II).

Foundrymen in die casting have successfully reaped the benefits offered by Unimate robots motivated by a highly competitive business atmosphere. Investment casting foundrymen have turned an art into a science through the use of robots with dramatic cost and quality benefits. Ferrous metal foundrymen are at the threshold of automated systems with similar potential benefits yet to be realized.

References

1. Steel Founders' Society of America Report No. 77-176: "Machine assisted finishing of steel castings - metal removal". (June 1977).

2. Holland, S.W.: "An approach to programmable computer vision". Society of Manufacturing Engineers Technical Paper MS77-747. (October 1977).

3. Rosen, C.A., Nitzan, D.: "Some developments in programmable automation". IEEE Intercon 75. (April, 1975).

4. Abraham, R.G., Stewart, R.J.S., Shum, L.Y.: "State-of-the-art in adaptable-programmable assembly systems". Society of Manufacturing Engineers Technical Paper MS77-757. (October, 1977).

5. Hill, J.W.: "Force controlled assembler". Society of Manufacturing Engineers Technical Paper MS77-749. (October, 1977).

APPENDIX I

Description of

UNIMATE^R Series 2005F

(with Velocity and Path Control)

Specification 2005F Series Unimate^R

Mode 1 - Point-to-Point

Programming: Record-Playback using Teach Control

Positioning Accuracy: $\pm$ 1 millimeter ($\pm$0.040 in.)

Max. Radial Velocity:	765mm/Sec (30 in./Sec)
Max. Vertical Velocity:	1270mm/Sec (50 in./Sec)
Max. Rotational Velocity:	110°/Sec
Max. Wrist Bend Velocity:	110°/Sec
Max. Wrist Yaw Velocity:	110°/Sec
Max. Wrist Swivel Velocity:	110°/Sec

This mode of operation is used for basic positioning movements.

Mode 2 - Velocity Control

Programming: Record-Playback using special marking tape

Working Speed: 102-3250mm/min (4 to 128 in./min) resultant electrode tip velocity

Speed Regulation: $\pm$ 2%

Accuracy: $\pm$ 1mm ($\pm$0.040 inches)

Automatic Speed Selection: 4 independently adjustable channels

Ambient Temperature: 0°C to 50°C

Power: 220/440 volt 60 HZ

                                                           11.5 KVA

                                      380 volts 50 HZ

Technical Description of Continuous Path Operation

The Principle of Linear Interpolation

Fig. 1 illustrates a typical convoluted path along which the tool should be moved. The tool is illustrated as being at a succession of points along the desired path. Note that the attitude of the tool must be changed as it moves along the path, in order to keep the tip normal to the work surface. This shows that the articulations of the Unimate^R arm must continually adjust (in relation to each other) to accomplish this in a satisfactory manner. In other words, each axis of the Unimate^R should move in synchronism with the others.

Assume that the exact positions illustrated are programmed into the Unimate's memory, but that only the positions showing the tool have been recorded - how can the Unimate^R

be controlled to follow the dotted line between the points?

The Unimate^R Control Computer accomplishes this feature by dividing the distance between two taught points into a multiplicity of small, equal increments. It then controls the Unimate^R arm to move sequentially over these incremental distances at a controlled rate. Since the number of incremental moves is the same for all axes of the arm, then if one articulation has to move further than another, each move will advance each articulation the same <u>proportionate part</u> of the full distance, thus synchronizing the total movement.

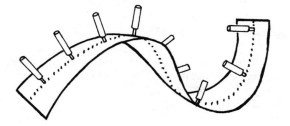

FIGURE 1.   LINEAR INTERPOLATION

Programming

The work is held in the desired position and especially marked tape applied exactly where the bead is to be placed. The tape is used as a teaching aid and greatly reduces programming time.

The point to point steps are taught first. These steps include those which approach the workpiece, retract from the workpiece, and perform any gross reposition motions when moving from one area of the work to another.

Velocity control steps are taught by positioning the tool tip over the tape markings. The marks provide convenient fixed points against which the tool tip can be placed to record steps of equal length.

Attention must be paid to the <u>Record Length</u> and <u>Tool Tip Velocity</u> switches when recording velocity control steps.

Record Length Selector

The Unimate acts as a continuous path machine via a process called Linear Interpolation. This means the distance between taught points is divided into smaller equal increments, which then appear as desired position from memory. The record length selector determines the number of small increments between recorded points. The switch position should be coordinated with the tape during the teaching process. When steps are recorded every eight tape divisions, the switch set to the "8" position etc.

A general set of rules for selecting how far apart to place recorded positions follows:
  1. Never place recorded points more than 8 divisions apart. Areas with little curvature lend themselves to this setting.

2. Where a curvature is present, such as traversing the circumference of a pipe, measure the radius with the special marking tape.
If the radius is:

   1 to 4 divisions select position 1/2
   5 to 8 divisions select position 1
   9 to 16 divisions select position 2
   17 to 32 divisions select position 4
   33 to 64 divisions select position 8

The tighter the curve the more closely the recorded points should be.

3. Teaching points should be kept as far apart as will permit following the contour properly. This minimizes teaching time and also uses memory most economically.

Tool Tip Velocity Selector

The speed the tool tip will move over the work when the Unimate^R is performing automatically is determined by this switch. Four ranges are available:

1. 4 to 16 inches per minute
2. 8 to 32 inches per minute
3. 16 to 64 inches per minute
4. 32 to 128 inches per minute.

There are four potentiometers on the Unimate^R next to the tip velocity switch, one potentiometer for each range. The potentiometer functions only in the Repeat mode, and selects the exact velocity performed within each range. The tool tip velocity will remain constant as long as coordination between number of divisions on the marking tape and the Record length selector is maintained.

Operation and Troubleshooting Indicators

Indicating devices on the Unimate's front panel provide useful information for the programmer and troubleshooter.

Memory and Encoder Display

There are two five digit displays which provides a decimal number readout of signals from the Unimate^R position indicator (encoder) and memory. When the Unimate^R reaches the desired position, both memory and encoder numbers will agree. This provides a powerful troubleshooting tool when needed.

Memory Input and Memory Output Display

Two groups of 16 Light Emitting Diodes (LED's) whose primary use is to display the auxiliary function command information. This display is especially useful to the programmer as it clearly shows on which program steps are taught signals which allow the Unimate^R to interact with other equipment. In addition, the display is useful to the troubleshooter as both auxiliary and positional information can be selected.

Unimate Robots in Foundry Mold Preparation

At Caterpillar Tractor Company, Mapleton, Illinois, U.S.A. iron foundry Unimate series 2105B industrial robots have been installed on several mold cope and drag lines. Fig. 1 illustrates pictorially a typical line. Some of the lines are indexing, others are continuously moving.

As shown some robots wield spray heads to spray refractory wash into the copes and drags. Following this additional Unimate robots play gas torch flames onto the molds to dry them.

Each robot has up to sixteen programs in its local memory. Identification of the molds as they come down the line triggers proper selection of the program developed for that mold configuration.

The flexibility of the robot provides a very cost effective means for performing these operations in a batch processing environment.

Spraying with the robots assures coverage of all critical surfaces without overspraying. Drying with the robots also assumes proper coverage, while reducing fuel consumption by 67% compared to oven drying.

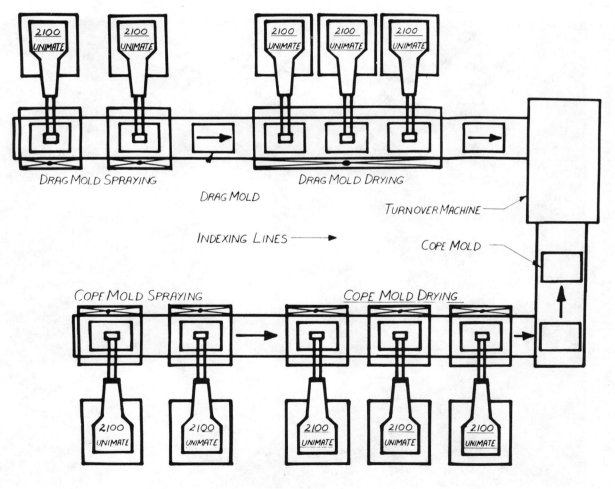

1. Figure 1 - Unimate Industrial Robots in Foundry Mold Preparation

Basic Operations
_____

Pattern Preparation
Cope and Drag Mold Manufacture
Mold Curing, Venting
Core Making and Setting
Metal Pouring
Shakeout
Casting Cleaning
Repair Welding
Finishing, Heat Treatment
Inspection

Cleaning Room Operations
_____

Shot Blast
Air Hammer Chipping
Sledgehammer Flogging
Flame, Plasma, Water Jet Cutting
Air-Arc Washing
Grinding
Saw Cutting
Repair Welding

Table 1. Ferrous Metal Foundry Operations

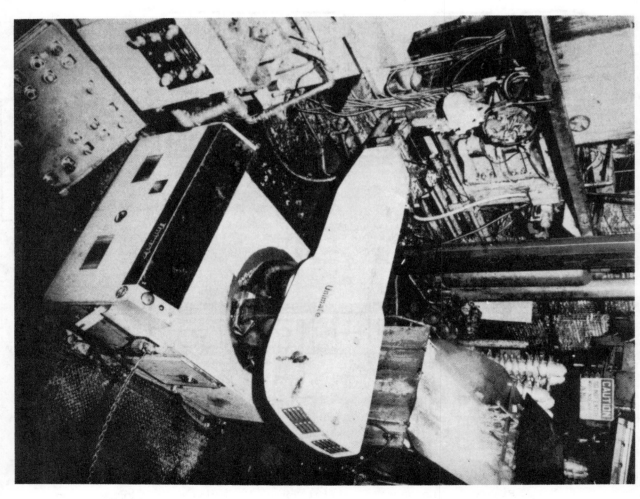

Figure 1 - Typical Unimate Industrial Robot Installation in Die Casting Foundry

SUMMARY OF CHARACTERISTICS OF STEEL CUTTING SYS.
TAKEN FROM SPECIAL REPORT #15 (S.F.S. OF A.)

CUTTING METHOD	THICKNESS (T)	CUTTING RATE (V) (FOR 1-IN. THICK MAT.)	WORKPIECE MATERIAL	ADVANTAGE	DISADVANTAGE
1) ABRASIVE WHEEL	$T < 2\frac{1}{2}$ INCHES	APPROX. 20 I.P.M.	ANY STEEL	1) LOW CLAMP FOR. 2) MIN. HEAT GEN.- DURING CUT. 3) SMOOTH CUT SURF. 4) EASY TO SENSE WHEN CUTTING. 5) CLOSE CUT TO CASTING CAN BE MADE.	1) STRAIGHT CUTS ONLY 2) LIMITED ACCESS
2) OXY-FUEL	0"-10" MOST ECONOM.	APPROX. 20 I.P.M.	CARBON STEEL LOW ALLOYS.	1) CAN CUT CONTOURS 2) GOOD ACCESS TO MOST AREAS OF CASTINGS. 3) SMOOTH CUT SURFACE IF AUTOMATIC.	1) FLAME DEFLEC. BY SAND. 2) CRACKS ON HIGH ALLOY STEEL. 3) MUST LEAVE SOME PAD.
3) PLASMA-ARC	0"-4"	APPROX. 30 I.P.M.	ANY STEEL	1) CAN CUT CONTOURS 2) CAN CUT HIGH ALLOY STEEL WITHOUT DAM-AGE (I.E., SHALLOW HEAT AFFECTED ZONE) 3) NO PREHEATING REQUIRED.	1) LIMITED ACCESS. 2) MORE COSTLY EQUIPMENT THAN OXY-FUEL
4) LASER	0"-$\frac{1}{4}$"	$\frac{1}{8}$" @ 30 I.P.M.	ANY STEEL	1) VERY ACCURATE CUT. 2) NARROW KERF. 3) SHALLOW HEAT AFFECTED ZONE 4) CAN CUT CON-TOURS.	1) HIGH CAPITAL COSTS.
5) BAND SAW	ANY	5 I.P.M.	ANY STEEL	1) CAN CUT CONTOURS.	1) VERY SLOW. 2) DIFFICULT TO REACH SOME AREAS 3) BLADE GUIDE ADJUSTMENTS FOR EACH SIZE PIECE REQUIRED.
6) CIRCULAR SAW	0"-3"	20 I.P.M.	ANY STEEL	NONE	1) EXPENSIVE TO REPLACE BLADES. 2) SAFETY HAZARDS FROM BROKEN BLADES AND TEETH. 3) HIGH CLAMP FORCES. 4) STRAIGHT CUTS ONLY.

Table 2. Summary of Steel Cutting Systems

Figure 2 - Typical Unimate Industrial Robot Installation in Investment Casting

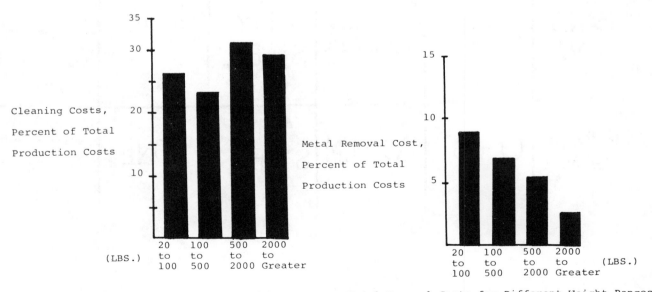

Cleaning Costs for Different Weight Ranges   Metal Removal Costs for Different Weight Ranges

Figure 3 - Cleaning and Metal Removing Costs for Different Weight Ranges

Figure 4 - Complete Gate for 10cm Steel Gate Valve Body Casting

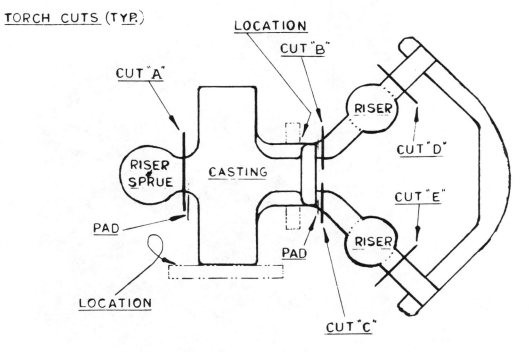

Figure 5 - Steel Valve Body Gate

Figure 6 - Oxygen-Fuel Control Panel

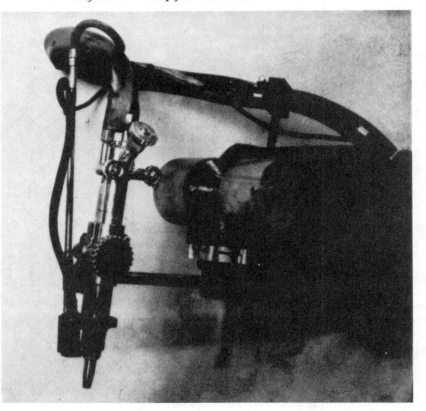

Figure 7 - Cutting Torch Showing Oxygen-Fuel Electric Igniter

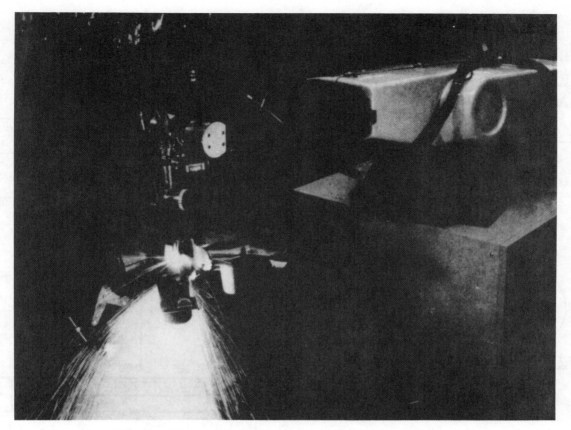

Figure 8 - Unimate Industrial Robot Flame Cutting Rigging from Casting

PATTERN #1

PRE HEAT TIME 35 TO 45 SEC.

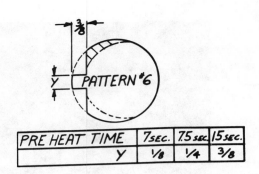

PATTERN #6

PRE HEAT TIME	7 SEC.	7.5 SEC.	15 SEC.
Y	1/8	1/4	3/8

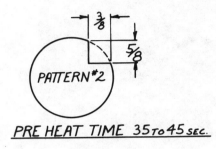

PATTERN #2

PRE HEAT TIME 35 TO 45 SEC.

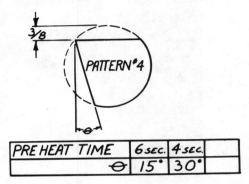

PATTERN #4

PRE HEAT TIME	6 SEC.	4 SEC.	
θ	15°	30°	

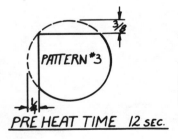

PATTERN #3

PRE HEAT TIME 12 SEC.

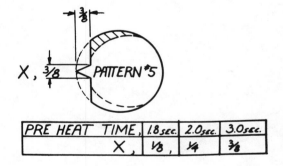

PATTERN #5

PRE HEAT TIME,	1.8 SEC.	2.0 SEC.	3.0 SEC.
X,	1/8,	1/4	3/8

Figure 10.   Experimental Cross Sections to Reduce Pre-Heat Times

Reprinted from Foundry Management & Technology, November 1977

SWEDEN PIONEERS FINISHING PROGRESS

Electro-hydraulically controlled manipulator grinds large steel castings efficiently. The operator controls the machine from within a cabin that can be raised from the ground. The pendulum grinder is suspended from a travelling crane capable of cross movements. The 610-mm grinding wheel has a peripheral speed of 60 m/sec, and grinding pressure ranges from 300 to 3,000 N. The casting rests on a turntable.

FOUNDRIES in Sweden have done a great deal of work in the attempt to increase productivity of their cleaning operations. Government regulations related to safety and health and environment, typically more stringent than those of our own OSHA and EPA, have caused their progress to be more rapid than ours. One technique that has been used with great success in Sweden has been to remove the operator from the heat, dust, noise, and danger of the cleaning room and to replace him with a manipulator or industrial robot. Development work in Sweden on the use of manipulators and industrial robots has been made by the Association of Swedish Foundries in cooperation with machine manufacturers and foundries.

Robots have proven their worth by grinding flash from castings. They also are capable of chipping and cutting off gates and risers. The technology is available to use a number of robots, in sequence, to perform successive operations.

Pendulum grinding always has been a heavy, dangerous foundry operation. In Sweden, remotely controlled manipulators have made it possible to remove the man from the work area and at the same time to increase productivity and reduce labor costs.

Removal of flash has been successfully performed on a high-production basis with hydraulically powered trim presses. For several years, this has been done with ductile iron castings. Flash also has been removed via the cryogenic process. Castings are cooled in a bath of liquid nitrogen to the point at which gates and risers, as well as flash, can be broken off easily.

Transfer machines have been used in Sweden to clean large quantities of identical castings—specifically cast iron engine blocks. The castings pass through machines that remove flash and also perform some surface grinding operations.

Handling large castings has been simplified through the use of a positioner akin to those used in high production welding. The actual cleaning operations then are performed by a remotely controlled manipulator with exchangeable tools for cutting, grinding, and chipping.

When it is necessary to clean castings manually, much attention has been paid to the workplace—especially in terms of safety. The mandate for foundries to control the threshold limit value for dust has caused many Swedish foundries to install individual cleaning booths. These units solve problems of dust, ventilation, and protection from flying sparks and metal, but are expensive to purchase. They also isolate the workers. These are viewed as temporary solutions until more sophisticated means of cleaning are discovered and brought onstream.

Lars Villner, director, Association of Swedish Foundries, suggests that the ultimate way to improve casting cleaning operations is by eliminating them. Rather than searching for better ways to clean castings, researchers should look for better ways to produce castings. Elimination of risers, more accurate patterns, and more accurate molds and cores could reduce drastically the amount of cleaning required.

Another solution being explored in Sweden is to allow the machine shop or the customer to remove gates and risers as a part of the machining operation.

Industrial robot is being programmed to grind off flash from a casting. Normal programming capacity is 250 position instructions, carried out according to the point-to-point principle. Changes, erasures, and additions to existing programs can be made easily. The robot has a lifting capacity of 60 kg.

Ingates and feeders from steel castings are being cut off by a robot. Because of the noise and dust associated with the process, the cutting is done in a separate room. The operator places castings on the turntable that rotates into the workroom. He monitors the operation in safety and comfort through a window.

Reprinted from Robotics Today, Fall 1979

Robots

Clean Up

at the Foundry

It is often difficult

to find the manpower

for the demanding work

of grinding and cleaning

iron castings.

OSHA requirements

make the search for labor

even more arduous.

An ASEA industrial robot

with electrical drive systems

is obtaining excellent

cost-effective grinding results

HANS SKOOG
and
HANS COLLEEN
Engineering Managers,
Industrial Robot Systems
ASEA Incorporated

1. The robot is shown here cutting ingots at Kohlswa Jernverk, Sweden.

In the foundry industry, manual grinding is often very difficult and performed in an undesirable environment. Regulatory agencies have dictated that conditions in the cleaning room be improved. One answer to the cleaning room problem is the implementation of robot casting cleaning. Industrial robots are cost-effective solutions when dealing with minimum lot sizes and two and three-shift operations.

ASEA's industrial robot model IRb-60 has been used with success in a number of applications in the grinding field where robotic precision, contouring capability, and force sensing are important attributes. Here are highlights from application studies of actual installations using the ASEA IRb-60 for grinding iron castings, *Figure* 1.

Grinding Characteristics. When grinding is performed by a robot, a certain pressure must be exerted against the casting. Any variation of fin contour must be removed without damaging the surface of the casting. Normally, a robot is programmed around a finished casting. As the robot follows the contour of the casting, with fins of various sizes along the parting line, there is a variation in pressure from the robot as it attempts to maintain the original programmed path. The larger or higher the fin on the parting line, the greater the pressure exerted by the robot. In addition, the force depends on the robot's rigidity and the distance of the programmed path from the casting.

A correlation can be seen in *Figure* 2 between the force built up from the variation in fin size and the various rigidities of the robot itself. It is assumed that this correlation increases proportionally to the distance from the preprogrammed path. The grinding force generally increases with the fin size. This has a progressive effect on the cutting volume, which is more pronounced with stiffer robots.

For a constant contact force, the cutting speed will be proportional to the thickness of the material (i.e., to the grinding pressure). The grinding pressure will vary greatly depending on the shape of the fins.

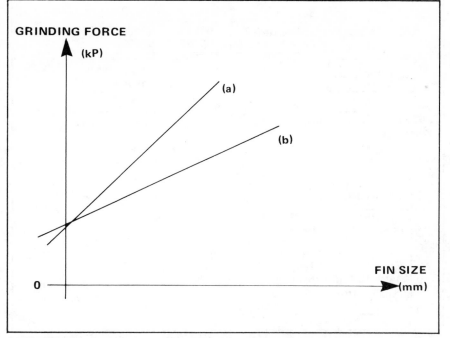

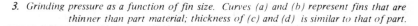

2. *Grinding force as a function of fin size. Curve (a) has double the stiffness of curve (b).*

There are two main types of fins: those with a thickness much smaller than that of the material of the part itself, and those with a thickness about equal to that of the material of the part. *Figure* 3 shows the grinding pressure achieved with a constant force according to *Figure* 2 for the two different types of fins. In the case of thin fins, the grinding pressure will decline radically when the material behind is contacted, which means that the change in cutting speed is significant.

In *Figure* 4, the removal rate is shown as a function of the original fin size for the different fin shapes and for the different stiffness of robots. In all cases, the contact force selected must remove the maximum fin size in order to have a constant cutting time.

For smaller fins, the grinding penetration into the casting will follow different curves. Curves *a* and *c* show the cutting depth for a constant force independent of fin size. Curves *b* and *d* show the cutting depth for forces increasing according to *Figure* 2. Curve *e* describes the ideal case when the desired cutting is achieved independent of original fin size.

Figure 4 shows that fins much thinner than the material behind can

be cut successfully. It is also preferred that the contact force increase in proportion to the size of the fins. On the contrary, it is more difficult to achieve acceptable results with thicker fins.

Other factors that make it difficult

to cut with acceptable results are differences in the shape of the castings, possible mistakes in gripping objects, wearing of the grinding wheel, and the robot's difficulty in following the desired path.

The negative effects of the above factors can be decreased by reducing the stiffness of the robot and/or by connecting force sensors to the robot for these applications. The reduced stiffness capability decreases the change in the desired contact force due to tolerance faults and wearing of the grinding wheel, for example. This gives consistently better finishing results for thinner fins. The force sensors can be used to sense the size of the fin or sense when the grinding of the fin has been completed. They can also be used to compensate for the wearing of the grinding disc itself.

It is preferable that the robot handle the casting from magazines than that it be fed by a worker. There may be several necessary operations on the casting requiring several grinding machines. It is possible to install a number of grinding machines within the working range of the robot for

3. *Grinding pressure as a function of fin size. Curves (a) and (b) represent fins that are thinner than part material; thickness of (c) and (d) is similar to that of part.*

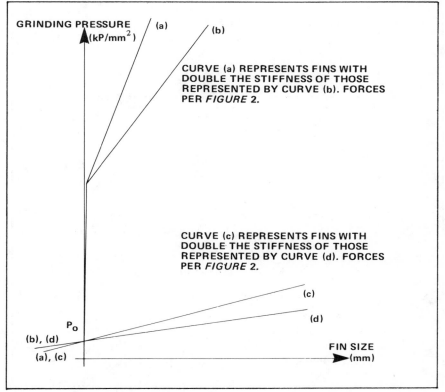

CURVE (a) REPRESENTS FINS WITH DOUBLE THE STIFFNESS OF THOSE REPRESENTED BY CURVE (b). FORCES PER *FIGURE* 2.

CURVE (c) REPRESENTS FINS WITH DOUBLE THE STIFFNESS OF THOSE REPRESENTED BY CURVE (d). FORCES PER *FIGURE* 2.

more than one application. In some cases the casting is too heavy to handle. The grinding unit is then attached to the robot and the casting placed in a fixture. This normally requires more manpower to feed the robot.

Faster cycle times per piece are possible using a robot than in a manual operation because the robot can achieve the higher grinding force. This is usually not true with small, lightweight parts that require very little metal removal.

The limitations of the systems are the total machine horsepower that can be carried by the robot and the characteristics of the grinding wheels themselves.

Automatic Grinding Installation. It is necessary to have the following equipment for cleaning of castings in an automatic grinding installation:

• Grinding equipment with sufficient horsepower.

• An industrial robot with good positional accuracy to perform the movements required and apply the grinding pressure needed for the application. The robot must be rigid but have the capability of adjusting to

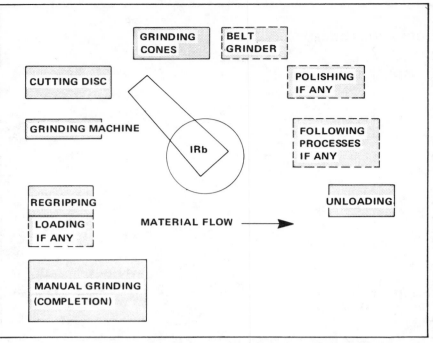

5. *Drawing of a practical layout for automatic grinding with the ASEA-IRb.*

the application, and must have force sensing which can determine when the grinding is completed for a variation of fin sizes. It must also be able to compensate for grinding wheel wear. It should be possible to read out the con-

tact force digitally if required.

The ASEA industrial robot system, model IRb-60, with its optional spring force servo, force sensing, and measuring equipment meets the demands of most grinding applications.

• The gripper should be attached to a well defined surface and removable fins located in accessible positions. Inside fins should be avoided unless they are on a radius. The automatic grinding installation is thereby simpler and more effective.

Grinding Equipment. There are four standard ways of grinding fins or parting lines from castings: breaking the large, brittle fins with a hammer; cutting with an air-operated chisel; grinding with cup or cone wheels; and cutting with cutoff saws. It is difficult to cut with a chisel automatically. Such operations should be replaced with trimming, cutting, or grinding with high machine power. Cutting of ingots should be made a separate operation as with an industrial robot.

Large and brittle fins must be broken manually before automatic grinding is done. Sand and other irregularities on the castings must also be removed manually.

The remaining grinding can be done

4. *Cutting as a function of fin size. Constant force is represented by (a) and (c), progressive force by (b) and (d). Curves (a) and (b) refer to thicker fins, (c) and (d) to thinner fins.*

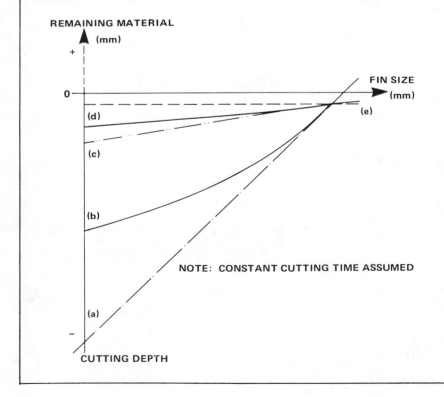

automatically. The grinding machine with the greatest horsepower should be used for most heavy grinding applications. The optimum solution for a grinding installation is to have the robot handle the casting and grind it on the largest stand grinder available.

The following list covers the horsepower used for various types of grinding tools:

Grinding wheels or cups: max power, 40 hp (29.8 kW)

Grinding cones: max power, 7½ hp (5.6 kW)

Cutting discs: max power, 75 hp (55.9 kW)

The Robot. The maximum contact force permitted on the ASEA IRb-60 depends on the load and torque that is seen by the wrist. A grinding installation is typically designed to counter normal loading of the robot arm itself. Grinding forces considered here are in the area of 30 to 60 kg, requiring 10 to 20 machine hp (7.5 to 15 kW).

With the optional spring force servo, the normal stiffness of the robot can be adjusted to reduce the values as required by the application. The electrical drive system makes this possible. By changing the speed controller to a proportional type, the return forces will be proportional to the position error. By using the outputs from the robot, the lower stiffness can be adjusted as required in the program. The spring force servo can be used to compensate for smaller variations in grinding disc size and small tolerance faults in the casting.

The robot can be programmed to stop or start as desired when the force sensors define a force as being above or below a certain level. A torque detector can be used on robots with electrical motors where the sensing is detected by the motor current, to avoid extra sensing equipment on the wrist. This arrangement generally offers a very good service life, since it is not in the area of the grinding itself.

In some installations, force sensors may have to be located on the robot's wrist. In such cases, the ASEA PRESS-DUCTOR system, seen in steel mills throughout the U.S. for weighing purposes, is used. This system can withstand great mechanical forces and survive in the worst environments, still maintaining very high accuracy.

The force sensing system can be used as a search function resulting in a transformation of the program. In practical use, the robot will constantly compensate for wearing of the grinding disc, detection of fin edges, and the casting surface.

The force sensing system can also be used to detect when the fin is cut and when it is time to continue to the next fin, especially in grinding ingots.

Figure 5 shows a practical layout of an automatic grinding installation. The station is designed to have a robot handling the workpiece. Several advantages are therefore realized. First, there is no limitation on machine horsepower. Too, tool changing is not required. Material handling is solved with minimum use of manpower. Finally, other in-process machining can be integrated with the automatic robot installation.

One operator working at several stations is responsible for the inspection and sorting of incoming castings; the removal of large fins; loading of castings in the magazine for orientation to robot; inspection, if required; and palletizing of finished casting, if required.

Advantages of Automatic Grinding. An automatic grinding installation made with an industrial robot offers the following advantages compared with manual metal removal:

• Higher and more uniform quality.

• Faster cutting times because of the utilization of greater horsepower and force which can reduce the finishing time by 50%.

• Better working environment, since the robot can be housed in a grinding enclosure.

• Safer working conditions for employees who are no longer exposed to exploding grinding wheels.

• Faster throughput of castings.

• Longer life of cutting disc and grinding wheels or cups.

Actual Installation. Located at Kohlswa Jernverk in Sweden, two

6. Robot holding the hydraulic grinding machine.

7. Milling segments before and after grinding operation.

ASEA IRb-60 robots are utilized in grinding operations. One robot is used for cutting ingots, *Figure* 1, with a hydraulic cutting machine of 55 kW, *Figure* 6. The disc has a diameter of approximately 300 mm, and the cutting speed is 0.5 sec per cm^2 cutting area.

The casting is manually placed on a turntable with two fixtures. While the cutting takes place, the operator removes the cut workpiece and loads the fixture with a new workpiece. When the cutting is completed, the robot signals the operator to index the table 180°.

The robot is placed in a large cage for effective protection from noise, dust, and possible disc breakup. To compensate for the wearing of the cutting disc, it is equipped with a torque detector connected to the search function.

second robot at Kohlswa Jernverk is used for grinding the milling segments, *Figure* 7, that have passed the ingot-cutting station earlier. The milling segments are made of stainless, acid-resistant steel, which has good dimensional characteristics.

In this case, the robot handles the mill segment. The gripper is hydraulically actuated and is equipped with a pressure switch controlling the gripping force to approximately two tons (17.8 kN). The station layout is shown in *Figure* 8. The overall robot station is housed in a large enclosure for environmental reasons. Grinding operations are accomplished by two electrical stand grinders.

Machine *3* has a power to 7½ hp (5.6 kW) with a disc diameter of approximately 500 mm. The robot can compensate automatically for the wearing of the disc up to 5 to 10 mm,

approximately the same way as for machine *3*. This machine grinds the narrow areas in the rear side of the milling segments.

The work sequence is as follows: The robot operator loads magazine *1* once per hour. The robot picks up the casting in a fixed position and moves it to fixture *2* for regripping to obtain an exact gripping position. The robot then moves to machine *3*, where the inner radius and one side are ground. The search program, which has a built-in torque detector to compensate for wearing of the disc, is used as the operation proceeds.

Small grinding operations are done on machine *4*. The casting is placed in fixture *2*, where it is regripped, and the outer radii on the other side are ground in machine *3*, as described earlier.

After the grinding operation is completed, the casting is put on the pallet in a programmed pattern. The total time cycle, including some manual grinding, is approximately six minutes per detail, compared with ten minutes for the complete manual system.

Today it is possible to accomplish many operations with a robot. More and more research is being done to develop the options required for optimizing future grinding systems. Because of OSHA and the problems of recruiting labor for the cleaning room, now is the time for introducing automatic grinding systems into the foundry cleaning room. ∎

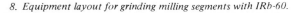

By using a robot for machining, the company achieves faster, more accurate cutting and better working conditions

By using a robot for cutting operations, the company achieves faster cutting (because higher force can be used), more accurate cutting (because there will be less material to remove in the later grinding operations), and better working conditions. In addition, the ASEA robot automatically slows down to cutting speed when the disc comes into contact with the workpiece; it maintains the same cycle time independent of the disc diameters.

Grinding of Milling Segments. The

corresponding to approximately 24 hours of operation. The stand grinder must occasionally be moved with a jackscrew to compensate for the 65-mm wear permissible on the grinding wheel. This machine grinds the outer and inner radii and the sides of the milling segments, shown in *Figure* 9.

Machine *4* is rated approximately 2½ hp (1.9 kW) and uses a smaller and thinner grinding disc. Wearing of the disc in this case is negligible. Compensation for the wear is achieved in

Adapted from "Grinding of Castings with Industrial Robots," MS78-679.

8. *Equipment layout for grinding milling segments with IRb-60.*

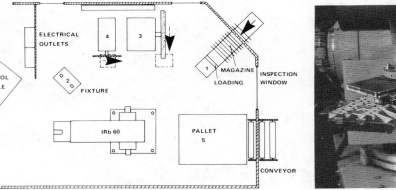

9. *Grinding of milling segments in initial laboratory test.*

CHAPTER 8

PLASTIC MOLDING

Commentary

Robots are used in the following plastic molding operations: injection molding, structural foam molding and compression molding.

The robots load inserts and compression molding charges, unload molding machines and presses and perform secondary operations such as degating, trimming, machining, assembling and packing. These secondary operations require die trimmers, gate cutters and fixtures, rather than the simple hand tools usually used by people.

Equipped with appropriate hand tooling, the robots can unload multiple-cavity molds in a single operation. They can reach into large machines which would require that an operator actually enter to unload.

Robots can improve quality and reduce scrap in plastic molding processes by operating at a consistent, optimum cycle rate.

Presented at the Fifth International Symposium on Industrial Robots, September 1975

Robots In Plastic Molding

By Bud Gregory
GTE Automatic Electric Company Inc.

Mechanical robots have been applied successfully to industrial applications, lifting them from the specialized and limited tasks of handling hazardous materials and working in unsafe or extremely fatiging environments. The layman is quite aware of robots being used to handle radioactive substances and working in spaces where extreme temperature variations are encountered.

The power and control systems developed for the early use of robots have been developed to a point that makes currently marketed machines highly reliable and places their cost within the range of many potential industrial users. Most robots are "powered" by hydraulic or pneumatic means using control systems that are principally of solid state construction. The control system is supplied with its command signals from a computer, rotary cams, punched "NC" tapes, and sequence "patch boards."

Mechanical robots have been developed to the advantageous position of being able to perform manual labor tasks without the adverse effects of fatigue, absenteeism, and possibility of injury. The power and control systems can be maintained by competent personnel after a few hours of instruction which is generally furnished by the robot manufacturer.

The above discussion is not implying the robots are to be considered as euphoria -- quite the contrary. There are far fewer applications for robots than for manual labor and automatic machinery. We must also keep in mind that manual labor offers the ultimate in versatility.

So that we may progress to the basic steps that are necessary (at least recommended) for a successful robot installation, we will define these steps and use two case histories from a plastic molding application as emphasis.

PRODUCT SELECTION

The money invested in a robot for use in tandem with an injection molding machine is considerable. The ratio is from one third to as much as the investment in the molding machine itself. To justify the need for the robot, it is desirable that the product occupy nearly the total production capacity of the machine. The product should represent a stable part of your total product line, preferably with a life of ten years or greater. The decision for investing in a robot based on product life is largely determined by your accounting procedures governing cash flow.

The on-going energy crisis has caused the cost of plastics to inflate to a level that drastically alters the ratio of processing cost to material cost, and in particular, the ratio of direct labor to material cost. It is common for direct labor costs to be one third and less of the finished cost of a product. Lowering of labor costs is one of the primary factors for installing a robot, even though its fractional size is decreasing. Product volume and life are the multipliers which provide you with one of the economic justifications.

In general, it may require several hours for a molding machine and mold to come to a stable thermal equilibrium level that produce products of satisfactory and uniform dimensions and appearance. Since material and labor costs have been incurred during this stabilization period with little if any part volume salvaged, these costs must be included as part of total cost for producing a given order quantity. To minimize the ratio of the start-up cost to total cost, long production runs are desirable. A robot can be used for part extraction during the warm-up and actual production run intervals. If the robot is replacing an operator, no direct labor costs are being incurred, but your variable overhead costs per hour are increased by virtue of the amortization effect of the investment for the robot. You have also eliminated "shift differential" costs if you have this in your wage policy.

PRODUCT EVALUATION

After a product(s) has been selected, we move to evaluating the product design and desired quality to insure that a robot will not adversely affect any of these features. This step is usually the most time consuming and requires input from the most segments of your organization. It is not uncommon to discover previously overlooked or unnoticed design deficiencies of a product in this phase.

Pickup points or grip points are a first factor to consider. If the surface of the part requires asthetic quality, grip points and grip devices are limited. The most commonly used grip points are part edges, round or rectangular openings, ledges, or smooth flat surfaces on the product. The grip points should remain clear of side actions, ejector pins, and extended cores in the mold to prevent damage in the event of a malfunction.

The part should have sufficient ribbing or geometry change at grip points to distribute gripping forces into the main walls of the parts. Grip points should be located as near the center of gravity as possible or spaced symmetrically about this center. Doing this minimizes the effect of dimension change caused by shrinkage (inherent in all thermoplastics), allows the part to self balance around one principal axis, and distributes extraction forces over the largest possible area.

Avoid placing grip points near areas of a product that may require development. This prevents having to modify grip tooling as well as the mold.

COST SAVINGS

To this point in our study we have selected our products to be automated using robot(s). We have evaluated the selected products from a design and quality stand point to insure that mechanized handling will not impair part integrity. At this point we have only achieved a cost savings by eliminating the need for a molding machine operator and volume is required to produce a savings that would amortize a robot in a reasonable time span. To develop additional savings, we turn to labor intensive operations such as secondary operations on the molded part itself and moving next higher subassembly operations internal to the product molding cycle.

A proficient molding machine operator can remove two parts from a molding machine up to approximately the 500 ton range as quickly as a robot can accomplish the same task. The operator can also degate, do light trimming, bag, and box the two parts during the next machine cycle. When the machine size increases, parts are automatically degated during ejection, and the number of parts molded per cycle increases beyond two, a robot is faster for part extraction and cycle restarting; again we are assuming that surface integrity is a prime consideration.

Time study analysis should be applied to the end product to determine which components requiring assembly at the next higher assembly level should be moved to your molding center. In general, the amount of assembly and packaging time should be 90% to 95% of the molding cycle time. The slack time allows for the assembly operator in the molding center to monitor the molding machine and robot and stop the cycle if a malfunction has occurred.

The type of subassembly method should be closely scrutinized. Since snap together assemblies usually require no equipment and only light fixturing, their application should be stressed. If a more rigid assembly method is required, ultrasonic welding and staking is a reliable alternative. It is even possible to use a single power source to control a number of ultrasonic welding stations. Ultrasonic welding equipment is relatively inexpensive. Ultrasonic welding operations are rapid, clean, and quiet.

If the above methods cannot be utilized on the end product, you will have to resort to the use of rivets or screws, with riveting being the least expensive of the two methods. These methods require more accurate fixturing, rivet machines, and power screw drivers. The biggest objection to these methods is the added cost to the end product itself as a result of the fastener.

You will undoubtedly discover many applications of fastening that offer more economical methods in the product lines you manufacture. You may just not have had the opportunity or been challenged to defend the method you are using. A product analysis of this type affords the opportunity.

A cost savings analysis of this type is more commonly performed by a captive or proprietary molder who is responsible for the end assembly. He enjoys the luxury of experimenting with alternate assembly methods on a product, testing any prototypes deemed necessary, and altering the type of plastic in joined components to accomplish the soundest assembly at minimum cost.

The custom molder does not have these luxuries at hand as a total responsibility. He does have one advantage though -- flexibility. The custom molder can try alternate methods of fastening, experiment with alternate plastics, and produce prototypes for his customers in a short time frame and a generally lower total cost. He may elect to add this type of service to his customers in addition to supplying parts.

As a result of offering this service, the custom molder is in a position to work hand in hand with his customers, engineers, and designers on cost savings developments. This type of teamwork between customer and supplier promotes full understanding of the intent (function) of the end product as well as individual components. The custom molder is suddenly in a position to offer some forms of subassembly of components to his customers. This will improve the productivity of his labor force with the reward being added profits.

The customer who receives the subassembled products can now determine whether he or his plastic components supplier can perform the required subassembly at a lower cost. If the component supplier offers the lower or equal cost, the customer may experience lower inventory cost due to fewer individual parts in his stockroom. This is particularly true if the subassembly is comprised of several plastic parts. If the customer is a multiplant complex, the inventory savings would result in each of his plants. If plastic parts are being supplied by several custom molders, this cost savings analysis may result in reassigning these components so that the custom molders will be able to perform the next higher assembly operation.

Another source of possible cost savings is to remove many parts from a mold that require careful handling if you are using a runnerless mold. If the parts were previously produced on a conventional sprue and runner mold, the molding machine operator could extract the parts from the mold by grasping the runner to which they were attached. At current plastic prices, this becomes an expensive handle. By using a runnerless mold, material costs can be reduced but the parts removal problem remains. The parts can be retained on the cores partially ejected so that the machine operator can free them from the mold. If more than two parts are molded per cycle the mold open time would be extended to allow clearing all the parts. If the parts cannot be partially retained on the cores, a robot extractor may be the only solution. The raw material savings is then retained and the labor cost of a molding machine operator provides additional savings.

Special tooling can be placed on the end of the robot arm to grip a multitude of shapes and as many parts as can be spaced in the mold. The limiting factor is tie bar clearance of the molding machine. The gripping action can be hydraulically or pneumatically actuated mechanical motions, vacuum cups, or simple "drop-in" receptacles. The advantage of hydraulic and air motions is that the source can be the robot power system itself.

After the parts are extracted from the mold, the robot can place the parts directly into shipping containers. If the containers are conveyorized past the molding machine, the robot control system can order the containers to the discharge station and return them to the conveyor when filled. This allows one machine operator to attend a greater number of machines than would be possible if he is performing a packing fuction in addition to other assigned duties.

This whole cost savings analysis leads to improved productivity, better utilization of floor space, effective inventory, quicker market response, and increased profits to the custom molder and his supplier. It will often be the only way that a custom molder can justify or even generate the required capital for automated equipment, particularly robots.

At this point in your evaluation, you have selected the product(s) to which to apply robots for various reasons, evaluated the design for continued integrity, and developed the economic reasons upon which your capital flow will depend. You are able to recommend what and why to your company management at this point. You may be wise in preparing a fact finding report now, in order to inform management of existing potential, and request approval to continue with the implementation program.

The next question to be answered is how.

MACHINERY SELECTION

We first make the assumption that we have the product but inadequate or insufficient machinery to produce the product. A package, even a "turn-key," purchase arrangement with the molding machine manufacturer is highly desirable. This is necessary since control interlocks have to be established between the molding machine control system and the robot control system. To affect the purchase agreement, you need to inform the molding machine manufacturer of your intention for a package purchase and furnish him with the names of robot manufacturers that are acceptable for your needs. Be sure that you select a robot with sufficient freedoms of movement to complete all the tasks required, but do not select more robot than is required to do the job. If you intend to adapt the robot to future tasks, have the control system designed for these tasks and retrofit the power system components when the future tasks are implemented.

A joint meeting with the robot manufacturer, molding machine manufacturer, and you is a very desirable step to review and finalize machinery specifications, establish delivery schedules of equipment, select the project managers for all functions and companys involved, and designate the point for and responsible personnel to perfrom the factory acceptance prior to shipment to the customer. It is also necessary that the molding machine and robot manufacturer include in the machinery cost an allowance for installation services in the customers plant. This service is most usually provided at a daily rate and will encompass three to five working days to complete. Make sure that all machinery is functioning to your specification and satisfaction. An additional several days of service man cost is a small investment for the complete and successful installation in your plant.

If you have selected any foreign manufacturers as an equipment source, be sure that electrical and hydraulic systems are compatible with American manufactured equipment and the systems in use in your plant. Of particular concern are electrical voltage, hertz, and phase standards, and hydraulic pressure and volume standards. Also, symbols appearing on system schematics should be compatible.

When making your machinery equipment selection, you have to place confidence in that choice. Select a reputable firm, with a proven record of prompt delivery, reliable operation, and competent service force. In some instances you may select suppliers of equipment already in use in your plant to minimize spare parts cost. In any event, there are numerous American and foreign manufacturers who can build equipment to your specifications and satisfaction.

SPACE ALLOCATION AND EQUIPMENT ARRANGEMENT

At this point our equipment has been selected. The physical dimensions required to make a plant floor layout are available from the selected suppliers. In addition to physical dimensions, the connection points for electricity, water, air, and hydraulic supply can be determined so that these services may be prepiped to a convenient connection point near the planned equipment location. Providing these points can speed your installation schedule by several days once the equipment is delivered in your plant. If you must break into existing facilities to get the services needed, you can schedule service interruptions so as to not hinder your normal plant operations. To try to schedule an interruption of services in a short time span after you receive your equipment is quite often difficult and expensive.

Several factors must be thoroughly considered in making your layout. Since most molding machines are constructed with the control console located on or near the stationary platen adjacent to the front sliding gate, this location must be kept clear so that personnel can control machine functions with safety. It is an added cost to the molding machine if this control console has to be mounted in other than this standard location on the machine. This limits robot entry into the molding machine to from above, the rear, or from under the molding machine platens. Entry from the rear and above the platens provide equal size openings since most horizontal molding presses have tie bars spaced in a square pattern; wide platen options are the exception, providing more clearance from above than from either side. Entry from below the machine is the most restricted and seldom used except for specialized products.

Placing the robot at the rear of the machine for tooling entry between the tie bars entails the least expense and the greatest versatility. In this location it is easily accessible for maintenance. The disadvantage is that the floor space area included within the swing radius with the robot's arm fully extended must be guarded to comply with O.S.H.A. safety regulations. It is permissible to locate machinery performing functions controlled by the robot within this guarded perimeter if personnel are not required in conjunction with this controlled function.

Placing the robot above the machine to permit entry through the machine tie bars from overhead conserves floor space. The only feature to be mounted on or near the floor is the control station. Overhead mounting is more costly because a frame work has to be erected. The supporting frame work can be extended so as to allow the robot to extract parts from a group of machines in a sequenced fashion. The most common scheme is to extract from two machines making similar or like parts allowing this robot to ride a rail between the two molding machines.

It is important to position the robot so that the extending arm moves perpendicular to the molding machine centerline and parallel to the horizontal centerline of the machine platens when extending into the mold opening to extract parts. Movement of the arm on this line will provide accurate position of gripping head at the take off position. To maintain this alignment it is good practice to anchor the robot support frame to a rigidly anchored framework or bolted to the plant floor.

When mounting the robot at floor level for rear entry into the molding machine, a frame constructed of steel channels or angles and bolted to the floor provides an inexpensive but functional platform on which to mount the robot frame. The frame is constructed by placing four channels to provide a pair of wheel guides and held in position by two or more cross members. Cross members should be placed on approximately five foot centers. The frame is then anchored to the floor by the cross members with expandable floor anchors. By using this method the robot can easily be rolled parallel to machine centerline on the rails for different mold openings, to provide additional room for mold setting, or clear of the machine for maintenance.

The wheel channels in the floor frame provide an access through which the robot frame can be anchored with pull-down bolts that are easily reached for loosening and retightening when robot repositioning is necessary. The floor frame also acts as an energy absorber to dampen the effect of positive or negative acceleration at the beginning and end of arm movement. The acceleration effect is significant enough to alter gripping head stability when it has reached the extract position. When positioning accuracy of approximately .050 inch or less is required, dampening capability is desirable and the floor mounted frame provides an inexpensive solution.

If the plan is to perform assembly on the plastic components you are molding using the robot, it is well to place the assembly stations at the molding machine utilizing the space adjacent to the fenced off safety enclosure or the space along the front of the machine after allowing clearance to electrical supply cabinets and control consoles. The use of light duty powered or gravity conveyors permits the orderly flow of parts and containers into and out of the assembly work stations that are positioned around the robot equipped molding machine. In our plant we have installed a 500 ton horizontal molding machine, robot, and assembly work station which occupies 950 square feet of floor space. Also included in this space is a dehumidifying dryer, four mold water temperature controllers, hot runner controller, and a plastic resin loading station. Another installation in our plant occupies 550 square feet including a 375 ton horizontal molding, robot, power conveyor, product packaging station, three mold water temperature controllers, dehumidifying dryer, hot runner controller, and a plastic resin loading station.   As a general rule, the length of the molding machine is the controlling multiplier for floor space.   This is due to the fact that machine widths increase at a smaller rate than machine length as the tonnage rating climbs.   Ancillary equipment is generally standardized in physical size and is required regardless of machine tonnage. The approximate width of the floor space is twenty feet for robot equipped molding machines from 100 to 500 ton in size.   Assembly operations performed at the molding machine generally utilize otherwise wasted floor space that is created by the "T" shaped arrangement of a floor mounted robot and molding machine.

If the robot is arranged for vertical entry from above the molding machine, it is feasible to transfer parts to a suspended mezzanine level for further assembly and packaging operations.   In plant sites where land values are at a premium it is sometimes more economical to expand up instead of out.   If overhead space already exists, converting it to usable work space is an alternative not to be overlooked.

EQUIPMENT INTERLOCKING

This is an imperative planning step that furnishes information to the machine supplier and people in your own organization.   The necessary interlocking is best planned on a sequence layout using bar chart format.   The elements forming the sequence layout is time in seconds plotted on a horizontal scale and the description of all functions occupying a time segment during the molding cycle ordered vertically.   The time required for the function is determined by an estimated timer setting or dividing the distance to perform the function by the rate of travel during the function.   The functions are then plotted on the sequence chart, the first function starting at "time zero" and starting the plot of a new function at the ending of the function immediately preceding it. By stepping the function plot vertically, the resultant plot forms a diagonal display.   When all functions are plotted, investigate any possibilities for overlap or coincidence of functions and adjust the elapsed time plot where necessary.

When your sequence layout is completed, it is a pictorial representation of your molding cycle.   The description of functions tell exactly what signals are required to initiate a function and the source that has to supply the signal.   When the signal is being supplied by the robot to a control device on the molding machine or vice versa, an interlock has to be established.   It is sometimes required that the initiation signal start two functions simultaneously; the sequence chart will alert you to this condition.   The sequence chart will assist greatly in determining if functions can be interchanged if the need arises.

Once you have developed your second sequence layout, you will most probably notice that the only apparent difference is the total elapsed time. This is not a coincidence since the function descriptions assigned are those that appear in the operating manual you received with the molding machine. The function descriptions you have added to the standard molding machine sequence chart are those inherent to the robot. If you intend to use a number of robots, you will design a standard chart for future use.

Your molding machine sequence chart will identify the electrical solenoids, limits switches, control relays, and timers that will be involved in the interlocking that you need. The sequence chart that you formulate should be included as part of your molding machine and robot operating manuals.

Functions that will be interlocked are cycle start, press opened, ejection start, and ejection complete. Some of the signals being transmitted with the above function signals will be "safe" signals to prevent machine and tool damage. Limits of clamp unit movements are set by limit switches; these switches should be mounted in a linear position indicator so that movement limits can be controlled to an accuracy of one eight inch. This option can be installed on a new machine for approximately $1,500. This feature is standard on some makes of machines.

The interlocking signals are completely electrical; if signals to be interlocked are not electrical they should be transformed so that they are electrical. The press manufacturer has to have the interlocking information at the time the molding machine is ordered. This allows him to alter his electrical system immediately and be able to deliver the molding machine on schedule. Since electrical signals are the only type being transmitted, other machine systems are unaffected.

The robot to molding machine interlock can be interrupted by means of a two position switch. This switch should be located on the molding machine operator console and of the style that is illuminated when in the "on" position. Placing the interlock switch in this console provides the capability to interrupt robot and molding machine operation from a safe vantage point, yet allowing visual scrutiny of equipment movement. With the switch in this location, there is a single position established from which to start the molding machine and robot as an interlocked unit. If an assembly station is part of the robot - molding machine center, an emergency stop switch for the robot will interrupt the cycle through the interlocking. It is then necessary for the cycle to be restarted from the main interlock switch on the molding machine console.

With the floor plan and interlock determination completed, you have determined __how__ in your analysis. If you have not ordered your equipment at this point - and it would have been premature to do so - your analysis will form the basis for the recommendation letter to company management. If they concur with your recommendation, you proceed to purchase order preparation if that was what you recommended.

SUMMARY

Your implementation plan has been formulated, the equipment ordered, and installation activity begun within your own plant. When it comes time to begin the use of your new equipment, the thoroughness of your plan will have resolved many of the last minute snags all of us have encountered while implementing a new method or beginning the manufacture of a new product. These snags result in time delays and added expense for which you had not planned. If you are beginning the manufacture of a new product that has an inflexible marketing date, a time delay is an unplanned event that cannot be tolerated.

why detail plan?

When forming your robot analysis team, select reliable and experienced people from your organization. If your application is on a molded product, you will need a manufacturing engineer from your molding division, an industrial engineer for plant layout and time study, the assembly engineer responsible for the "finished" product, and the design engineer for the product. This "basic" team may have to be augmented by people with other skills from time to time during the analysis. Designate one member of the team as the analysis (project) coordinator. All information dealing with the analysis should be routed to him in original or duplicate copy form. This leader should set all → *PERT ?!* schedules required to complete the analysis and implementation. The schedules will inform management of equipment availability so that production schedules can be formulated.

The analysis team may be informed of successful installations in other plants. It is often valuable to visit these installations to gather information. During these visits pay particular attention to the type of product being manufactured, arrangement of machinery, flow of material around the equipment, and the number of people required to produce the product and support the equipment operation. Do __not__ cut your visit short; a few more hours on the site to answer a few questions or gather more data can eliminate the need for return visits and make your planning more complete. It is better to have more information than required than to lack the answer to one seemingly insignificant point; it may become an important point in your analysis.

You have now completed all tasks necessary to have a successful robot - molding application. By appointing well qualified people to the implementation team and ensuring that they complete the analysis steps in proper order, you will attain the highest degree of success. Do not "short cut" any step in the procedure that has been outlined above. Depending on the size of equipment you require, you will find the study phase will take one to three months, delivery of equipment four to ten months, and installation and debugging less than one month. In total, six to twelve months lead to a successful implementation program.

Industrial Robot Automates Transfer Molding

Labor savings, increased output, and reduced cycle times were the objectives when this firm installed industrial robots.

GTE Sylvania Inc., Warren, Pa., has initiated a program to investigate the feasibility of automating their transfer molding machines, with the multipurpose in mind of upgrading part quality, increasing output, decreasing press cycle times, and reducing labor costs. The firm both injection molds and transfer molds a proprietary line of computer connectors, electrical connectors, photographic flash cubes, and cathode ray tube yokes.

They have installed a Series 2100 Unimate industrial robot which operates two 50-ton transfer molding presses. The robot manufactured by Unimation Inc., Danbury, Conn. has an articulated arm with a gripper which is controlled by electronic memory. In addition to unloading the two presses alternately, the Unimate also performs secondary operations during the cycle times of the presses.

There were three phases of the program for automatic transfer molding: 1) the pill (molding compound) must be preheated and placed; 2) the part removed from the mold; and 3) secondary operations performed when practicable.

Extruders, which prepare the molding compound, preheat it, and supply a molding pill, were selected because of their convenience in changing pill weight during production runs, and the consistent pill temperature throughout. Although an extruder represents approximately three times the investment needed for a customary preheater, one extruder can serve two transfer molding presses even though different parts are run in each press. Thus it was decided that the automated operation would have one extruder for each pair of transfer molding presses.

A special pill feeder was devised by the Research and Engineering Dept. to supply pills from one extruder to the two presses alternately. The feeder consists of a pair of hydraulic-cylinder operated arms, one arm for each press. Each arm has two cylinders; one to sweep the arm in an arc, and the other to lengthen radius of the swept arc. Each arm transfers a pill in a cup from the extruder to the press and loads the pill in through the rear. Another feature of the feeder is an automatic mechanism that permits pill weight for each transfer molding machine to be adjusted and pre-set independently.

To satisfy their objectives, the industrial robot was included in the system to automatically unload the parts from the molds, synchronize the operation of the presses and feeder, and transfer parts to secondary operations which could be performed automatically.

The left hand transfer press molds printed-circuit-card connector parts while the right hand press molds "split

Unimate Industrial robot in foreground is unloading one of the two transfer molding presses.

block" connector parts. Both parts are molded from glass-filled black phenolic, with a left hand pill weight of 72 grams and a right hand pill weight of 60 grams. Cycles of each press are approximately 80 seconds, representing a 5 second savings over manual operation.

While the right-hand press is closed, the Unimate robot reaches into the two-cavity mold of the left-hand press, extracts the parts, moves them past a feeler-operated limit switch to sense whether the parts have been extracted, and deposits the parts in a box. The extruder and pill feeder load the left-hand press which closes as the industrial robot turns to the right-hand press.

As soon as the right-hand press opens, the robot extracts those parts from the two-cavity mold, presents the parts to the limit-switch sensor on that press, and loads the parts into special secondary-operation equipment. Following this, the robot initiates the cycle of the secondary operation equipment, and re-sets the equipment upon completion of the right-hand secondaries as the transfer molding press closes. Next, the Unimate swings back to the left-hand press, poised to extract the next parts and repeat the entire cycle again.

Secondary operations performed on parts from the right-hand press are cleaning contact holes (25 in each of the two parts) and gate grinding. Both operations are performed in the same equipment. First, the industrial robot

Here the Unimate robot is removing a part from the right hand press. The automatic pill feeder can be seen behind and between the two presses.

places the parts on a sliding fixture and pushes the fixture slide inwards to position the parts for hole cleaning. On a signal from the Unimate, the holes are cleaned and the robot pushes the fixture slide inwards again to advance the parts slowly past built-in gate grinders with silicon carbide grinding wheels. Continuing to advance the slide until the parts fall out and into a box, the robot then pulls the sliding fixture outwards and back to the starting position before turning to the left-hand transfer press.

Industrial Robot

Unimate industrial robots are general-purpose, programmable manipulators designed to operate unattended and duplicate the actions of a worker. An articulated arm with gripper, controlled by an electronic memory, it loads and unloads workpieces. The arm can be equipped with a variety of production tools to suit the application. Interlocks such as limit switches synchronize operation of the industrial robot with the equipment it tends. The Unimate receives synchronizing signals from the equipment and provides signals that will control equipment operation. Unimate robots are available with from two to six articulations. Memory capacity can go as high as 1024 sequential steps. Weight-handling capacity ranges up to as much as 500 pounds.

This particular Unimate has a memory capacity of four 32-step programs and a "random program selection" capacity. At present, a 12-step program controls operations at the left-hand press, and a 29-step program is used for the right-hand press including the secondary operations. The remaining two programs are reserved.

All four programs will soon be employed when two more presses are added for a total of four grouped around the robot. A second extrusion feeder will be added for the second pair of presses.

By means of the random program selection capability, the Unimate can automatically skip one or more of the transfer molding presses while continuing to operate the remaining presses. The feature is a convenience that permits a press to be taken out of service for adjustment without shutting down the entire operation.

The firm anticipates that even with a two-press operation, the industrial robot will pay for itself out of labor savings in less than a year and a half.

This program to investigate the automation of transfer molding operations is providing information which will give guidance for future decisions on applications for additional Unimate robots at this facility.

After removing the part from the press, the Unimate robot delivers it to a secondary station where it will be automatically cleaned.

Reprinted from Plastics Engineering, September 1980

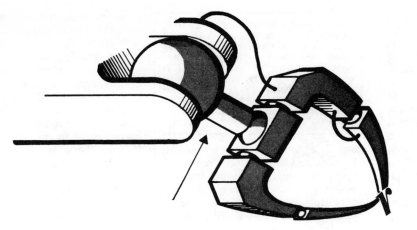

R U Ready for obots?

Ugly and dumb, robots don't want to take away anyone's job. But give these industrial tools a job that no one else wants—one that is hazardous or dirty, repetitive or boring—and they will work, work, work, three shifts a day, day after day, after day, after day. . . .

William R. Tanner, President
Tanner Associates
Farmington, Mich.

The term robot, from the Czech "robota" meaning to work, was introduced into our language in 1921. Czech writer Karel Capek used the term in his play, *R.U.R.* (Rossum's Universal Robots) to describe artificial beings manufactured to replace workers. Today, industrial robots are found in many manufacturing industries doing just that. But unlike Capek's mechanical workers, these robots will not revolt and turn against their creators. Instead they will take on any task, no matter how repetitive, no matter how dirty, no matter how hazardous, and they will work and work and work.

Sound like a Utopia? Rest assured there is no true Utopia for any manufacturing environment, but if you are ready for this sophisticated form of automation, the robot has something unique to offer and may be an effective tool for your operation. As a matter of fact, in many injection molding shops around the world, robots are at work, unloading parts, degating, trimming, and handling simple assembly operations. And their users report solid gains in productivity and profitability, and sizeable reductions in waste.

Robots in plastics

The concept of the programmable automation device, which has become known as the industrial robot, was developed in the 1950s. The first units were put into production in the early 1960s and, since their introduction, several thousand industrial robots have been installed in factories and foundries in the US. There are more than 14,000 robots in use in Japan and nearly 4000 are at work in Europe.

Currently, robots are performing a broad spectrum of tasks in the plastics industry: unloading parts from injection molding machines; spraying paint, stain, and plastics resins; handling hot, cold, fragile, large, small, light, and heavy parts, quickly, safely, and reliably.

Robots also perform many secondary operations: trimming, deflashing, degating, drilling, palletizing, and packing. They handle SMC compression-molded parts and charge molds that are producing parts too large for a man to handle. They are used to unload large structural foam molding machines, and in glass-fiber layup operations, they are at work spraying chopped glass and resin, and rolling out layup.

As universal machines, robots should be able to perform almost any operation. And today's robots actually can. But, because humans will forever offer the greatest flexibility of motion, they will always have the edge. And, because they possess complex decision-making abilities that can never be totally imparted to robots, human workers can never really be completely replaced.

In certain applications, however, robots solve a number of manufacturing problems in the most cost-effective and efficient way. They are especially suitable for those operations that humans are unavailable to do and for those operations that must be performed in inhospitable environments that humans simply shouldn't occupy. Today we are faced with manpower shortages bound to continue for the near term; robots are available now to fill these gaps. Robots promise increases in productivity of 15 to 20 percent. They yield better parts because they handle all items uniformly. A robot can be on its way into the mold of an injection molding machine before the mold is completely open; it can remove the part while it is still too hot for a human to handle; and the mold can close onto the next part, for a shorter cycle, more stability in temperatures, minimal cooldown and reheat, and increased savings in energy. Robots provide for better utilization of capital equipment, prolonging the lifetime of existing equipment by around 20 percent.

One robot can replace one, two, or more people and can work for three shifts each day. They can serve multiple stations for even greater gains in productivity, and they can be retrained to adapt to new operations should product lines change. And, most importantly, they can perform hazardous operations in harsh environments dangerous to human health and well-being, often exempting workplaces from OSHA and EPA censure by their presence.

What is a robot?

In November 1979, the Robot Institute of America (RIA) adopted the following definition of industrial robot:

"A robot is a reprogrammable multifunctional manipulator designed to move material, parts, tools, or specialized devices, through variable programmed motions for the performance of a variety of tasks."

Essentially, the terms "reprogrammable," "multifunctional," and "variable programmed motions" all refer to the robot's abilities to perform tasks that require some degree of dexterity or flexible motion, and to their adaptability to a variety of tasks. These are the primary keys to understanding the differences between robots and other forms of fixed or "hard" automation.

In the past, automation of industrial processes meant replacing a worker with a machine that performed one or more operations in its own machine-like way. The technology of hard automation was mechanical in nature and the motion used to perform the task differed greatly from the motions of human workers.

Robotic technology, on the other hand, closely imitates human performance, moving away from specialized, single-application machines

This is Part 1 of a three-part series on industrial robots in which we will take a look at these remarkable devices, see how they work, and analyze their operation and potential in the plastics industry. We'll also tour the robots market today and speculate on its future, and we'll survey some robot manufacturers for an overview of product lines and costs.

toward a more universal approach. Where fixed automation meant designing one machine to do one thing in one industry, the robot is a machine for many industries—you describe the task and we will teach this machine to do it—regardless of whether the job is welding, spray-painting, moving materials, unloading parts from dissimilar machines, or some combination of all of the above. In short, robotic devices are those that are sufficiently dextrous or flexible to perform relatively complex jobs in more than one industrial environment.

Robotic dexterity depends upon the ability to move the tool through different points in space, and, when at the proper location, to actuate it for task performance. To achieve maximum dexterity, the robot must be able to follow a particular sequence of events, memorize different sequences, and communicate with the outside world to modify its sequential operation. Finally, and most important, a robot must be able to sense and adapt to changes in its environment.

A machine that approximates human behavior and fits into almost any manufacturing scheme is, of course, likely to find many markets. The robotic approach to automation is, thus, aimed at exploiting these advantages, advantages that become even greater as the dexterity of the machine increases. Today we sit only at the tip of the iceberg in robotic technology; its greatest achievements are yet to be realized.

How do robots work?

Today's industrial robots, while they vary widely in shape, size, and capabilities, generally consist of three basic components: the manipulator, the control, and the power supply that drives the manipulator.

Manipulators are the mechanical devices that do the work and provide the dexterity by pneumatically, hydraulically, or electrically driven jointed mechanisms that can perform as many as seven independent coordinated motions (*Fig. 1*). In robotic

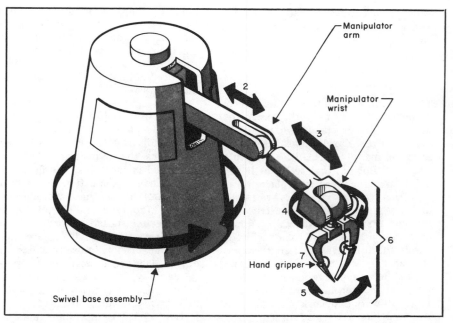

Figure 1. Machine arm of modern robots can perform coordinated motions in as many as seven axes, or degrees of freedom: (1) Rotation of swivel-base assembly. (2) Movement of manipulator up or down, in or out (shown), depending on the manufacturer's design. (3) Swivel motion side-to-side or vertical (shown) of lower manipulator arm. (4) Swivel motion of wrist (roll). (5) Swing motion of hand gripper (pitch). (6) Combined motions of roll and pitch (yaw). (7) Gripping motion of hand gripper.

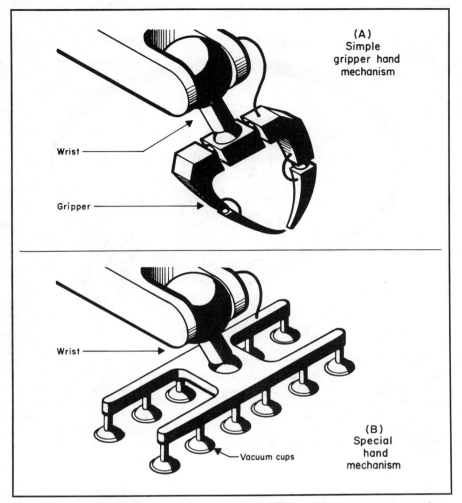

Figure 2. The robot hand, the tooling that customizes the robot to the task, may be a simple gripper (A), or a complex array of suction devices (B).

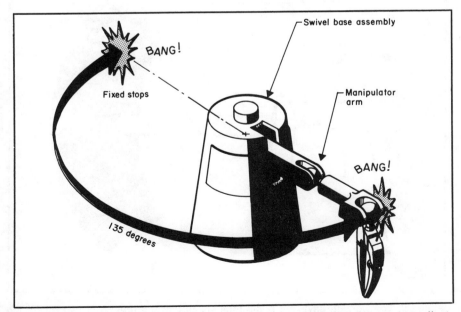

Figure 3. The range of motion for nonservo robots is limited by two fixed, but adjustable, stops for each axis. The robot arm moves from one to the other to perform its operation.

technology, these jointed motions are referred to as axes or degrees of freedom. Simple robots offer as few as three axes and complex robots as many as seven.

Feedback devices on the manipulator's joints send data to the control regarding motions or positions. A gripping device or tool, often called the "hand," is mounted at the outermost joint of the manipulator. It is this tool that customizes the robot for a given application since it is usually designed specifically for the task to be performed (*Fig. 2*).

The various motions of the robot and their sequence are stored in the memory of the control. It is the control that interacts with the machines, conveyors, or tools with which the robot works. Controls vary in complexity; they may be as simple as stepping switches or as sophisticated as computers.

Hydraulically driven robots use an electrically driven pump, control valves, a reservoir, and a heat exchanger in a system that provides fluid flow and pressure to actuate the manipulator. Pneumatically driven robots usually connect to a resident compressed-air system in the factory. Electrical systems are directly coupled to manipulators.

Currently industrial robots fall into one of two categories: nonservo or servo-controlled. Nonservo robots, named for their limited motions, are sometimes called "end point," "pick and place," "bang bang," or "limited sequence" robots. Fixed, but adjustable, stops for each axis by which the end-positions of movement for the mechanical arm are set, define the range of motion for nonservo robots (*Fig. 3*). Since nonservos are limited to alternate between two positions of movement in each axis, they are necessarily consigned to simple operations. Some, however, do have movable stops, inserted or withdrawn automatically, that allow for more than two-position motion.

Nonservo robots are relatively inexpensive, simple to program, and require little maintenance. The typical sequencing controls used can execute single programs of some 24 consecutive steps.

Servo-controlled robots incorporate feedback devices that continuously measure the position of each axis. The manipulator arm can stop at any point within its total stroke (*Fig. 4*). Some ultra-sophisticated 7-axes robots can stop at any one of 4000 points in space. Since they are able to position the tool or gripper anywhere within this total space, or working envelope (*Fig. 5*), servo-controlled robots have considerably more capability than non-servos. But they also cost from 2 to 5 times as much as nonservos of a similar size.

Many servo-controlled robots use minicomputers or microprocessors in their control systems and can execute more than one program; each program may contain several hundred sequential steps.

Using robotic technology

Japan leads the way in applying robots to its industrial operations (see *Table*). In 1977 (the latest figures available), there were more than 10,000 plants involved in plastic molding in Japan, most of them relatively small operations. There were about 45,000 injection-molding machines in use, and of these, about 33,000 employed some sort of automatic part removal system, including 17,000 robots or simple fixed-sequence manipulators. By 1985, it is projected that approximately 95 percent of all the injection-molding machines in Japan will use some type of automatic part removal system, with robots playing a major role in this growth.

Europe, too, is rapidly gaining on the US in raw productivity statistics, primarily because they are eagerly adopting these kinds of advanced automation tools. Robots have received wide acceptance in many industrialized European countries, as is evidenced by the *Table*.

By comparison, statistics for the US

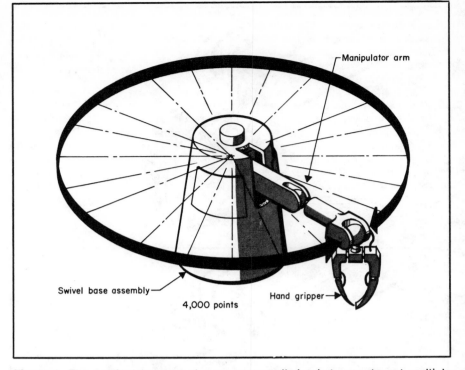

Figure 4. The manipulator arm of a servo-controlled robot can stop at multiple points within its total stroke. For some sophisticated units this means access capability at as many as 4000 points in space. Given programs select and sequence activity points for a particular operating scheme. Programs can be varied to maintain the scheme while changing the activity points. You can, of course, change the program to change the entire scheme.

The robots-at-work population by geographic concentration.[a]		
Japan		14,000
Europe[b]		3,750
Belgium	20	
England	190	
Finland	110	
France	400	
Italy	450	
Norway	170	
Poland	360	
Sweden	1200	
W. Germany	850	
United States		3,250

[a] Data supplied by RIA, based on industry estimates and results of their survey.
[b] While actual statistics are unavailable, it is known that robots are at work throughout Eastern Europe in significant numbers as well.

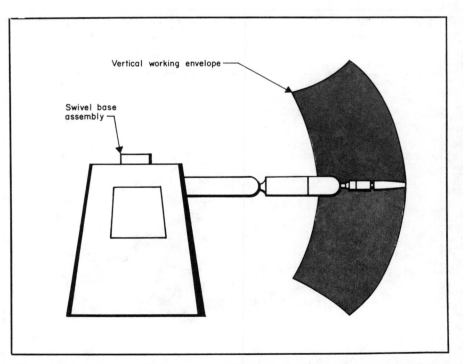

Figure 5. The working envelope is the total space that the robot arm can reach: up, out, down, and side-to-side. The envelope shown is highly idealized. Numerous options are available to the manufacturer to tailor robots to specific tasks.

are not so impressive. In 1980, the plastics industry accounts for only an estimated 3 to 5 percent of the total robot population. While this is expected to grow to around 12 percent by 1985, with sales to plastics processors estimated at around $44 million, the plastics industry will still have a long way to go to catch up with its international competitors. The plastics industry is seated at the top of the growth industries in the US for the decade of the 80s. If this industry is to maintain the competitive edge it has enjoyed by virtue of its

materials costs advantage, it must become more productive, through increased automation, and through leading-edge technologies like robotics.

But there are times when robots are simply out of the question. The robot, to be cost effective, must displace at least two people in a typical plastics operation. While robots can be relocated and retrained to perform other jobs, most users report that this is not likely to happen because putting a robot on a line means retiming the entire line. The robot paces the operation, not the reverse.

And although robots cannot now, and can never, totally duplicate human performance, they are rapidly increasing in capability at the same time as they are decreasing in cost. A single robot ranges from $10,000 for a simple nonservo to around $60,000 for a servo-controlled robot with a broad range of positioning capability. Clearly, the initial investment in a robot is not insignificant. But labor costs are also rising, and replacing one, two, or three salaries with one robot, and enjoying increased productivity as a result, sounds like good business. □

Part 2 of this series will analyze current experience with robots in plastics operations and discuss their payback potential.
Associate Editor
Pat Richter

Reprinted from Plastics Engineering, December 1980

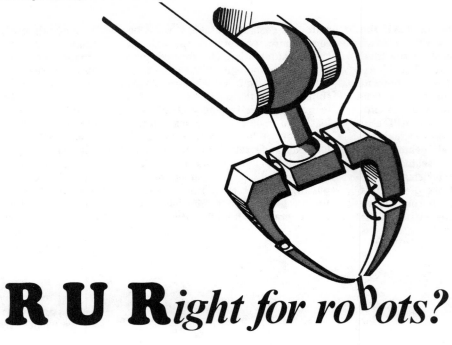

R U Right for robots?

Whatever your motive for purchasing a robot—higher productivity, improved workplace safety, less employee turnover—you're ultimately going to have to justify the purchase in economic terms. With robot prices going down and performance going up, maybe you can.

William R. Tanner, President
Tanner Associates
Farmington, Mich.

and

Patricia J. Richter
Associate editor

In the first of this series of three articles on robots (PLASTICS ENGINEERING, September 1980), we defined a robot, explained how robots work, and talked about who uses them. In Part II, we will try to justify the application of robots to the plastics industry.

Annual productivity gains in the US, once the hallmark of industrial might, have been slowing dramatically over the past decade and the decline seems to be continuing unabated. Wages simultaneously have been increasing steadily, leaving industry to search for ways to improve productivity while holding the line on costs. Robotic tools offer one solution to this di-lemma, and as the following report illustrates, using examples from the plastics molding industry, robots can be a cost-effective approach, generally, to increased production.

Are robots for you?

Robots have been successfully used in industry for many years. In their early applications, they were used primarily for handling hazardous materials and for performing dangerous or fatiguing and repetitive tasks. But recent advancements in robotic technology, control systems, and computers have vastly broadened the applicability of robots. Where robots were once used where men feared to tread, they are now being used where it is too expensive for men to tread.

The new attitude toward robots is keyed to the dramatic reduction in computer costs over the past 25 years even as the industry simultaneously witnessed an equally dramatic increase in processing power (*Fig. 1*). Much of this gain has resulted from the introduction of faster, cheaper, and more reliable electronics, such as integrated circuitry. With lower-cost yet more powerful electronic controllers spun off from the computer industry, robot unit costs are dropping at the same time that the robot's ability to handle more complex operations and to be reprogrammed to handle a wider range of operations is on the upswing. Add to this the upward trend in labor costs, and robots become an attractive alternative to the manual or traditional fixed automation that are the rule in industry.

Yet the investment in a robot, despite decreasing costs, can be considerable. Robots are not yet for impulse-buying. A robot can cost anywhere from one-third to as much as the total cost of an injection-molding machine. Careful cost analysis should precede placing a robot in a plastics operation.

For instance, a proficient molding-machine operator can remove two parts from a molding machine up to approximately the 500-ton range as quickly as a robot. He can also degate, do light trimming, bag, and box those two parts during the next machine cycle. The robot cannot. But should the machine size grow, parts be automatically degated during ejection, and the number of parts molded per cycle increase beyond two, then a robot is faster.

Speed is not everything, however. In plastics, the run of parts to be removed from the molding machine, then degated, inspected, transferred, or assembled by a robot should be a significant portion of the shop's overall production. The kinds of parts that can justify placing a robot in a plastics shop must not only represent a stable percentage of the total product line, but the line should be one that is apt to have a life of 10 years or more in your production scheme.

Parts that require careful handling and removal, such as those made in runnerless molds, could use robots to advantage. Parts produced in molds fed by conventional sprues and runners can be readily extracted from molds by an operator, who simply grasps the runner to which the parts are attached. At current resin prices, however, this is expensive handling.

In some cases, parts can be retained on cores—the parts are partially ejected, and the operator can free them from the mold. But if more than two parts are molded per cycle, the mold-open time must be extended to allow clearing of all the parts. For parts that cannot be partially retained on the cores, a robot extractor may be the only solution.

Parts of all shapes and sizes, and as many as can be spaced in the mold, can be gripped and removed by special tooling placed at the end of the robot's arm. (The clearance of the tie bar in the molding machine is the only limitation.) The robot can subsequently place extracted parts directly into shipping containers. If the containers are conveyorized, for example, the robot's control system can order the containers to the discharge station and return them to the conveyor when they are filled. Coordinated operation such as this relieves the machine operator of packing, enabling him to attend a greater number of machines.

Why you'd like one

But even when robots cannot be justified in straight terms, they might well be the answer to the productivity problem that appears to be plaguing the nation. Robots increase productivity not so much by working faster as by their constancy of pace and consistency of operation. Increases in output attributable to robots can be

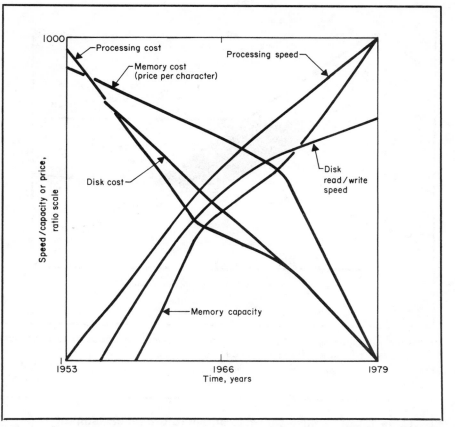

Figure 1. Decreasing processing costs and increasing product speed and capacity over the past 26 years have revolutionized control. The spinoff has helped make today's robots more capable and more affordable. (Data courtesy of IBM.)

measured by the larger number of parts produced without losses from downtime, shift changes, absenteeism, or other human causes.

Operating consistency generally means better part quality and reduced scrap. By introducing robots to plastics injection molding, you achieve constant and often faster cycle times, which allow temperatures to stabilize and remain stabilized during continuous operation. In coating applications, robots attain greater consistency in coating thickness and make more effective use of material.

Beyond consistent operation, robots offer manufacturers a degree of flexibility lacking in fixed-automation devices. Should their function at one station become obsolete, robots can be reprogrammed or moved to another work-station where they can continue to be productive. Fixed-automation devices must often be retired to the graveyard of early obsolescence.

Can you afford one?

No matter how desirable owning a robot might seem, you should be able to justify buying a robot strictly in

terms of dollars and cents. A robotic cost-study analysis should include: *cost avoidance* and *cost savings*.

Cost avoidance. The less stringent of the two economic objectives, cost-avoidance economic analysis is made simply to determine the least costly of several alternatives. A discussion of new production equipment for the following year's model in an automotive plant, for example, might turn on the comparative advantages and shortcomings of a labor-intensive approach, the use of special-purpose fixed automation, and the use of robots (*Fig. 2*). The choice to be made is not between "do we or don't we retool for this job?" But rather it focuses on "which approach to retooling will result in the lowest lifetime cost?" Intensifying labor (in effect, a "tooling" decision) would mean less capital outlay and a relatively low obsolescence cost, but operating expenses would remain high. Special-purpose fixed-automation equipment involves a relatively high capital investment and operating expenses are moderate to low; but obsolescence and changeover costs are high. Robots

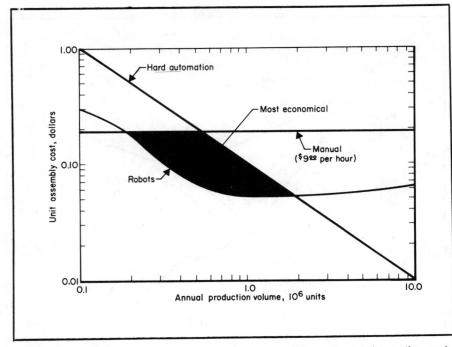

Figure 2. A comparison of the costs of three assembly methods shows the cost-effective range (shaded area) of robotic tools.

Table 1. Cost-analysis form to obtain return on investment.

Expenditures

1.	Robot and accessories cost	$	45,000[a]
2.	Installation costs	$	2,000
3.	Related rearrangements costs	$	5,000
4.	Subtotal	$	52,000
5.	Special tooling costs	$	3,000
6.	Total expenditures	$	55,000

Annual operating savings[b]

7.	Direct labor, man-hours per day 16.5		
8.	Direct labor (about $12.55 per hour)	$	51,800
9.	Indirect labor, man-hours Estimate		
10.	Indirect labor, dollars	$	(Estimate)
11.	Operating supplies	$ (	0)
12.	Maintenance and repair	$ (	2,800)
13.	Launching costs	$ (	500)
14.	Taxes and insurance	$ (	1,175)
15.	Special-tooling amortization	$ (	0)
16.	Depreciation	$ (	5,500)
17a.	Other savings	$ (	1,675)
17b.	Other costs	$ (	0)
18.	Profit before taxes[c]	$ Calculate	
19.	Profit after taxes (50 percent of Line 6)	$ Calculate	
20.	Investment base $27,500		
21.	Return on investment (18 or 19 divided by 20)	Result %	

[a] Sample cost approximates a typical high-technology robot.
[b] Long-term average dollars—10-year life; amounts in parentheses are costs.
[c] Calculate as follows: Line 8 minus Lines (10 to 16) plus Line 17a minus Line 17b. Take 52 percent.

would likewise require a relatively high capital investment and operating expenses would again be low to moderate. The big difference is that obsolescence and changeover costs would be considerably lower.

The escalating cost of plastics has drastically altered the ratio of processing cost to material cost, and, in particular, the ratio of direct-labor to material cost. It is common for the cost of direct labor to be one-third and less of the finished cost of a product. And even though its slice of the pie is becoming thinner, labor costs remain an important element of the cost equation for determining whether to install a robot.

Yet labor costs (and material costs) are frequently concealed during the often unproductive startup periods, before satisfactory parts are produced. If the true production picture were drawn, both would have to be included. Only during long production runs can they be minimized.

On the other hand, a robot used for part extraction during the warmup and actual production-run intervals can make a difference. If it is replacing an operator, direct-labor costs are, of course, being avoided (although the variable overhead costs are increased by virtue of the amortization effect on the investment in the robot). Shift differential costs, should you have them in your wage policy, are also avoided.

As you would expect, cost-avoidance economics can be used to determine the best method of meeting safety requirements or reducing workplace hazards. In a press-loading operation, for instance, the cost of installing safety devices on the press must be compared with that of a robot installation. Here the production rate of the robot approach should be plotted against the manual-with-guarding approach. The cost of part orienters for the robot should likewise be plotted, as should the potential reduction in labor costs.

Cost savings. Whereas cost avoidance requires an outlay of capital, cost-savings economics can be used in situations where one option is to do nothing. Thus the potential economic gain of an investment, as measured either by return on investment or by payback period, must be weighed against the advantages of standing pat, particularly since the demand for discretionary capital usually exceeds the supply. Establishing a potential target ROI helps not only in the initial screening of requests for these funds, but supplements your decision with an additional measure of perspective.

To calculate your ROI, you can use a simplified sample cost analysis method (see *Table 1*). Just fill in the blanks. The completed form will yield a fairly accurate estimated long-term (10-year) average return on investment after taxes.

If you want to calculate your ROI on the basis of depreciated cash flow, apply the data developed on the form shown in *Table 1* to the form shown in *A, Table 2*. The data given in *Table 1* are applied to the form shown in *B, Table 2*, to calculate the payback period.

You might want to develop a simple rule-of-thumb for calculating your ROI. You work backwards through the cost analysis, starting with your required ROI and the already-fixed direct-labor rates. After accounting for launching costs, you will have to assume some average operating costs for indirect labor, operating supplies, and maintenance and repair. Your objective should be to project expenditure levels, supporting them with the direct-labor saving you expect at the ROI rate you seek. Once established, this rule-of-thumb applies until direct-labor rates change.

Here's how it works. Let's assume that for an ROI of 40 percent, the elimination of two men per day will support an initial expenditure of $80,-000. Use this rule when estimating the potential saving in manpower, the cost of the robot and the end-of-arm device, and the cost of installation and rearrangement. When tabulated, the data should give you a quick clarification of your position. If your ROI is less, you can summarily eliminate uneconomical applications without doing the more-detailed cost analysis.

Productivity revisited

It is time to reconsider the noneconomic justifications for robots discussed earlier, this time quantifying them. If an increase in productivity can be projected with the use of a robot, the number of extra parts produced should be estimated, and, based on their value, the annual saving determined. Similarly, if improvement in quality and/or in scrap reduction can be anticipated, the potential annual reduction in scrap or rework costs and the material savings should be calculated and applied to production costs. When a robot is used for

an undesirable operation, one that is prone to high labor turnover, a saving equivalent to the annual cost of hiring, training, and reaching optimum performance levels for a new employee should also be credited to the robot, as should the elimination of overtime costs to make up for production losses.

More difficult to quantify is the saving in costs that are direct extensions of labor costs and thereby an integral part of any analysis that attempts to equate productivity to real costs and the destiny of a given manufacturing process. Such expenses include the costs of protective clothing, safety equipment, lighting and ventilation systems, parking, dining, washroom and locker-room facilities, and

supervisory needs. Although the elimination of each of these may constitute only a small part of the overall saving, the total may be sufficient to make a marginal project financially attractive or to move a potential application from the unacceptable to the acceptable side of the required ROI account.

A robot in your future?

A market study by Frost and Sullivan, Inc., New York, projects a period of solid growth for the robot market. The industrial market, which stood at only $26 million in 1977 and reached $79.5 million in 1979, will, according to the survey, "soar to $438 million by 1985, propelled by a rapidly enhancing technology." And the report goes on to say that most

---

Table 2. Calculating return on investment by depreciated cash-flow method (A) and calculating payback period (B).

Data from Table 1. Numbered lines refer to Table 1.

Annual labor savings (Line 8)	$51,800
Other savings (Line 17a)	1,675
A. Total annual savings	$53,475
Annual operating costs:	
Maintenance (Line 12)	$ 2,800
Launching (Line 13)	500
Taxes and insurance (Line 14)	1,175
B. Total annual costs	$ 4,475
C. Net annual savings (Line A minus Line B)	$49,000
D. Net annual savings after taxes (52 percent of Line C)	$25,500
E. Depreciation tax credit (48 percent of Line 16)	$ 2,640
F. Total annual net future savings (Line D plus Line E)	$28,140

(A) Calculating return on investment.

$$\text{Current value of savings} = \sum_{1}^{n} \frac{\text{Total annual net future savings}}{(1 + r)^n}$$

$$\text{Investment level} = \text{Total annual net future savings} \sum_{1}^{n} \frac{1}{(1 + r)^n}$$

Assuming these savings are constant over the life of the robot (10 years), for a discount rate of r, and solving for r for the current value equal to the investment of $55,000,

$$\frac{55,000 \text{ (Line 6)}}{28,140 \text{ (Line F)}} = \sum_{n=1}^{10} \frac{1}{(1 + r)^n}$$

$$r = .51 \text{ or } 51 \text{ percent ROI}$$

(B) Payback period.

$$\text{Payback} = \frac{\text{Total expenditures}}{\text{Total annual net future savings}} = \frac{55,000 \text{ (Line 6)}}{28,140 \text{ (Line F)}}$$

$$= 1.95 \text{ years}$$

robots will be purchased to reduce production costs or help raise production levels.

Unquestionably, insufficient ROI has held the market back from fulfilling past optimistic projections. But those were not the only factors. Prospective buyers of robots were cautious about accepting the new concept. Robots faced stiff competition from other methods of automation. Robot technology for many applications had not reached present levels.

But rising labor and materials costs may just tilt the scale, making the forecasters of a robotic invasion right after all. While until recently basic robots have cost too much (compared with the costs of human labor or fixed automation), today's high-technology robots are starting to sell in the $40,000 range, two or three times lower than their earlier price tags. As robot production increases, unit cost should drop even more—all against a backdrop of mounting labor costs.

In 1963, the hourly labor costs of a simple robot were as high as a human's. Today, a far more sophisticated robot is available for the same cost. A 6-axis, servo-controlled, computer-driven robot, amortized over 8 years, costs only about $4 an hour, whereas automotive human labor costs over $14 an hour on the average. By 1990, while labor costs could go to $30 an hour, the same robots, if made in very high volume, could, by some estimates, cost only around $10,000, or about $1 per hour.

The bottom line

Economics remains the most significant and widely used justification of robots. Even when robots are proposed under the guise of altruism or reasons other than pure economics, the underlying reason for using a robot is usually profit-oriented. Whatever your reasons for thinking robot, in the final analysis you're going to have to convince your boss or your banker of its economic justification. You're going to have to itemize all of the potential costs and cost benefits of a robot installation, quantifying, wherever you can, the economics of direct-labor replacement and those other cost reductions related to the labor replacement itself.

A word to the wise: try to make original estimates as accurate as you can. Many of the cost factors that can only be estimated during the justification preparation can usually be easily and accurately checked out once the robot is installed. Remember, it is far better to be complimented when the robot's performance exceeds original projections than to be criticized when it falls short of them. □

CHAPTER 9

WELDING

Commentary

Resistance (spot) welding is one of the most extensive applications of robots and one which often involves a number of robots, fixturing and transfer devices integrated into a total system.

Robots are usually not quite as fast as human workers in handling spot welding guns; however, the robots' consistent placement of welds may permit a reduction in the number of spot welds required, for the same net output.

Gas-Metal Arc Welding (GMAW) is a robot applications area with great potential. Productivity of robot arc welding operations may be two to three times as high as with people. Robot arc welding requires, however, more consistent, repeatable joint fit-up and positioning than does manual arc welding.

The availability of practical, reliable joint-tracking devices, which are currently under development, will greatly increase the use of robots for these operations. This utilization will not only significantly improve arc welding productivity, but will also relieve people from exposure to potentially hazardous fumes and from tasks which require a high level of concentration and skill.

Reprinted from Automotive Industries, February 1, 1977

Peter J. Mullins

ROBOT WELDING
of car bodies

One of the largest and most sophisticated fully-automatic welding lines anywhere in the world has recently been commissioned at Volvo's Torslanda plant in Gothenburg, Sweden. The line, costing over $6 million replaces an existing manual welding line. It is designed for a maximum production of 50 Volvo 242 car bodies (2 and 4-door) an hour.

Full production is being maintained while the changeover takes place and the flow will feed one complete 242 car assembly track.

The former work force of 67 people will be replaced by only seven on the new line. Volvo expects to recover the capital outlay on the installation in a little over two years. Main benefits are a considerably improved working environment, more consistent work cycles and better quality welds.

The line was designed and built to Volvo's specification by the Italian company, MST of Turin—part of the Comau group. MST also undertook complete responsibility for the installation. The Italian company has over the past few years gained considerable experience engineering similar robot welding lines for the Fiat 132 and 131 lines in Fiat's Mirafiori plant. Currently Fiat is the biggest user of robots in Europe—with Volvo a close second.

The new Volvo welding line—which contains a total of 27 Unimate robots—is a technical breakthrough from several points.

First, the installation is fully automatic, including loading and unloading stations, and intermediate assembly of all body parts. Second, the bodies are tacked at the second station without the use of the traditional multi-spot welding jigs. Lining up and clamping is by a system of swinging gates at both sides and two from the top for the front and rear windscreens. These provide 36 centering and clamping points and ensure correct positioning, while seven robots perform 128 tack welds at structurally important locations. This system is claimed to be more versatile and economical than multi-spot welding jigs.

The third novel feature is the use of four roller welding guns mounted on pendulum-type arms (two on each side of the body) to weld the roof to the drip channel. These welding guns provide overlapping spot welds to form a continuous seam along the line of the roof and down the windshield pillars. The guns are guided by a cam having the same shape as the outline of the top of the car. Formerly, this work has been carried out by hand due to the difficulties of automation.

The line comprises ten stations, the first and last of which are used for loading and unloading. The bodies are mounted on pallets at all times and are conveyed from station to station by a sophisticated conveyor system. The pallets revolve on two levels. The upper used for welding—the lower for returning the pallets to the starting point.

The pallets on the upper (working) story are moved step-by-step by a central pull bar which hooks on to the pallets to move them to succeeding stations where they are positioned for welding and held firmly by special devices. The pallets are returned by a chain rotating continuously at 49 fpm (15 mpm). Special spring arrangements on the chain hook up the empty pallets one at a time.

This system allows pallets to build up one against the other. They are hooked up and released automatically and this provides for great flexibility at the body loading station.

Two special fork lifts at the first and last (loading and unloading) stations convey the pallets between the two levels.

Running through each station:

1. Loading. Body is automatically loaded onto pallets for lifting.

2. Tacking. Body held in position by swinging gates and tacked. Six horizontal (Unimate 4000B) and one vertical (Unimate 2100) robots perform 128 welds.

3. Respot. Three robots (2-horizontal 4000B and 1-vertical 2100C) perform 82 spot welds.

4. Respot. Five robots (4-horizontal 4000B and 1-vertical 2100C) perform 138 spot welds.

5. Automatic roof loading. Two types of roof: normal and hatch. These are stacked on either side of line. Controller selects desired roof and loader picks it up and places it precisely on car body. One robot (vertical 2100C) makes 14 spot welds.

6. Position checking and securing of roof. Roof fixing device clamps roof and one robot spots in drip channel to fix the position. One man loads tie plates into automatic

A Unimate robot (top) welds the lower section of a Volvo body on a completely automated welding line.

Volvo used this pilot line to gain experience for their present major automatic installation.

transfer machine. These are then automatically placed and held in position for welding. Six robots (4-horiz 4000B and 2-vert 2100C) weld 176 spots.

7. Automatic roll welding of roof. Four pneumatic roll welding guns (guided by templates shaped according to the roof configuration) overlap spots to form a seam. Each gun performs 60 spot welds at 0.20-in. (5 mm) intervals. Speed 6.5 fpm (2 mpm). No robots.

8. Final respot. Five robots (4-horiz 4000B and 1-vert 2100C) make 156 welds.

9. Back up station. This is used for checking and left free in case of robot trouble. Manual back-up can be provided.

10. Unloading. Body unloaded and pallet descends.

All robot operations are coordinated by 2-8k programmable controllers. A main supervision panel gives full indication of working-start/stop, cycle time, overload, bodies in position, etc. For easy fault finding, this central panel indicates which station is in trouble. The maintenance operator then goes to one of six local panels pro-

vided to find the precise fault.

Volvo intends to install a (number 11) station equipped with two small robots manufactured by ASEA, of Vasteras, Sweden. These six-axis machines will ride along a rail and be hooked into the mini-computer. They will be used in a checking mode for providing a complete report on each body produced.

The present installation brings the number of robots in use at Volvo's Torslanda works to over fifty.

"Robots are now well accepted on the Swedish automotive scene," says Rolf Johaneson, head of R & D at the Torslanda body factory, "They are essential in our attempt to improve productivity—and they are welcomed by the unions who recognize that they eliminate an awkward, dirty, unpleasant job. The workers displaced by robots are all being assigned easier and more interesting tasks."

Reprinted from Material Handling Engineer, October 1976

Pass the bodies, the robots have to be fed

Putting in a line of robot welders is one thing, but keeping five seconds ahead of them is something else. Chrysler picked handling equipment that could take the one-a-minute, three-shifts-a-day punishment and gave it some touches of their own.

The automotive industry can make jelly out of the beefiest piece of handling equipment. You realize this quickly when you watch good handling devices running through their paces at 60 to 70 cycles an hour on a two- or three-shift basis. All of a sudden you know why the auto industry does what it does.

At the Newark, Delaware, assembly plant of Chrysler Corp., you see major material handling equipment suppliers pace their equipment to the 1000-cars-per-day-plus output and do some remarkable things. You also see Chrysler engineers mix and blend handling and production technologies and add their own "it's gotta run" factor, and see what has to be done to insure a three-shift operation. There are lessons here aplenty.

This plant builds Plymouth Volares and Dodge Aspens. These are basically frameless automobiles, and their structural integrity depends upon their being put together correctly. This is why Chrysler sought a way to automatically spot weld the body panels so that each body would get all good welds in all the right places each and every time.

At the beginning of the framing line, punched steel parts for the sides are put into gates. The right and left hand gates suspended from overhead conveyors meet up with a towline-propelled framing truck carrying the floor pan and other parts. The top is added and all the parts are tacked with manual spot welders to hold them in position.

The old system was all bullwork

The original scene (more than a year) had welders with water-cooled cumbersome suspended spot welding machines placing welds in up to 500 spots around each body as it moved on the build truck down the towline path.

The increased speed of the line made the job tedious and unpleasant.

To eliminate these problems, Chrysler designated the Newark plant for design, development and proving of an automatic system of spot welding the bodies. A main segment of the system is a robot welding line which had to be built off of the main towline conveyor. A tough handling challenge was getting the tacked 500-lb bodies from the framing line trucks to the robot welding line fixtures and back onto the towline trucks when the robots had finished the 500 welds.

Chrysler engineers at Newark conceptualized the system. Unimation, Inc., put their robots and handling experience together to create the automatic respot line. American Monorail worked with Chrysler to design a stacker-like handling system to get the bodies from the towline build trucks to the respot line and back to the build truck line and position them within .035 inch.

At first glance it doesn't seem like much of a challenge. It is not the first automatic welding line which combines robots and spot welding guns. It is the speed of the line (62 per hour) and its use of pre-programmed controls and its tough interfaces with different kinds of conveying, lifting and positioning equipment that points the way for future systems.

It also points out that material handling equipment can be engineered to meet the grueling pace of a machine tool when the situation warrants it. You also can't help but note the exceptions that Chrysler took to some aspects of design that didn't permit quick, on-line maintenance to keep the line running on a three-shift basis.

As pretacked bodies approach the automatic respot line on the body build trucks, the latches that hold them to the trucks are manually released. Here an American Monorail stacker crane waits to pick up the body, move it about 34 feet and position it on a respot car. The cars which move and stop every 58 seconds as the robots weld the bodies are part of an SI Cartrac conveyor within the respot line.

The Load System stacker crane moves in all planes. It waits as the pretacked body approaches, a floor-mounted pusher synchronizes the stacker to the body truck speed to a lift point while automatic controls extend the rack-and-pinion-driven forks under the body. The body is lifted, traversed to the respot car and positioned on it precisely. The forks are retracted and the stacker returns to its waiting position for the next body. This cycle happens once every 58 seconds and the stacker cycle takes about 53 seconds. It can cycle faster or slower to meet production demands.

The respot line: like a small factory

The welding or "respot" line is on a mezzanine level. The center of this line is an SI Cartrac conveyor which receives the car bodies and advances them through the five automatic spot welding stations and then positions them for pickup by the unloaded stacker at the output end of the line. When the body has been removed from the car, a hydraulic lift table will lower the car and return it beneath the respot line to the beginning for positioning of another body by the stacker.

At the beginning of the respot line a hydraulic lift table will lower with its section of track and receive the Cartrac car. It will then raise and align with the respot line at the mezzanine level. When the load stacker has positioned a body onto the cart, and withdrawn, the car will be powered into the first station of the respot line.

The robots wait in a "pounce" position which is low and a few quick motions away from their first weld point. When the car has positioned the body precisely (within ±.035 inch) and the Unimates change programs according to the body to be welded (either 2-door or 4-door), the Unimate welders will go through their complete program of welds. The body is advanced by the Cartrac to the next two welders for a repeat of the performance making a completely different set of welds. At one station an overhead robot will perform the spot welds around the rear window.

When all welds have been completed, the body will be moved out onto a hydraulic lift table and await transfer by the American Monorail unload stacker.

One thing you have to keep in mind is that the bodies are transferred about one every minute. Bodies are produced as long as the machines work. When the machines in the respot system, including the load and unload stackers, stop for any reason, the alarm sounds and the downtime clock starts running.

Aiming at maximum uptime, you can see a number of important changes that were made in the equipment to make maintenance super fast.

Where power cords flex and move as the stackers go through their paces, pin-type connectors and receptacles were installed so that a broken cord can be quickly replaced.

Where cords connect motors and switches to power panels but no flexing is involved, simple locking plug and receptacle connectors are used on the cords so the electrical components can be mechanically unbolted and electrically disconnected in a hurry. A full complement of spares is kept on hand at all times.

Because the solid state speed controls for the motors that power the stackers are responsive to motor current draw, a bad motor can slow down the stacker response. A panel of ammeters has been added to indicate the draw on each motor so a faulty one can be spotted quickly.

Key parts of components that would be inaccessible during any part of the cycle are being modified so that they will be accessible. For example: a limit switch might be hidden under a lift table when it is lowered, slowing maintenance until the table is raised. So that limit switch will be relocated outside of the lift table framework. Coiled cords that rise and fall with the lift table will also be equipped with pin-type connectors so that they can be replaced quickly.

On-line maintenance is key.

You can tell that the yellow lines that enclose the approximately 25 by 140 foot area mark off more than an area of automatic handling and high technology. Electrical engineer Don O'Shea and his crew of electronic technicians have an office/workshop beneath the controls mezzanine a few feet from the sparking, buzzing robot spot-welding line.

There are about 190 conditions that can shut the

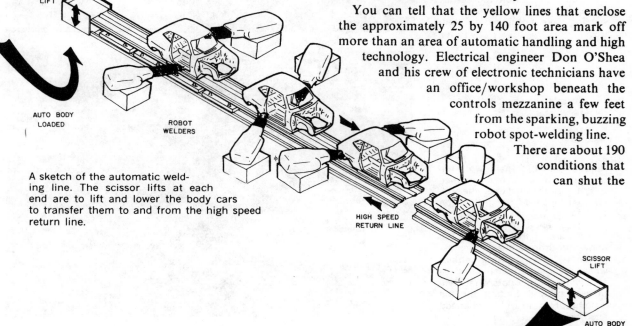

SCISSOR LIFT

AUTO BODY LOADED

ROBOT WELDERS

HIGH SPEED RETURN LINE

SCISSOR LIFT

AUTO BODY UNLOADED

A sketch of the automatic welding line. The scissor lifts at each end are to lift and lower the body cars to transfer them to and from the high speed return line.

The unload stacker picks up a completed body from the Cartrac car at the end of the respot line and prepares to replace it on the same build truck it started out on. The trucks continue at their same pace past the automatic respot line indicating that transfer and welding take nothing from line travel time.

An auto body in transfer while the Unimate robots wait in the pounce position for the next one. The bodies are propelled from one station to the next in eight seconds stopping precisely each time. The rotating tube in the center of the conveyor propels the cars independently by engaging a contact wheel mounted on each car. Sensors will tell the robots what style of auto body is waiting and the welding program will be changed accordingly.

line down, and they are there to make certain that the conditions occur as infrequently as possible and, when they do occur, to get the line working again as soon as possible. Toward this end, on-site engineering was undertaken to design and produce a kind of scoreboard for the line. The scoreboard will show the machine and function at fault, the kind of fault and show the accumulated downtime as the minutes and seconds pass.

Because of the responsibility for the line centered in this area, it operates more like a factory within a factory than as a segment of a production line. The area encloses more than a dozen programmable controllers (PC). There is one PC on each of the 13 robot welders plus the master PC which controls the stackers and the shuttle car line and another used for testing and standby uses.

The Newark plant points out how challenges to continuous handling and productivity are being met with good hardware and software on the one hand and user savvy on the other. ■

—GENE SCHWIND, editor

Looking toward the input end of the respot line you see the lead body with all welds completed. You can also see the lower return line for the Cartrac cars.

At the unload end a stacker deposits an auto body on the body build truck which will carry it on to cleaning, painting and final assembly. The two cover panels marked "Danger" enclose the rack-and-pinion-driven forks which will retract once the body is in position on the truck. Here you also see the floor-mounted chain-driven pusher which will pace the stacker to the build truck exactly. Above the emergency manual control pendant are meters for monitoring motor performance on the stacker. A second bank of meters, suspended from the bridge, monitors bridge function motors.

Detail of the unload system stacker about to deposit a body on the input end of the respot line. Note the oil-tight plugs on all power and control cords which makes maintenance faster. An Electrolift double drum wire rope hoist is used for the lift and lower motions of the stacker mast.

Presented at the Robot III Conference, November 1978

Robotic MIG Welding of Tubular Frames

By Marvin H. Johnson
Unarco Commercial Products

Welding light gauge tubing with automatic machines is not an easy task. Tubular shopping cart frames are especially difficult to weld because of the variety of weld positions, thickness differences and part variations due to bending tolerances.

Conventional automatic welding machines are complex, difficult to maintain, and, most importantly, lack versatility to change from one configuration to another.

Programmable robots offer the versatility needed to change from one style frame to another. The problem of part variation is handled by making program "touch ups" as required.

This paper describes the approach taken to weld tubular frames with robots. Methods, tooling, manpower and problems encountered are outlined.

INTRODUCTION

For many years shopping cart frames have been welded manually. Formed tubes, caster plates and caster horns are positioned and hand clamped into a two station weld fixture mounted on a trunnion and turn table. The operator loads and unloads the first station while the weldor welds the frame in the second station. Welding tubular frames manually is difficult and costly. Weld quality is dependent on operator skill and is difficult to maintain.

Machine welding tubular frames is also difficult because of the variety of weld positions and material thickness combinations which vary from 1/4" plate to 16 ga. tubing. Several years of research were spent evaluating different types of automatic welding equipment. The requirements were as follows:

1. Capability to change from one style frame to another in a reasonably short changeover time.

2. Ability to adjust for part variations.

3. Simple operation and maintenance procedures.

4. Attractive economic justification.

The programmable robot was the only machine that met the project criteria.

EQUIPMENT

Unimate 2000B, Five Axis, Continuous Path Robots were chosen. Welding equipment consists of Standard Hobart 300 AMP Power Supplies with 105 Control and H45 Feed Heads. The welding gun is a standard unit with a special bracket for mounting to the robot. Welding wire is purchased in 250 lb. reels. Shielding gas is straight CO_2. Fixturing consists of a two station fixture mounted on an indexer. Air logic controls the sequence of operations and provides safety interlocks to protect the operator and equipment.

TOOLING

The first generation weld fixture was constructed entirely from steel. Two station fixtures are mounted on an 180 Deg. rotation indexer. Clamping is done by air cylinders and air toggle clamps. Clamping and unclamping functions are controlled by air logic and are interfaced with the robot, indexer and welder.

Second generation fixtures are being made out of aluminum. The decreased weight enables a faster indexing time. Other changes have been made to simplify finished part removal.

METHODS

The fixtures and robots are arranged in a side by side location, which enables one operator to load and unload two weld fixtures. Frame parts are arranged in the work center to minimize loading time. This arrangement makes it possible for one operator to keep up with two welding robots. Robots, welders and weld fixtures are interfaced with adequate interlocks to insure proper sequence of operations.

PROGRAMMING

Programming is accomplished by guiding the robot weld gun under manual control to each position by means of a teach control. Step-by-step, each successive position of the arm and weld gun is recorded in a digital electromagnetic memory. Weld gun positions, velocity, timed waiting periods, etc., are combined in the program. Weld gun position and speed of travel are critical in welding light gauge tubing. Therefore, it is essential that the programmer be knowledgeable in welding procedures. Program "touch ups" are required occasionally to compensate for part variations. Touch ups usually consist of changing the weld gun position relative to the weld joint.

Completed programs are transferred to cassette tapes for future use. Once the program is on tape, the robot memory can be utilized for other programs. Programs on tapes can be transferred back into the robot memory at a later time.

TOOLING CHANGEOVER PROCEDURE

The weld fixtures are designed to be changed over from one frame to another in a short period of time. Changeovers are accomplished one of two ways:

1. Locators and clamps are changed by removing eight bolts and installing new locators. Several minor product design changes were made to simplify tooling changeovers.

2. The entire weld fixture is removed from the indexer by removing two bolts and one air line. Guide pins locate the fixture accurately on the indexer.

Weld program changes are made by re-programming or by transferring existing programs from the tape deck. After entering the new programs, the robot is stepped through the cycle for verification. Minor changes are occasionally required. As stated earlier, these program touch ups usually consist of slight gun position changes.

PROBLEMS

Many problems were encountered during the initial months of operation. Some of the more serious problems were as follows:

1. Inconsistent Parts

    Part tolerances were more critical than anticipated. Tolerances were tightened up on both piece parts and fixtures.

2. Inability to change weld parameters during the weld cycle

    The only welding parameter that can be changed during the welding cycle is the speed of travel. It is difficult to establish voltage and amperage settings that will handle the several material thickness combinations and joint configurations. The problem is being handled by varying the speed of travel and weld gun position. Future applications will include welding controls that will permit voltage and amperage changes during the welding cycle. A control such as

the Hobart 123 Solid State Tri-Schedule Motor Control
would provide the additional parameters needed to
simplify the welding operation.

3. Repeatability

    Burn through was a problem during the initial start
    up because the robot did not repeat weld positions
    accurately. Weld gun position is critical when weld-
    ing light gauge materials at high heats. One problem
    was "overshoot", which was corrected by removing the
    wire feeder from the boom and mounting it on a separate
    platform. Various changes were made on the robot
    and the problem was corrected.

4. Cleaning of welding gun tip

    Spatter build up on the contact tip and nozzle is
    always a problem in MIG welding. In a short time
    the spatter build up chokes off shielding gas, which
    causes porosity, weld contamination and nesting of
    the wire in the feeders. A simple blow nozzle,
    coupled to an in-line-lubricator, blows a fog of
    anti-spatter agent on the weld gun nozzle. The two
    second blast delivers the anti-spatter agent and
    removes existing spatter build up. The robot is
    programmed to clean the weld gun nozzle at regular
    intervals. This device keeps the weld gun nozzle
    and contact tip clean for several hours.

5. Protection of Air Lines

    The first generation weld fixtures were equipped with
    thermoplastic type air lines which were covered as much
    as possible with guards. Weld spatter constantly
    bounced up under guards and burned holes into the
    plastic tubing.

    All stationary air lines were replaced with steel
    braided, hydraulic type lines or were covered with
    wire springs. This change has been effective and
    has reduced down time considerably.

SUMMARY

After long months of de-bugging, the equipment is performing
well. Weld cycle time and quality standards are up to requirements outlined in the justification. Additional applications
are being considered in both resistance and MIG welding
operations.

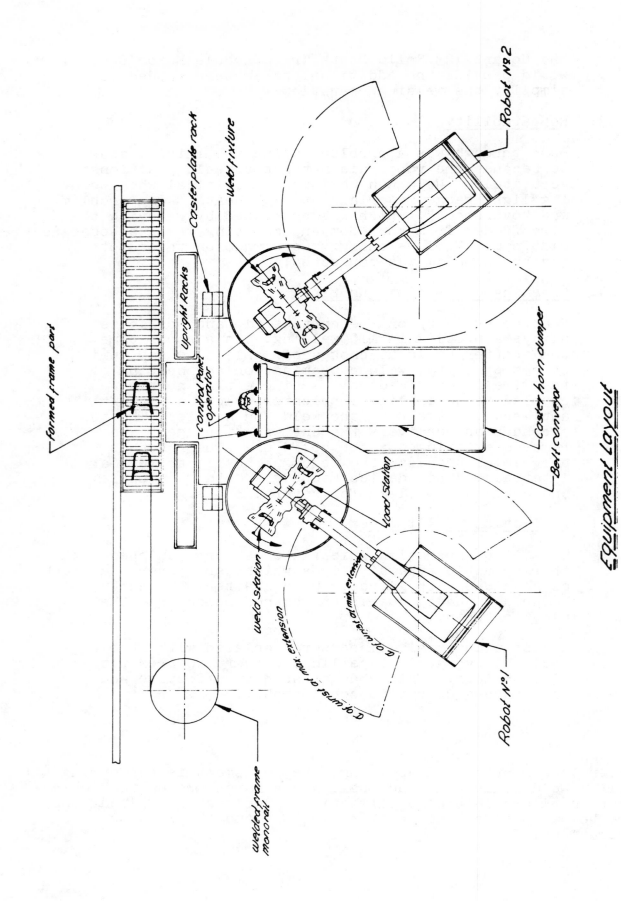

Equipment Layout

Robot Nº2

Robot Nº1

Welded frame
monorail

Formed frame part

Caster-plate rack

Weld fixture

Upright Racks

Control Panel
Operator

weld station

C or wrists at max extension

C or wrists at min extension

Load Station

Caster horn dumper

Belt conveyor

Presented at the Robot IV Conference, October/November 1979

A Flexible Robot Arc Welding System

By John G. Holmes and Brian J. Resnick
Cincinnati Milacron, Inc.

Automation of the arc welding process results in process and equipment requirements which were totally irrelevant with manual welding. With a human controlling the welding function, all process and machine control was interfaced through him and dependant on his skill. Now sophisticated electronic equipment interfacing and special sensors for process monitoring must be developed in order to have a successful automatic system. Most of today's automatic systems are dedicated to a specific type of welding task and are extremely difficult to alter. Future automatic welding systems must be designed to be more flexible in their operation. In fact, their flexibility and decision-making capability must begin to approach the typical human welder. A special type of flexible system is presented in this paper. The system uses a computer controlled robot as the weld gun manipulator and for master control of the system's operation. This self contained cell has the robot as the prime decision-maker. The advantages and problems of such a system will be explored. Special emphasis is placed on software which aids not only in teaching and in automatic operation, but also in the system's response to sensors which contributes significantly to the system's flexibility. Typical productivity increases and economic justification will also be discussed. Finally, a discussion will center around the future research efforts required in automated arc welding.

INTRODUCTION

Arc welding is one of the foremost processes used in industry for joining metals. Today, it is estimated that over 850,000 welders are working in industries which are heavily involved in welding. Welding is the third largest job category after assemblers and machinists. The total dollar volume for the welding equipment and consumables market in 1978 was approximately 1.54 billion dollars, and this value is expected to triple by 1986. Of the total 1978 welding market, arc welding accounted for 74%, or 1.14 billion dollars. Almost sixty percent of this amount is the cost of consumable electrodes, the remaining amount is for the welding machines and accessories. The consumables are used in two primary classes of arc welding: stick electrodes and those using continuous wire (electrode) feed. The use of stick electrode is primarily a manual welding process which would be extremely difficult to automate. The wire-fed methods, such as Gas Tungsten Arc, Submerged Arc, and in particular Gas Metal Arc are well suited for automation. These processes account for approximately 37% of the welding market. Considering the size of the welding market, arc welding still remains basically a manual operation.

The primary reasons relatively few flexible automated arc welding systems exist today are the dexterity and adaptive control required for the welding process. These requirements have usually led to a large initial investment in systems which was difficult to justify. Thus, the majority of automated systems available today are dedicated to a single task such as welding pipe or long straight welds in plates. These systems have mechanized welding heads which move along rigid tracks parallel to the joints being welded. Any options such as seam trackers or weaving devices are merely add-on self-contained modules. This adds significantly to the cost of these welding systems. These dedicated systems do, however, function well for their limited scope of operation.

A truly automated arc welding system for the manufacturing industries in general must be designed for inherent flexibility. The system must be capable of handling a variety of parts. This requires the welding equipment to be flexible or programmable. A computer controlled industrial robot is both programmable and flexible in its operation (Figure 1). The

Figure 1: T3 ARC WELDING ROBOT

Figure 2: SEAM WELDING WITH THE T3

computer controlled robot adds a new dimension to arc welding. Software features can be added which minimize programming time, monitor and control the process parameters, and extend the robot's capabilities in areas such as seam tracking and arc weaving. The computer controlled robot plus a welding software package minimize the need for external add-on hardware once necessary to perform arc welding (Figure 2). The computer controlled robot coupled with the appropriate welding equipment promises to do for arc welding what NC machine tools did for machining.

Welding Automation

There are basically two levels of automated welding systems: fully automated and semi-automated systems. A fully automated system will be the ultimate type of welding system. There will be no human intervention in the welding process other than loading and unloading the parts and consumables. A fully automated system will have both adaptive position control and adaptive weld process parameter control. Adaptive position control is available in some systems today in the form of commercially available add-on seam trackers which correct for deviations in the programmed weld path. Here a position sensor, just ahead of the weld gun, monitors the position of the weld seam. The positional information is used to drive the slides on which the weld gun is mounted. This is present state of the art technology. Adaptive process control is a totally different problem. Very little has been accomplished to date to successfully implement adaptive process control. One of the main reasons is the difficulty in measuring such parameters as penetration, bead geometry, arc stability, deposition rate and weld quality in real time. Also, a control strategy relating these weld quality parameters to the controlled variables is required. However, the controlled arc welding variables such as weld gun travel speed, arc voltage and wire feed are relatively easy to measure and control. This allows for the possibility of open-loop process control instead of adaptive process control.

The lack of adaptive process control does not mean that a totally automated system is not possible. Knowing the geometric characteristics of a weld seam and using the proper preset weld parameters result in acceptable welds a high percentage of the time. Consistency in the weld seam can be obtained through better quality control in the manufacturing processes upstream from the welding, and/or better fixturing prior to welding. New and better measuring techniques, such as computer vision, may in the future be implemented to study the weld seam.

If this still does not suffice for a given application, the alternative would be to have a semi-automated system with a welding system operator. The operator would be responsible for accurately fixturing parts on a weld positioner and selecting the proper robot welding program for each part. The resulting semi-automated arc welding cell would be controlled primarily by the robot, with the operator having the capability of real time adjustments of the gun position during weld cycle.

Benefits of a Robot Arc Welding System

There are two reasons why a manual welding operation should be converted to a robot arc welding system; safety and productivity. The most obvious is the removal of the welder from the vicinity of the arc. The environmental conditions resulting from the welding process are potential hazards to the welder. In order to eliminate these severe working conditions, OSHA may soon impose more stringent limitations and safety standards on the welding industry. To achieve these standards, costly safety equipment will be required. Once the expensive safety equipment is installed and operational, there is no guarantee that the productivity of this particular welding operation will increase or even be maintained after the welding area is "cleaned up". The use of a robot arc welding cell not only minimizes these safety problems involved with the welding process but also increases the productivity of the particular operation. When the robot is programmed to do a certain weld pattern, it continues to reproduce these instructions independent of the environmental conditions. Smoke, temperature or radiation from the arc are no longer constraints on the welding process.   The use of the robot arc welding system results in higher arc-on time, lower costs and higher quality welds than are available with manual welding.

Increased productivity in robotic arc welding results primarily from the fact that the robot minimizes non-productive time in the welding cell. If either two weld positioning tables (Figure 3) or an indexing device with multiple fixturing (Figure 8) is employed in the system, a welding cell operator can be loading and fixturing new parts while the robot is welding in the other location. This increases the flow of parts through the welding system. The result is reduced lead times, higher outputs and more efficient production scheduling. The net result is a reduction in the overall floor to floor time.

A robotic arc welding system also allows for better process control. The optimization of welding parameters can be determined in advance by qualified welding engineers and stored in the welding software in the robot's control. The repeatable positions and velocities of the robot coupled with the proper weld parameters allow for more consistent and higher quality welds than possible with manual welding. A desirable feature of a robot arc welding system is that the

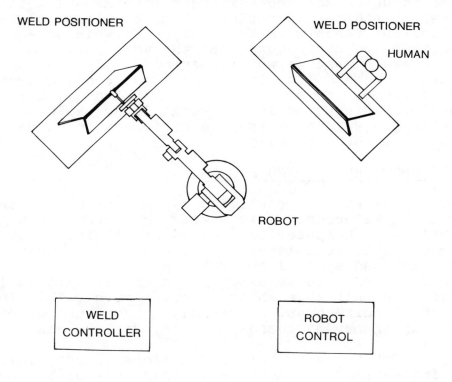

Figure 3: DOUBLE FIXURED WELDING

weld parameters be controlled by the robot and become part of the robot's program. The data is thereby reproducible from part to part.

A robot arc welding system should not be thought of as a replacement of a human, but merely removing the human from the vicinity of the arc, and allowing him to perform functions other than the actual weld gun manipulation. The operator would load and unload parts, monitor the cycle and perform quality control duties. The function of a weld cell operator would be similar to those performed by a CNC machine tool operator. A robot arc welding cell not only tends to upgrade the work of the welder but also minimizes such problem areas as OSHA requirements, poor working environment, job satisfaction, and the lack of trained welders. It is anticipated that a weld cell operator may be able to attend several robot welding cells further increasing the productivity of the specific operation.

REQUIREMENTS FOR AUTOMATIC WELDING

It is now an accepted fact that a computer controlled industrial robot has the manipulating and logic capabilities for automating the arc welding process; however, many of the requirements which were non-existent with manual welding must now be taken into consideration. Some problems relate to interfacing the robot with the peripheral welding equipment such as the weld power supply and the weld positioner. Special software must be written specifically for controlling the welding process. Special sensors and sensing techniques must be developed for process control. Finally, all of the above must be integrated into a workable arc welding system. The robotic arc welding system to be described addresses and solves many of these problems.

FLEXIBLE ROBOT ARC WELDING SYSTEM

A flexible robot arc welding system has been introduced by Cincinnati Milacron. This system consists of a high repeatibility T3 Computer Controlled Industrial Robot, a Programmable Weld Positioner (PWP), a wire feed unit, a weld power supply, and special arc welding software. All the components of this arc welding system are integrated into a single working unit through the robot control. Distinct advantages of this computer controlled robot arc welding system, both in hardware and software, will be discussed.

The T3 Robot has two prime functions in the flexible arc welding system. It must accurately control the position, orientation and velocity of the weld gun. Then, as the control center for the flexible arc welding system, it must communicate with and command all the peripheral equipment in the cell.

POSITION, ORIENTATION, AND VELOCITY CONTROL

The arc welding process requires the robot to accurately position and orient the weld gun in relation to the seam. Then while maintaining these coordinate relationships, the robot must control the velocity of the weld gun along the weld seam at the desired welding speed. Since the welding speed is very slow in comparison to typical robot applications, the robot has been modified with smaller servo valves on the positioning axes. This modification results both in smoother low speed performance and improved positioning repeatability (+.025 inch) required for arc welding. In addition to this hardware change, a special software package is provided to simplify the programming task and expand the capabilities of the robot in arc welding applications.

Velocity Function

Since it is conceivable that each segment of a weldment could require a different velocity for optimum welding conditions, the T3 arc welding robot provides a wide range of velocity selections: from 1.0 ipm to 127.5 ipm in 0.5 ipm increments. As each segment of the weld is taught, the operator assigns a velocity table entry number to it. The operator then loads the desired velocity value into the proper position in the velocity table. This is accomplished by using the Velocity Function. If examination of the trial weld indicates that a different velocity is required, it is a simple matter to edit the program and change the Velocity Function table entry to the desired value. The major advantages of this method of velocity selection are:

   a. It associates velocities with welds.
   b. It provides greater editing and programming posibilities.
   c. It reduces the velocity selection time.

High Speed Replay

After a program has been taught to the robot, it is desirable to replay the program prior to welding to observe the relative motions and general flow of the program. The very slow welding velocities would cause this procedure to be excessively time consuming. To facilitate this program verification, a high speed replay feature has been created. The operator enters a velocity clipping value via the keyboard and all velocities below this value are then raised automatically to the clipping value. The obvious advantage of this feature is the time-saving possibilities associated with it.

Multipass Welding

The computer in the T3 robot also assists in multipass welding. The operator teaches the robot a root pass using the portable teach pendant, and then via the keyboard creates as many copies of the root pass as required to fill the seam. The coordinates of the additional passes are then "offset" and welding parameters changed, if required, via the editing and modification capabilities of the robot control. The entire multipass welding sequence can be created with the robot actually being taught only the root pass.

Weave Function

If the weld joint width becomes excessive, it is often desirable to weave the weld gun from side to side in the seam. Weaving can now be accomplished using the T3 robot without the use of a mechanical oscillator. The weave function can be programmed between any two taught points. The operator teaches a point which includes the Weave Function at the beginning of the weld seam where the weaving will start, and then teaches the end point of the seam where the weaving will stop. Five parameters are involved in specifying the Weave Function:

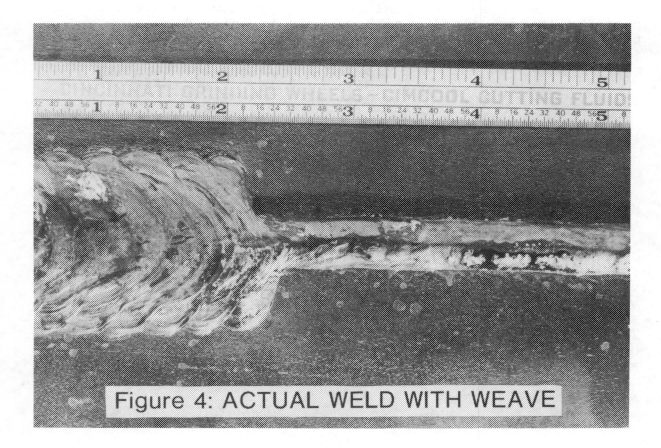

Figure 4: ACTUAL WELD WITH WEAVE

1)   Number of cycles (weaves) per inch
2)   Magnitude on the right (inches)
3)   Magnitude on the left (inches)
4)   Percent dwell on the right
5)   Percent dwell on the left

Figure 4 demonstrates the weaving capability of the T3 robot. The flat weld was made at 24 inches per minute, then the robot was programmed to weave the remainder of the weld. This particular weave pattern was programmed at six inches per minute. The oscillations were at three cycles per inch with a right and left amplitude of .5 inch. The weld was also programmed with ten percent dwell at both peak amplitudes. The various combinations of the five weave input parameters result in a variety of symmetric and asymmetric triangular and trapezoidal weave patterns as shown in Figure 5. The weave patterns are made perpendicular to the axis of a flat weld. Figure 6 illustrates types of seams that can and cannot be welded with weave patterns. Both flat welds, whether perfectly flat or on a slight incline, are possible; however, weaving is not possible on the horizontal weld.

PERIPHERAL EQUIPMENT

The second principal function of the robot in a flexible arc welding system is the control of all the peripheral equipment. Some of the simpler tasks the robot performs use the Input and Output contacts: turning on the weld gun, the cooling water, the shielding gas, the power to the weld controller and other equipment. The robot then waits for inputs from the peripheral equipment indicating that everything is ready prior to start the welding cycle.

Weld Schedule Function

A Weld Schedule Function has been developed which allows the operator to enter the Wire Feed Speed and Arc Voltage into a table. The advantage of the Weld Schedule Function is that both wire feed speed and arc voltage can be changed simultaneously at one programmed point by programming a different table entry number. Now each weld or segment of weld can have its own unique weld schedule. The values in this table are entered via the keyboard similar to the Velocity Function. The weld parameter data is stored in the robot's control and is thereby reproducible from part to part.

Wire Speed and Arc Voltage (Figure 7)

The T3 robot uses two digital to analog signal converters (DACS) to control the Wire Speed and Arc Voltage. Each DAC provides a zero to ten volt DC output and is interfaced to a weld controller that has been designed or modified to accept these control voltages. The wire feed speed and arc voltage

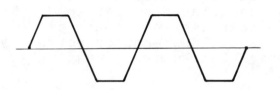
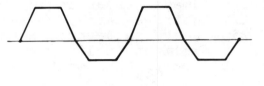

TRIANGULAR WEAVE PATTERNS: VARIOUS   AMPLITUDES

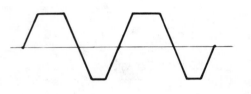
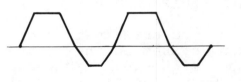

TRAPEZOIDAL WEAVE PATTERNS: VARIOUS AMPLITUDES AND DWELLS

SINGLE SIDED WEAVE PATTERNS

Figure 5: PROGRAMMABLE WEAVE PATTERNS

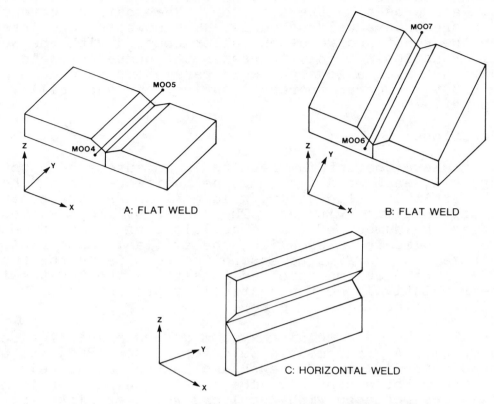

Figure 6: SEAM POSITIONS FOR WEAVE

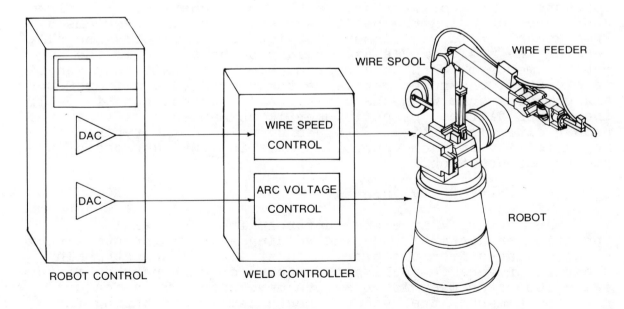

Figure 7: ROBOT CONTROLS PROCESS PARAMETERS

have been traditionally controlled through a bank of potentiometers, each preset by the operator. Each pair of potentiometers represents a weld schedule. The powerful DAC interface
between the robot and weld controller coupled with the Weld
Schedule Function not only increases the number of weld
schedules but also allows the weld parameters to become part
of the robot's program thereby insuring reproductibility from
part to part.

Weld Positioner

To achieve the optimum results in an automated robot
welding cell, each weld seam must be positioned in the best
possible attitude. Typically, weld positioners are controlled
by cam driven limit switches. The number of limit switches
determines the number of unique positions of the weld positioner. For complicated assemblies, the available number of limit
switches may not suffice to assure the seam is in the best
attitude. The problems involved in setting the limit switches
and their relatively poor repeatability presents a high
possibility of positioning error.

The T3 robot arc welding system offers a solution to
this problem. A Programmable Weld Positioner (PWP) has been
developed which is easy to teach and control accurately
through the robot's control. The cam-driven limit switches on
each axis are replaced with resolvers which are interfaced to
the robot control. The weld positioner's axis positions are
measured by the resolvers. The weld positioner's motor
controls are interfaced to the robot's Input and Output
contacts. The operator drives the positioner to the desired
configuration using the positioner's pendant and programs a
Programmable Weld Positioner Function in the robot's control,
causing the positioner's axis values to be stored in the
robot's memory. When the Programmable Weld Positioner Function
is executed, the robot retrieves the stored values from memory
and compares them with the positioner's current position. If
the values differ, the positioner is commanded to move in the
proper direction until the values coincide. The typical
repeatability on each controlled axis is $\pm.045$ degrees of the
programmed position.

Operator Override (Figure 8)

Manufacturing processes upstream from the welding
operation or distortion during welding at times create a seam
which deviates from the programmed straight line path. In
order to prevent reteaching the robot for each part, a feature
called Operator Override has been developed. This feature
uses the human as the feedback mechanism for adjusting for
deviations in the weld seam. The operator, through a set of
switches, can cause the robot to deviate from or shift its
straight-line path. The deviations are made in a direction

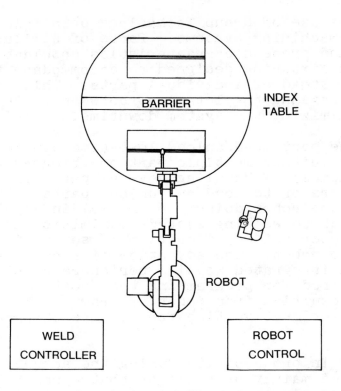

BARRIER

INDEX
TABLE

ROBOT

WELD
CONTROLLER

ROBOT
CONTROL

Figure 8: OPERATOR USING 'OPERATOR OVERRIDE'

normal to the line of the weld in the X-Y plane, and also
vertically in the absolute Z direction of the robot. The
maximum deviation is limited by software to one-half inch in
all directions. At the end of the move, the robot will automatically return to the programmed end point cancelling out
the path deviations input by the operator.

PRODUCTIVITY BENEFITS

It was emphasized above that the main reason for implementing a flexible robot arc welding system was to remove the
operator from the immediate vicinity of the arc. However,
this welding system, like most automation, has substantial
productivity benefits associated with it. Although specific
examples on productivity increases are not as yet available
for this particular system, areas where significant increases
over manual welding can be expected will be discussed. The
major areas which affect productivity are set-up, arc-on time
and rework time. The remainder of this section will discuss
how each of these areas are influenced by a flexible robot arc
welding system.

There are three basic areas which must be resolved during
set-up to significantly increase the productivity of a welding system. The first is the fixturing changeover required
from part to part. This problem can be minimized to some

extent through the use of Group Technology principles similar to those used in machining systems. Parts of similar physical characteristics and process characteristics should be grouped into families and fixturing designed to accommodate this family of parts instead of individual parts. This will have the effect of increasing the lot sizes between fixture change-overs thereby minimizing the system downtime.

Secondly, the part load/unload lost time can be minimized through the use of either multiple part positioners or an indexing table capable of handling several part fixtures. This allows an operator to load and unload parts without interfering with the robot doing the arc welding cycle. This type of an approach to welding systems maximizes the arc on time. Also, an important consideration in set-up is the programming of the robot. The advantage of a computer controlled robot and integrated welding peripherals is the reduced time required for programming the robot. Non-productive time resulting from robot programming is just as meaningful as non-productive time for fixture changeover and part loading.

The increase in productivity during the actual welding process is achieved mainly by the fact that a robot is capable of welding continuously on a part without a break or change of pace in a hot and smoke-filled environment. In most applications, the robot doesn't increase the deposition rate of the weld metal since this is dictated by the part design. However, the robot does perform the non-welding operations, such as gun positioning, in a minimum amount of time. The net result is an overall increase in the arc-on time. Some investigations have also begun in the use of the robot arc welding systems to weld at energy levels beyond those tolerable by human welders. If the part design, process and welding equipment are capable of this additional energy input, then productivity gains beyond those described are possible.

Finally, welding productivity is increased by the consistant quality of the weld made by the robot welding system. With proper fixturing, accurate control of position and velocity by the robot, and adaptive position control, welding engineers can specify the appropriate weld schedules. With all the parameters properly specified, the probability of an acceptable weld is very high. This results in a minimum amount of rework.

Most robot welding systems today have experienced productivity increases in excess of twice manual welding. If the productivity doubles, a basic economic evaluation can be performed assuming that the added throughput is equivalent to one operator. On a two shift basis, for a two and one half year payback period, and a labor rate of $11.00 per hour including fringes, the robot and extra peripheral equipment can cost in the vicinity of $110,000. As labor rates, government regulations, and manpower shortages increase, justification of such a system becomes much easier.

CONCLUSIONS

The benefits of a flexible robot arc-welding system have been discussed. Removal of the welder from the vicinity of the arc, increased productivity and higher quality welds are benefits worth mentioning again. It should, however, be obvious that there is much more involved in creating a robot arc welding system than merely taking a robot and bolting a weld gun to it. The use of a turnkey computer controlled "flexible robot arc welding system" insures that the peripheral equipment, robot, and software have been designed to perform welding in a most efficient and productive manner. It also has the advantages of being upward compatible with the continuously advancing field of automated arc welding.

Future developments in automated arc welding will also be of significance. Automatic seam tracking using a sensor leading the weld gun in the seam has been demonstrated. The T3 is capable of accepting information from a contacting sensor and altering its path. There are obvious limitations in the use of contacting sensors on certain weldments, but it is expected that non-contacting sensors will be available in the future greatly enhancing the attractiveness of sensors to arc welding.

Adaptive process control in some form is one of the major areas yet to be attacked. With the aid of computer vision, weld seams can be studied and analyzed before, during, and after welding. Various universities and research institutes are investigating this particular problem. Seam tracking will also advance with the implementation of vision. All of these devices can be integrated into a system with a computer controlled robot. All of these advances, along with automatic material handling and fixturing, will some day enable the welder to be removed completely from a truly hazardous environment with the creation of better quality welds at lower welding costs.

Presented at the Robot IV Conference, October/November 1979

An Electric, Continuous Path Portable Robot for Arc Welding

By Mortimer J. Sullivan
Unimation Inc.

A study of welding in industry indicated a requirement for a portable robot which could hold a torch, follow variable paths and do out-of-position welds, with poorly fitted material, in batch manufacturing processes.

The robot manufacturer in conjunction with a large shipbuilder developed a prototype and tested it in actual ship constructions.

Results were satisfactory, resulting in a new lightweight, easily programmed industrial robot, called the "Apprentice®".

This paper describes the contacts made with the industry, the initial design goals, the basic difficulties encountered in early designs, how these difficulties are overcome, the prototype test and operational problems, and finally the satisfactory portable Apprentice® robot.

INTRODUCTION

Arc welding is often a very unhealthy job. The smoke and fumes can cause many respiratory problems. Some of the fumes are toxic or carcinogenic. A precaution is even necessary in welding aluminum with small amounts of magnesium. Magnesium heated by the arc radiates harmful ultraviolet light. More and more pressure is upon management to protect people in welding as required by the Occupational Safety Hazards Act.

Unlike many unpleasant jobs, welding requires skilled personnel. To make an out-of-position weld takes training and some will never master the art. For example, welding of pressure vessels, nuclear power plants and aircraft parts-the welder has to pass tests to prove his qualifications and obtain certification.

Consequently, the industry is worried about the future of welding and interest in automation of arc welding is growing rapidly.

The time a welder is actually joining metal is called "arc time". In the United States the average arc time is 30% in industrial manufacturing. This means that during a 40 hour week only 12 hours are actually productive. It is the goal of robotic welding to increase this arc time and by so doing a robot justifies its existence.

BACKGROUND

During a tour of a large shipyard in Pascagoula, Mississippi, an engineering vice president was shown the problems in the production welding of the inner bottom section. There were a number of men doing arc welding in cramped quarters and this particular operation was definitely in need of some assistance in the area of automatic equipment. Visits to other shipyards indicated that this problem was universal.

During a visit to a shipyard in Malmo, Sweden, UNIMATION INC., the manufacturer of robots, was informed that the shipbuilder would be willing to jointly develop a robot capable of handling the torch for welding of ships. A contract was entered into between the two companies.

ANALYSIS OF PROBLEM

The first step in the contract was an analysis of the problem. The problem could be broken down into two areas; human factors and technical factors. In regard to the human factors, much of the work was done outdoors where temperatures varied widely. The operator of the welding torch was exposed to these temperatures. The quarters which the man had to work in were normally cramped, causing him to suffer muscular pain within a short period of time. Inner bottoms and other areas of ships under construction are very confining. Ventilation of these areas with removal of smoke and heat is necessary. Workers exposed to smoke and heat would fatigue very quickly, as a result of this, their arc time in welding was considerable shorter than what would be expected under other more favorable conditions.

The technical factors which had to be considered in analyzing the problem were that 1) heavy welds require high amperage, generating high heat in close proximity to the welder, 2) ship construction requires laying down many vertical welds using a weaving pattern, a very time consuming process,

3) large ships cannot be practically positioned for down welding, and
4) to make the technical matters even more complex, fits between pieces
in a ship vary considerably and cannot be held to the accurate tolerances.

ESTABLISHING THE SPECIFICATION

The type of power to be utilized to actuate the robot was an early consideration. Electric, hydraulic and pneumatic power were considered and
electric power was chosen for a number of reasons. First of all, it was
available in large quantity on the site. A totally electric robot could,
therefore, be provided power with no difficulty. While hydraulic power
was considered, due to the fact that the work was done outdoors in a sometimes cold environment, warm-up time of a hydraulic robot would be extensive. In addition to this, arc welding and hydraulic fluid would constitute a potential fire hazard. For these reasons hydraulic power was
discarded. Pneumatic power was not always available on site in early
construction, therefore, was not seriously considered.

The shipyard asked that the welding power supply be located at least 10
meters from the welder, providing a 20 meter radius of movement. This
would allow an operator to work with the portable robot without moving
the welding power supply. This would add to the flexibility of usage,
especially in limited access areas. It was hoped that a single operator
would be able to handle the welding robot so that labor would not be
increased.

The shipbuilder also indicated that since they already had large quantities
of arc welding equipment in inventory, they did not want a robot which
would require special torches. The configuration of the torch holder to
match the existing equipment was, therefore, established.

Perhaps the most difficult part in the establishment of the specification
was that of programming. It was fairly obvious from studying the work to
be done that the welding paths were not identical to each other, that fits
were poor and frequent re-programming would be required. While the paths
were different, the material to be welded remained fairly constant throughout many areas of the ship. Therefore, the weld parameters, the wire speed,

sequence of welding, weave patterns and dwell times did not have to be changed as often as the path programming. Weave patterns in the robot could be produced electronically rather than have the weave paths described by the welder as he programmed the path of the robot. This would be a significant labor saving.

Arc welding requires continuous path movement. Since the weld paths would vary considerably, a method had to be devised to quickly program the path points in the robot's memory. This requirement led to the development of the teaching head with a wheel which is perforated and interrupts a signal light beam as it is rolled along the seam to be welded (Figure 1).

DESIGN GOALS

The design goals were established from the specification. The first goal was for a lightweight robot and it seemed logical that it should be in two separate sections, one a control console (Figure 2) which would control the movements of the robot, its speed and its interface with the welding power supply; and the second (Figure 3), the robot itself which would manipulate the torch holder. The second design goal was a totally electric powered programmable robot. The third goal was a very quick method of programming the path by the operator. The fourth and more subtle design goal was that this robot, once developed, would be adaptable to industrial batch manufacturing processes, so that its use would not be limited to marine construction.

The design team on the project was separated into two groups, one in charge of electromechanical design and the other the electronic design. The two groups working together selected stepping motors as the ideal method of movement control. Initial mechanical design was not successful. Too much emphasis was placed on light weight and size, and the first choice of gear ratios was not stiff enough. Movement of the entire robot was too loose and therefore repeatability of its movements by electronic control was not accurate.

The initial design also utilized chains in combination with gears to produce some of the movement desired. Some of the gear design was exposed and this, in combination with the chains, produced problems in initial testing. These problems were overcome by going to improved gear ratios, removal of the chain type movement, and by enclosing exposed gear areas.

The welding torch is held in a wrist which is mounted to the robot central arm. The guidance of this arm proved difficult due to the fact that it was acting with gravity in one direction and against gravity in the opposite direction. It became obvious that while the motors were powerful enough to move the arm during the operating process, it was difficult for the welder to manually move it during the path programming procedure. This problem was overcome by adding a negator spring which balanced the arm of the robot and allowed for precise movement.

The initial control concept placed in the hands of the welder all the controls necessary to set the path of the robot and also the parameters of the weld. The human engineering in regard to performing these functions needed improvement. The most important decision made in the program was to separate from each other the controls for the robot path, and weld parameters. The former were placed on a small control panel (Figure 4) located on the robot itself; the latter were placed in the control console.

This decision to separate the controls had outstanding ramifications later in the program, when it was realized that an experienced welder could set the parameters of the weld on the control console and then lock the console, while a far less experienced man could easily be taught to program the path of the robot and have it perform welds of excellent quality.

While making modifications to the early engineering model it was noted that there was difficulty in repair due to the configuration. The unit was then redesigned with simplified mechanisms using spur gear drives, simple bearings and mounts. A modular motor drive was established so that motors could be interchanged in most areas. This allowed for easy removal for replacement or repair and a minimum of motors to be held in spare parts stock.

Perhaps the most difficult design area was the wrist since it incorporates an electric motor, electronics and mechanical functions. The mechanical clutch, latch for sliding movement of the torch, and electronic sensing

all had to be incorporated in something the size of a man's wrist
(Figure 5), and yet be rugged enough to meet the challenges of shipyard
use. Of all the electromechanical parts, the wrist went through the
greatest number of evolutions. The initial design placed within the
housing the electromechanical clutch and had the motor in the arm. When
this unit went into field test there were complaints regarding the size
of the wrist package and requests that it be reduced substantially. The
wrist was then re-designed using a mechanical clutch. This mechanical
clutch went through a series of design changes. It must spring open with
the proper load and yet provide for lateral movement without backlash.
This lateral movement was important in the operation of the machine and
backlash had to be eliminated in order to have proper repeatability of
the robot.

ELECTRONIC DESIGN

The five electric motors used to power each robot axis are stepping motors
with built-in encoders. During the teaching process when the path of the
unit is being fed into the memory, the motors are de-energized. As the
movements of the robot are made, the encoders spin and actually count
positions relating to the five axes movement. De-energizing the motors
puts the entire robot in a more or less "free-wheeling" position. On
playback, the motors are energized and the encoders allow movement until
they match the count in the memory for that particular axis at a particular
time. Since the speed of the robot movements is set on the control console, the programming time bears no relation to the actual welding time.
The programming of a three-foot seam to be welded can be accomplished in
say, seven seconds, while the actual time to weld might be as long as ten
minutes, depending on the quantity of metal to be deposited by the welding
torch.

In initial design it was decided to have no fan cooling, due to the dirty
environment in which the unit would operate and instead use convection
cooling. Two oil and dust tight enclosures (Figure 6) with no interior
cooling were placed back to back. Between these two enclosures are
externally mounted power transistors and ballast resistors so that a
chimney effect of cooling is created. With all heat producers in one of

the enclosures, the control electronics in the other enclosure never see the heat generated. The electronics are on printed circuit boards and everything is modular for quick replacement or repair.

PROTOTYPE TESTS

Prototype tests were conducted in the shipyard under actual working conditions. Basically they were very successful. Some buttons and switches which were on the teaching head had to be increased in size since a welder using gloves could not feel the movement of them. It was also found that the perforated wheel in the teaching head was too thin and in some cases it would get caught in poor fits. As a result, fatter wheels were developed and supplied which were very successful.

FINAL PRODUCT

A quantity of robots was ordered by the shipyard and put into actual production use. Results were quite amazing. It was found that one operator could operate more than one robot at a time, therefore increasing his arc time substantially. When working with only one robot welds which previously took 50 minutes to accomplish were now accomplished in a total time of 26 minutes by use of the robot. This latter figure included movement and programming in addition to the welding time.

As a result there is now available an electric, portable Apprentice® Robot weighing under 100 pounds. With its 200 pound control console, the robot is capable of arc welding within the envelope parameters shown on Figure 7.

Studies of its operation show that it can extend the arc time of a welder sufficiently to provide a return on investment in less than one year.

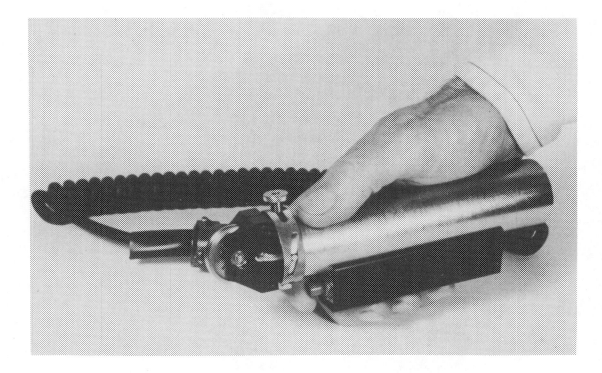

FIGURE 1 – TEACHING HEAD

FIGURE 2 – CONTROL PANEL, CONSOLE

FIGURE 3 – ROBOT, SHOWING ARM AND CONTROL PANEL

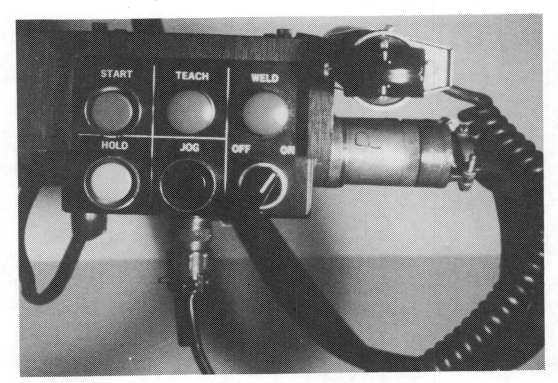

FIGURE 4 – ROBOT CONTROL PANEL SHOWING TEACHING HEAD
IN HOLDER

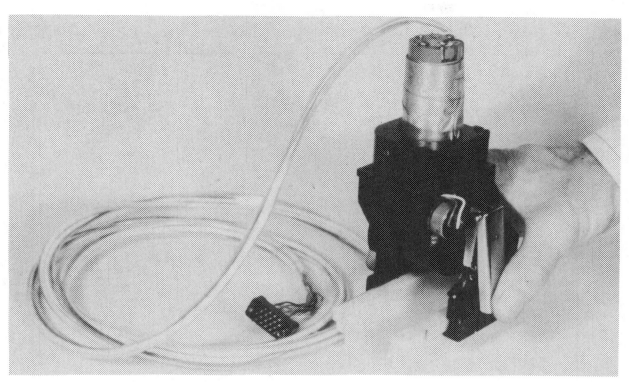

FIGURE 5 – WRIST ASSEMBLY

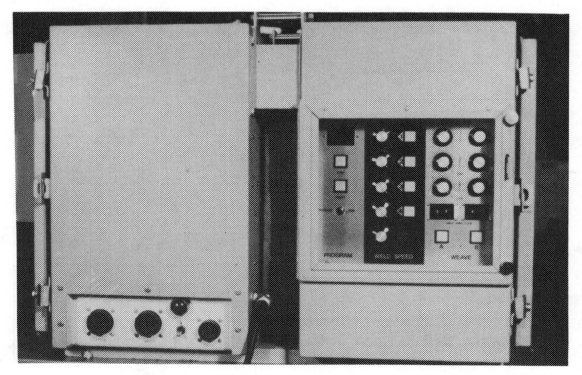

FIGURE 6 – CONTROL CONSOLE

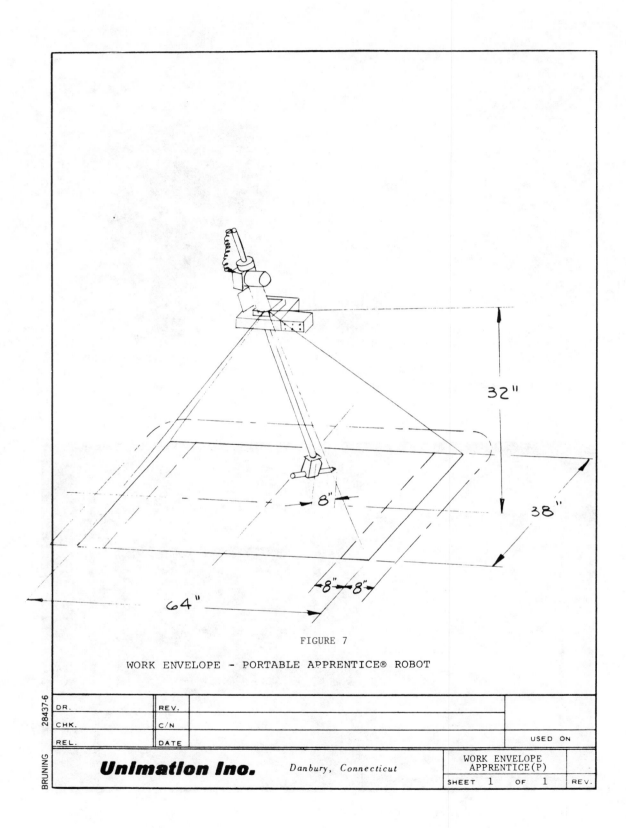

32"

38"

8"

8" 8"

64"

FIGURE 7

WORK ENVELOPE - PORTABLE APPRENTICE® ROBOT

28437-6

DR.		REV.			
CHK.		C/N			
REL.		DATE			USED ON

Unimation Inc.   *Danbury, Connecticut*

WORK ENVELOPE
APPRENTICE(P)

SHEET 1 OF 1   REV.

BRUNING

The Robot System at AiResearch Manufacturing Company of California: A Case History

By William A. Hosier
AiResearch Manufacturing Company of California

Two high-production manufacturing operations at AiResearch Manufacturing Company of California, welding motor housings and routing aluminum plenums, were selected for automation as a means of reducing operator fatigue and increasing productivity. Introduction of the Unimate 2000 industrial robot not only achieved the desired goal, but also decreased welding and routing time by 40 and 75 percent, respectively. This paper discusses the need for this innovation, the problems encountered in establishing a robot welding and machining facility, the programming techniques developed, the results obtained, and the future applications of robots at AiResearch.

INTRODUCTION

The introduction of a programmable industrial robot into AiResearch Manufacturing Company of California was based primarily upon economic and social needs. The promise of increased productivity, reduced costs per part, and greater profitability was considered in justifying the purchase of a new Unimate 2000 industrial robot. No less important a consideration was the need to improve the quality of working life and advance personnel into more challenging and rewarding positions. AiResearch Manufacturing Company seeks to eliminate jobs that are harmful, potentially dangerous, strenuous, unpleasant, or dull.

While all areas of manufacturing at AiResearch are periodically being advanced by the addition of new equipment and the latest technological developments, there were two high-production operations that appeared to offer the greatest return in the immediate future by the addition of programmable automation. Specifically, the extensive amount of GMAW (MIG) welding required to assemble large motor stator housings for various transit systems and the routing of a large-diameter hole in moderately thick aluminum plenums.

In the case of welding motor housings, operator fatigue had increased to a significant level. The steady handling of a heavy GMAW gun and cables plus the heat radiation from the massive welded assembly had a definite effect on the productivity of the welder. With hundreds of these housings to weld, this operation became a logical candidate for an automation process that would not only minimize operator fatigue, but also increase production rates. In the case of metal routing, a different set of problems existed. The machine operator was exposed to oil vapors, flying chips, and a high noise level for long periods of time. Operator fatigue was significant and contributed to low production output. Therefore, metal routing of aluminum plenums was a logical candidate for automation.

Based upon internal studies and discussions with Unimation, Inc. personnel, it was concluded that in addition to the benefits of reduced worker fatigue and improved product quality, a significant cost savings of one to three man-years was possible, permitting experienced personnel to be transferred to more challenging operations.

PURPOSE

The purpose of this paper is to present the results of AiResearch experience with the Unimate 2000B industrial robot, specifically problems associated with establishing a welding and machining facility, programming techniques developed, quality of the products produced, magnitude of cost savings achieved, and operating problems encountered.

EQUIPMENT

To facilitate MIG welding, the major function of the robot, an entire welding system was purchased from Unimation, Inc., of Danbury, Connecticut. This consisted of the following components:

(a) Unimate industrial robot Model 2000B with continuous path and five degrees of freedom (out/in, up/down, rotary, wrist bend, and wrist yaw)

(b) Unimate cassette tape recorder

(c) Hobart welding power supply, Model RC-300

(d) Hobart motor control, Model 123

(e) Hobart wire feeder, Model H4S

(f) Hobart MIG torch, Model GA-250H

Accessory equipment required to move and support stator housings or to make minor weld repairs included:

(a) AiResearch wire feeder dereeler system

(b) Two AiResearch rotary/tilt table positioners, each capable of handling 1500 lb

(c) A 14-ft jib boom crane and 30-ft-min. air-driven hoist, capable of handling 2000 lb

(d) Tungsten-inert gas (TIG) welding power supply with TIG torch

The equipment was located within a specified welding and machining facility shown in Figures 1 and 2. The Hobart wire feeder was mounted just above the wrist on the end of the twin booms, as shown in Figure 3.

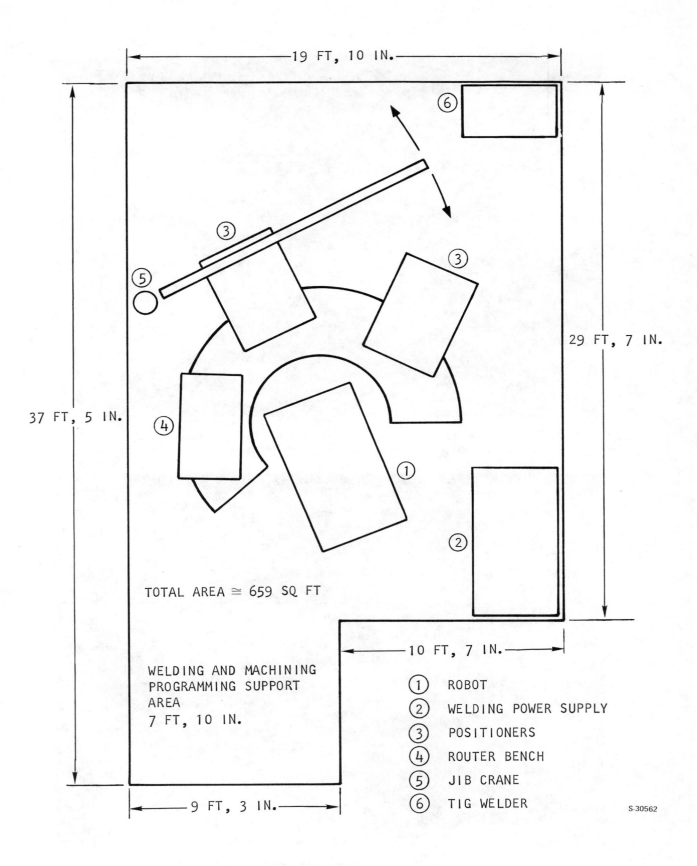

TOTAL AREA ≅ 659 SQ FT

WELDING AND MACHINING
PROGRAMMING SUPPORT
AREA
7 FT, 10 IN.

① ROBOT
② WELDING POWER SUPPLY
③ POSITIONERS
④ ROUTER BENCH
⑤ JIB CRANE
⑥ TIG WELDER

S-30562

Figure 1. Initial Welding/Machining Facility Layout

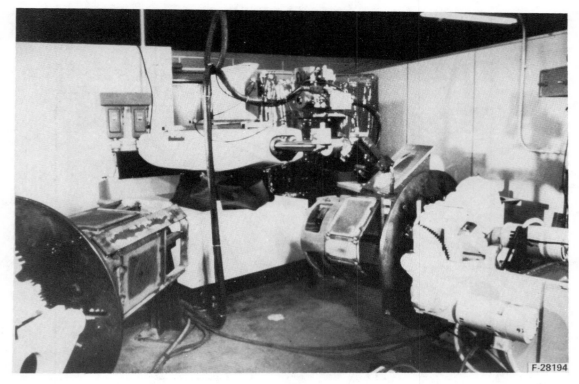

Figure 2. Robot Welding and Machining Facility Showing Motor Housings on Positioners and Router Bench with Plenum in Fixture

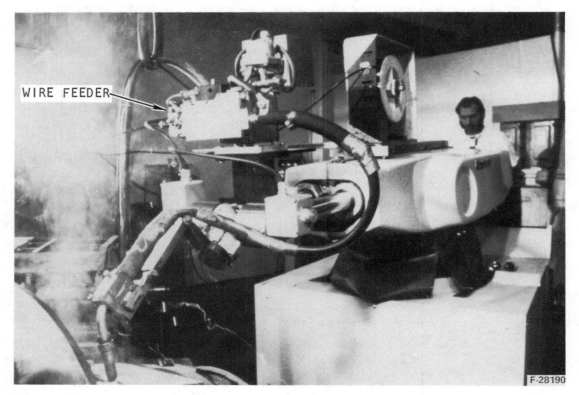

Figure 3. Wire Feeder Mounted Above Wrist at End of Robot Twin Booms

Also housed in the welding/machining facility were sufficient tools and accessory equipment to perform routing of aluminum plenums. Included were such items as:

(a) A 3-hp, 10,000-rpm, air-driven router motor

(b) High-speed router bits, 1/2-in. dia.

(c) AiResearch router guide

(d) AiResearch plenum holding fixture

(e) AiResearch lubricant application and recovery system

(f) A 3 by 5 ft steel bench to support aluminum plenums

It should be noted that selection of the 3 by 5 ft work bench was based upon calculations determining the maximum permissible rectangular area within the annulus of the minimum and maximum radii described by the robot arm (see the Appendix to this paper).

In order to maintain a constant positional relationship between the robot and other equipment, some of the equipment was fastened to the floor. The robot itself has two mounting brackets along the front; they are approximately 43 in. apart with 1.750-in. bushings (inside diameter). Two stainless steel pins with corresponding bushings were fabricated for machine locating; the bushings along with threaded inserts were embedded and bonded into the floor. Once the bushings, pins, and anchor bolts were installed, the robot was leveled and secured.

The two rotary/tilt table positioners and the steel router bench were also fastened to the floor, using very closely located embedded threaded inserts with standard 5/16-in. to 3/8-in. bolts. This arrangement made it possible to move the equipment, if necessary, without loss of position. This capability is absolutely necessary to relocate programmed points in relation to any device that may be moved and returned. Since excessive cummulative positional tolerances of accessory equipment and tooling are the most damaging factor to robot repeatability, they must always be kept to an absolute minimum.

DEVELOPMENT OF PROGRAMMING TECHNIQUES

Welding

Development of programming techniques for welding mild steel stator housings occurred in two phases. Phase I was a learning experience wherein all development was performed on mild steel standpipe test hardware shown in Figure 4. This practice avoided the possibility of scrapping expensive hardware during the learning phase. During this phase, not only were preliminary welding parameters developed, but development of the wire feeding-dereeling mechanism was initiated.

Figure 4. Standpipe Practice Piece

The wire feeder occupied the extreme position of the robot booms as shown in Figure 3. To minimize the weight on the booms, the wire feeder dereeler, which has to move freely regardless of the robot's position, was placed at the fulcrum of the moving portion of the robot. This placed the weight of the mechanism and its wire (about 55 lb) on the columnar support of the robot. The dereeler's adjustable friction or drag design allowed the wire to be pulled off the reel at a constant tension, thus minimizing any backlash that might otherwise occur when the feeder stopped. When the boom was fully retracted, the cable's radius of curvature was greatly reduced, frequently causing the wire to bind. This resulted in a slight interruption in the feed, which in turn would interrupt the arc and generate a skip in the weld. This problem was corrected by a cable control arm (see Figure 5) and a simple wicking lubricant application system (compatible with welding requirements) that applied lubricant to the wire just before it entered the cable. This dereeler design was subsequently replaced with a lighter unit that could be located beyond the fulcrum closer to the rear

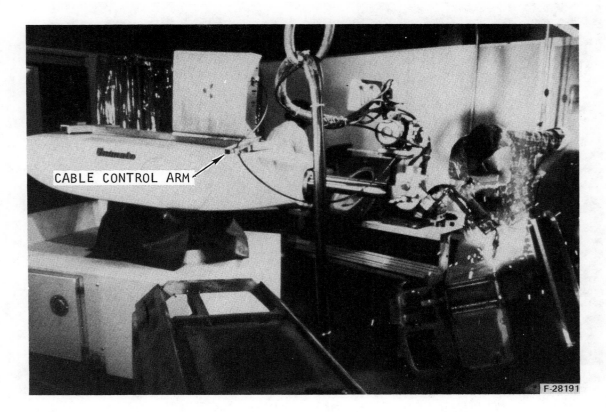

CABLE CONTROL ARM

F-28191

Figure 5. Wire Dereeler with Cable Control Arm

of the movable portion of the robot. This new arrangement effectively increased
the radius of curvature of the cable during motion of the boom, thus allowing
the wire to flow in a smooth, unrestricted manner.

Phase II development was conducted on stator housings using welding parameters established in Phase I. Motor housings were constructed by assembling
detail parts at a remote station adjacent to the welding/machining facility,
TIG tacking parts to hold them in place during defixturing, and transferring
them to a rotary positioner for robot welding. The alloy used in fabricating
housings is mild steel, ranging in thickness from 1/4 in. to as much as 1-1/4
in. The base flange and upper casting (shown in Figure 6) had cross sections
approaching 1-1/2 in. in thickness. Housings were positioned on the rotary/
tilt table with the aid of locating pins. In addition, an orientation detail
on the housing was aligned with a locating mark (similar to a timing mark) on
a platform fixture that had been permanently mounted on the rotary positioner
(see Figure 6). Loading was accomplished with the positioner table maintained
in a horizontal plane. After securing the housing, the table could be tilted
and/or rotated into the various welding attitudes dictated by a predetermined
program that consisted of signals from the robot's memory bank.

337

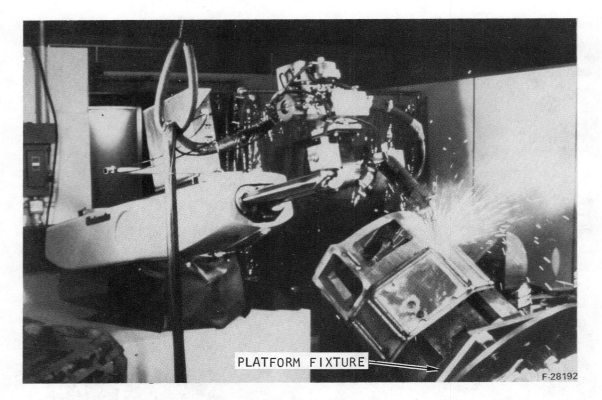

PLATFORM FIXTURE

F-28192

Figure 6.   Motor Housing with Upper Casting Being Welded in Place
While Secured to Permanently Mounted Platform Fixture

The welding operations called for eight panels to be welded in place utilizing all five of the robot motions. The boom out/in and rotary motions were the major movements; they were further refined at the arc by the simultaneous movement of the wrist bend, wrist yaw, and down/up motions. As each panel was welded in place, the workpiece was rotated and/or tilted into the next position by the program. This was accomplished through interlocks on the robot that allowed the positioner to be energized to another position as a programmed step. When the welding was complete, the positioner table returned to the horizontal position and the housing was picked up and returned to the assembly area for subsequent inspection. Inspection was made of 280 to 1024 inches of weld, depending upon the housing involved. The inspection showed that the appearance and structural integrity of the robot welds were far superior to average manual welds. Their appearance and penetration into the parent materials were much more uniform, due to the ability of the robot to maintain the torch at a constant velocity and position relative to the workpiece. The robot's stability in torch handling afforded torch tip velocities of 10 in./min for the thicker materials to 20 in./min for thinner sections. Quantitatively, the increased velocities decreased actual weld time as much as 40 percent.

The success of this type of welding lies to a great extent in the programming techniques. The Unimate 2000 is programmed by a hand-held control that simultaneously drives and programs. The operator or programmer drives the robot-held torch tip to points on a tape that has been placed along the joints to be welded.

As each point is reached, the operator records it in the computer memory simply by pushing a button on the hand-held control. The number of points used and their spacing is critical along curved welds or at corner welds, and the ability to ascertain these factors along with the proper torch angulation and distance from the joint is essential to the overall success of the welding task.

Important considerations in joint design are (1) to keep joints as accessible as possible (for example, to avoid welding under other components of the assembly), (2) to round off or angulate component corners wherever a change of weld direction is required, and (3) to keep welds in a horizontal plane (if economically justifiable) so that the force of gravity is exerted perpendicular to the weld. The same weld joint fit-up and dimensional tolerances used for other automatic shielded metal arc welding (SMAW) applications apply to this process.

Routing

The routing task consisted of routing a 10.17 ±0.03 in. dia. hole through 1/4-in. thick 6061-T6 aluminum alloy 55 degress off normal to a slanted surface (see Figure 7). This task was accomplished by a combination of programming and guiding. The programming was performed with the router, router guide, and knee attached to the robot (see Figure 8). The robot was driven to the holding fixture so that the router guide had a slight tangential interference with the fixture guide, as shown in Figures 7 and 9. The interference was required to produce a slight constant pressure by the router guide onto the fixture guide, assuring a continually smooth curved surface as the robot directed the router through the plenum. Without this type of processing, the generated surface would have a series of steps caused when the robot advanced from one taught point to another via coordinates instead of in a direct path. An alternative approach would have been to use a fixture track to contain the router guide completely throughout its travel, but this would have added considerable cost to the fixture with no visible improvement in the process and no additional hardware life. The preferred interference-to-guide method also prevents chip engagement between the router guide and the fixture guide. These chips would prevent the robot from attaining its programmed point(s), resulting in an unacceptable routed configuration.

The total effect of the interference stresses on the robot is not known, but to minimize that effect, AiResearch is considering placing a flexible link into a system that is otherwise rigid (except for the hydraulics). The contemplated flexible member is to be of resilient material such as rubber or plastic. It will probably be placed at the knee and at the robot wrist housing interface, as indicated in Figure 8.

Much work remains to be done to protect the robot from the stresses caused by interference and also from the transmission of vibration into its sensitive areas. Much of the operation vibration has been removed simply by weighting the workpiece with sandbags, but there are still unacceptable levels of vibration in the system since the encoders* need readjustment more often during the routing operations than during welding.

---

*The robot's encoders are primarily responsible for the accuracy of positioning and repositioning of the robot's arm.

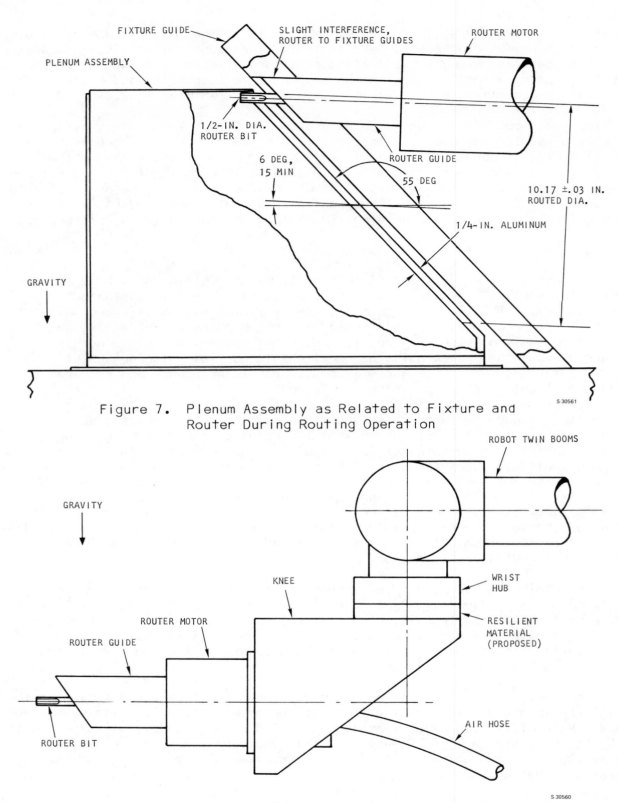

FIXTURE GUIDE

PLENUM ASSEMBLY

SLIGHT INTERFERENCE,
ROUTER TO FIXTURE GUIDES

ROUTER MOTOR

1/2-IN. DIA.
ROUTER BIT

6 DEG,
15 MIN

55 DEG

ROUTER GUIDE

10.17 ±.03 IN.
ROUTED DIA.

1/4-IN. ALUMINUM

GRAVITY

S-30561

Figure 7. Plenum Assembly as Related to Fixture and
Router During Routing Operation

GRAVITY

ROBOT TWIN BOOMS

KNEE

WRIST
HUB

RESILIENT
MATERIAL
(PROPOSED)

ROUTER MOTOR

ROUTER GUIDE

AIR HOSE

ROUTER BIT

S-30560

Figure 8. Router Mounting onto Robot with Contemplated
Addition of Resilient Material Shown

F-28193

Figure 9. Plenum Assembly Being Routed

To minimize tooling cost, and to verify the interference-to-guide method of routing, 7075-T6 aluminum alloy was chosen for the fixture guide. The alloy is easily machined and, with the proper lubricants, has sufficient wear resistance to last through several hundred cycles of this type of operation. The router guide, which is the moving member of the interference-to-guide method, was fabricated from oil-penetrated bronze; the bronze has good wear resistance and machining properties, and was readily available.

The final components of the routing system are a lubricating system and a router. The flood type lubricating system has a filtered recovery and uses good quality aluminum cutting oil, which lubricates both the router bit and the guides. A two-flute, high-speed router minimizes cutting tool chatter.

The routing system not only relieved the operator from the discomforts of oil vapor, chips, and noise, but also has reduced production time by about 75 percent.

DISCUSSION

The development of a successful program to weld stator housings came about only after considerable effort was expended in establishing suitable welding parameters for test hardware aud subsequently applying these parameters to production hardware. Parameters derived from test hardware had to be modified appreciably to compensate for large variations in section thickness of each housing. Once a program was established and recorded, it was usually repeatable. Nonrepeatability occurred infrequently and was primarily associated with malfunction in either the printed circuit boards or the encoders. A relatively simple readjustment eliminated the malfunction.

Programming the specific routing task described in this report was simpler than programming welding of stator housings, but simultaneously was more time consuming since many more steps were required per unit length to assure a smooth elliptical shape at the end of the operation. The greater the number of steps when programmng curved surfaces, the truer the curve reproduced. Although routing is a relatively simple operation for the robot, it proved to be more severe than welding since an unacceptable intensity of vibration was transmitted through the arm to such sensitive components as the encoders, resulting in the need for frequent readjustment. A means of absorbing most of the vibrations will have to be developed if the robot is to be used for routing operations in the future.

Performance of the robot to date has shown that both repeatably high quality and quantity are achievable when necessary preparations have been made. For example, the routing of aluminum plenums required development of substantial fixturing and tooling. Selection of the heat router bit configuration, an efficient lubrication system, and the establishment of the most practical router bit travel speed were all part of the necessary preparations that preceded programming. Once programming of the specific task or operation was completed, the robot continued to repeat with excellent accuracy. Experience with welding housings proved that end product quality will be as good as the accuracy of the positioning fixture and the dimensional fit-up of the parts to be welded; thus good preparations prior to any programmed operation are critical.

To date, the relatively limited experience with the Unimate robot has resulted in a production time reduction of 75 percent for routing aluminum plenum assemblies and a 40 percent reduction in time required to weld motor housings. These figures can be expected to increase significantly as the fit-up of components is improved prior to welding.

CONCLUSIONS

(a) Results to date have proven that programmable gas metal arc welding (MIG) of complex assemblies is not only a reality but a daily occurrence in a production facility. Using a remote teach-control unit allows a variety of programs to be stored in the robot's memory bank. This, coupled with accurate tooling and good fit-up of parts, gives excellent repeatability of quality hardware.

(b) The success achieved with automatic routing of aluminum plenums
     demonstrates conclusively that the robot may be adapted to vari-
     ous other machining functions that may be carried out whenever
     demand for welding stator assemblies has diminished.

(c) The robot has been proven capable of handling jobs considered to
     be hazardous and/or fatiguing to the human operator with little
     or no effect on the machine and with a substantial cost saving.
     The robot not only increased productivity in two manufacturing
     operations, but also allowed highly skilled operators to use
     their talents on jobs that are not readily adaptable to auto-
     mation or that are too small in quantity to make automation
     economically feasible.

FUTURE APPLICATIONS

The experience and knowledge gained from the welding and routing experiments are expected to serve as the basis for future investigation into the
following manufacturing operations:

(a) Punch press die loading and unloading of workpieces

(b) A brazing alloy deposition process similar to spray gun painting

(c) Mechanical assembly of a heat exchanger requiring tube placement

(d) Loading and unloading workpieces during shearing operations

(e) Loading, directing, and unloading workpieces during seam welding
     operations

(f) Loading a fixture, directing it through spot welding, and unloading
     the fixture during fixtured type spot welding

APPENDIX A

Selection of the 3 by 5 ft work bench for the routing task was based on calculations to determine the maximum permissible rectangular area within the annulus of the minimum and maximum radii described by the robot arm (see Figures A-1 and A-2).

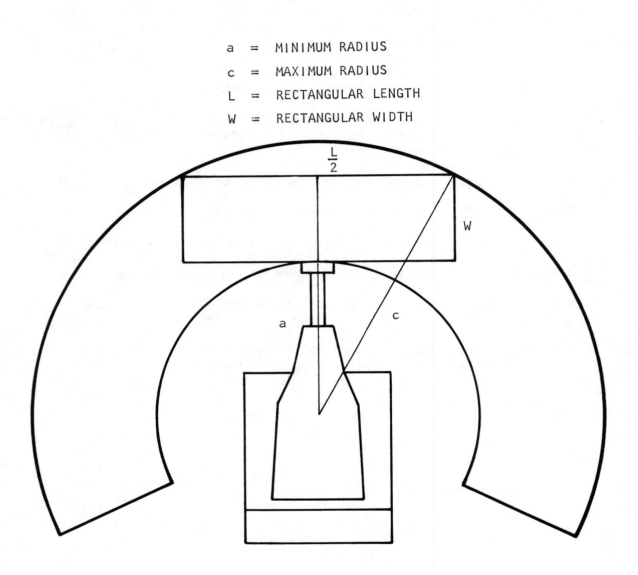

a = MINIMUM RADIUS

c = MAXIMUM RADIUS

L = RECTANGULAR LENGTH

W = RECTANGULAR WIDTH

S-30563

Figure A-1. Annulus Described by Robot's Minimum and Maximum Radii

$$(a + W)^2 + \left(\frac{L}{2}\right)^2 = c^2 \qquad \text{(Equation 1)}$$

and

$$A = LW, \quad L = \frac{A}{W}, \quad L^2 = \frac{A^2}{W^2} \qquad \text{(Equation 2)}$$

therefore

$$a^2 + 2aW + W^2 + \frac{A^2}{4W^2} = c^2$$

Differentiating with respect to W:

$$\frac{d}{dW}\left(W^2 a^2 + 2aW^3 + W^4 + \frac{A^2}{4}\right) = \frac{d}{dW}\left(c^2 W^2\right)$$

$$2a^2 W + 6aW^2 + 4W^3 = 2c^2 W$$

(At maximum, (A) is a constant. Therefore, $\frac{dA^2}{dW} = 0$)

Dividing by 2W,

$$a^2 + 3aW + 2W^2 = c^2$$

$$2W^2 + 3aW + a^2 - c^2 = 0$$

Solving for W,

$$W = \frac{-3a \pm \sqrt{9a^2 - 4(2)(a^2 - c^2)}}{2(2)}$$

$$W = \frac{-3a \pm \sqrt{8c^2 + a^2}}{4} \qquad \text{(Equation 3)}$$

Solving for W from Equation 3 when a = 38.25 in. and c = 79.25 in.,

$$W = 28.16 \text{ in.}$$

Solving for L from Equation 1 when a = 38.25 in., c = 79.25 in., and W = 28.16 in.,

$$L = 2\left|c^2 - (a + W)^2\right|^{1/2}$$

$$L = 86.49 \text{ in.}$$

Solving for A from Equation 2 when W = 28.16 in. and L = 86.49 in.,

$$A = 2435.56 \text{ sq in.}$$

<u>Note</u>: To minimize bench top deflection and vibration during the routing operations and to limit congestion within the facility, the smaller 3 x 5 ft bench was chosen as opposed to a 3 x 6 ft bench. This was the only commercially available size that would afford approximately the calculated maximum area at nearly the same length and width without excessively diminishing accessibility to itself and the other equipment.

S-30690

Figure A-2. Calculations Used to Determine the Maximum Permissible Rectangular Area Within the Annulus Depicted in Figure A-1.

CHAPTER 10

MACHINING

Commentary

Machining operations with robots are a relatively new applications area. Production operations include deburring of machined parts and removal of flash from parting lines of forged and cast parts.

Pneumatically or electrically driven tools are used and the tools are usually compliantly mounted on the robot. In some cases, the robot may employ force-feedback, as well.

Other machining operations under investigation involve drilling, profiling and riveting of aircraft panels. At present, these operations require the use of drill fixtures and routing templates. However, the long-range objective is to develop hardware and software capabilities which will do away with such machining aids.

Robotic machining has the potential to eliminate what have primarily been hand operations and, eventually, to eliminate costly tooling aids, as well.

Presented at the Robots III Conference, November 1978

Some Special Applications For ASEA Robots— Deburring Of Metal Parts In Production

By Bjorn Weichbrodt and Lennart Beckman
ASEA, Inc.

A significant number of installations made with ASEA robots have been of the type where the robot, holding a hand tool, performs some operations other than materials handling. Such installations include grinding, polishing, deburring, measuring, and many others.

This paper will attempt to describe some applications concerned with deburring of metal parts.

Deburring after a mechanical operation is a common and expensive operation which has earlier been difficult to automatize with flexibility. An industrial robot with the precision and contouring capability of the ASEA robot here offers a general and economical alternative.

INTRODUCTION

Almost always when machining is performed on metal parts, burrs are generated. The removal of those burrs are usually an expensive operation. Automation of this has earlier normally been possible only in special applications (using sandblasting, explosive shocks, etc). Most deburring has therefore been done by hand which is a monotonous and, with time, more and more expensive task.

One of the first generally usable and economical methods for automation of deburring uses industrial robots. By closely resembling the manual method the industial robot can solve most deburring problems.

AN EXAMPLE OF ONE INSTALLATION

The task is to remove the burrs generated from machining of a cast iron detail. The detail is part of a break regulator used in trucks and other heavy vehicles. The burrs are generated inside the detail at the intersections between holes drilled from different directions. The size of the burrs varies between approximately one and ten millimeters. Two contours are to be deburred, one rectangular and one with continuous curves.

Figure 1 shows the installation which uses an ASEA robot of type IRb-6. The robot works together with a parts magazine for the parts which are to be deburred, two deburring tools, and a conveyor for the finished parts.

In this installation the robot handles the part itself since the part is relatively light weight. The part is picked from the magazine. It is then moved by the robot between the two tools and manipulated in such a way that those very accurately and with controlled speed follows the contours which are to be deburred. At the end of the cycle, the part is placed on the conveyor belt and the robot continues the operation. The cycle time is approximately 40 seconds. The corresponding cycle time for manual operation was earlier approximately twice as long.

The magazine is arranged as a rotating index table and holds approximately one hour of production (this is later intended to be expanded). This allows the robot to work continuously even with very limited supervision.

Two deburring tools are used in this application. One is a rotating hard metal tool and the other is a rectangular reciprocating file which is used especially to reach the sharp corners at parts of the contour. Since the robot with very simple accessories performs the same task as two operators, and can work in two-shift-operation, it is easy to see that installations of this type are very profitable.

A further advantage is the very repeatable quality of the job performed.

Figure 1 Robot deburring installation at SAB Industry, Landskrona, Sweden.

DEBURRING WITH ROBOT

In the following some general views on robot deburring are
presented. In order to define the scope, some general
properties of burrs are discussed, thereafter some of the
requirements which have to be put on the robot are discussed.
The choice of deburring tools are also essential as well as
the method for fastening the tools to the robot or the work
table. Those questions are also discussed.

BURRS

Many different types of burrs occur, depending on how they
have been generated, where they are located on the detail etc.

Burr generation

- Machining burrs

  Such burrs are generally generated through drilling, milling,
  turning, grinding, punching, and cutting operations. Most of
  the applications for robots have concerned this type of burrs
  which will be discussed more in detail in the following.

- Fine casting

  Burrs from die-cast metal parts, and also some moulded
  plastic parts, are quite similar to machining burrs and can
  be treated in the same way.

- Coarse casting burrs

  Excess material generated for example on sand castings are
  often considerably heavier than those above. Such burrs are
  also well suited for robot removal but require other methods
  and are not discussed in detail in this paper.

Burrs locations

The burrs can be generated in a number of different places on
the workpiece, for example:

- Between a machined and a raw surface. In this case the
  position of the burr can vary from one detail to another.

- At the intersection between two machined surfaces.

- At each side of a drilled hole or at inner intersections
  between holes.

Burr properties

There are major differences between different burrs even when the burr generation and location are disregarded. Considerable variation has been noted concerning:

- Size

  Size can vary between fractions of millimeters and several centimeters.

- Shape

  For different operations the thickness of burrs at the base shows major differences.

- Brittleness

  Depends especially on the material.

When making a production installation, it is important to recognize that the individual variations can be quite considerable. One batch of details can differ from others depending on differences in material, tool wear, etc.

Most of the different types of burrs which have been mentioned above are suitable for robot deburring. However, it is necessary to choose the deburring tool and deburring pattern according to the situation at hand.

Figure 2   Robot deburring of machined gear box housings at Volvo AB, Köping, Sweden.

REQUIREMENTS ON THE ROBOT

A successful robot deburring installation puts high requirements on the robot. It has to generate complicated motions with very highly controlled speed but also be simple to program and maintain.

- Contouring

  This is a necessity. It must be possible to follow an arbitrary contour with high accuracy in order to obtain satisfactory quality.

- Repeatability

  Very high repeatability is necessary in connection with the contouring. It is especially important that the robot can move without noticeable vibrations. Deviations larger than a few tenths of a millimeter usually decreases the quality of the result.

- Speed

  Since a short cycle time is important for the profitability, it is necessary for the robot to be controlled within a wide speed range, from high speed for the materials handling to very low and accurate speeds for some of the deburring operations.

- Speed Programmability

  During a deburring operation, different speeds will be required for different parts of the contour.

- Servo Stability

  The servo system of the robot must be able quickly to compensate for varying loads which often are obtained during the deburring depending on the different sizes of the burrs.

- Simple Programming

  One of the most important requirements is the possibility for simple programming of the robot. This concerns not so much the original programming as the possibilities to optimize the program in order to achieve a good quality in combination with a short cycle time. Thus, it must be possible to change the location of the contouring path as well as the speed at different parts of the path easily. Such editing of the contouring program, similar to the editing capability of modern CNC-numerical control systems, is a necessity.

In order to meet the requirements above the robot will need to have a computerized control system, an accurate servo system, (so far we have achieved the best results with electrical servos), and a robust but accurate mechanical system.

The ASEA industrial robot IRb-6 is characterized by the following properties:

The contouring is obtained through the control of a micro-computer which achieves 3-6 axis interpolation of the contour, and gives a continuous control of the path velocity.

The positioning repeatability is $\pm$ 0.2 mm. This low value has been obtained through a very robust design of mechanical parts and the exclusive use of rotating ball- or roller-bearings.

Through the micro-computer all forms of editing of the robot program can easily be made. Changes of the contouring path as well as changes of the speed at different parts of the path can be easily inserted without changing the other parts of the program.

TOOL CHOICE

Experience has shown that the choice of the right tool is essential for success of the installation.

In most applications a rotating high-speed hard metal tool is used. In some applications, especially where it is necessary to reach into sharp corners, a reciprocating tool is used. In some applications band sanding machines and similar are being used.

The support of the tool is also very important, whether it is held by the robot or mounted on the work table. A machine with a rotating tool must have a well damped resilient mounting.

Considerable amounts of application laboratory work has been required to determine the characteristics of the tool support required for different types of installations.

Figure 3 shows support system for a rotating tool using support elements of rubber with a variable spring constant.

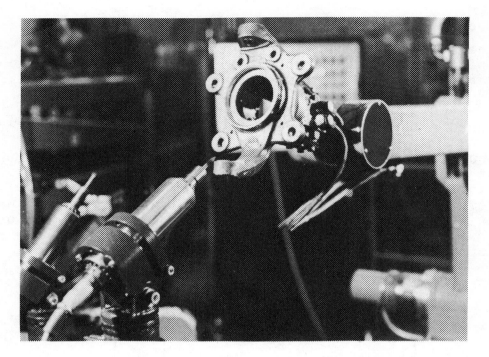

Figure 3   Robot deburring of forged metal details, using
           resilient mounting of the deburring tool

DEBURRING TOOL APPLICATION

It is essential that the deburring tool is used correctly
in order to obtain acceptable quality and to minimize the
cycle time.

This implies:

- Correct contact point

  Errors often lead to vibrations

- Correct contact angle

  Correct angle results in the avoidance of secondary burrs

- Correct path direction

  Incorrect direction of the tool motion relative to the
  rotation of the tool often results in vibrations.

- Correct path velocity

  Depends on type of burrs, desired quality etc.

- Correct resilient mounting

  The resilient mounting compensates for variations in
  position between different details, for example, the
  different location of raw cast or forged surfaces.

Different obtainable shapes of the rotating tools have
to be used especially in order to reach unaccessable burrs.

The deburring of small holes (Ø 15 mm) in a gear box housing
of cast iron (Volvo, Sweden) is a good example of the
accessibility problem. Both ends of the holes are to be
deburred. This is done in a production installation by using
a rotating spherical file with diameter 12 mm, and the use of
the forward edge for the outer side of the hole and the rear
edge for the inner side. See figure 4.

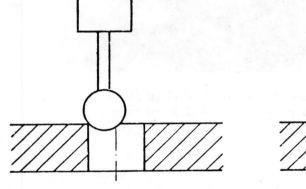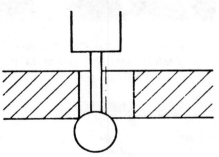

Figure 4   Robot deburring of holes using spherical
           rotating tool

CONCLUSIONS

A robot deburring installation is very often the only
automation alternative to manual deburring. It is usually
a rather simple installation because few and easily maintained
parts are included. However, difficult requirements are put on
the robot. The correct tools must be used and they must be
used in the correct way.

This paper mentions a number of the parameters which are
essential to control in order to obtain a good result.
Experience has shown that it is very difficult to calculate
theoretically many of these parameters. It has been possible,
however, to generalize experimental findings in such a way
that the specific application work for many new installations
has been significantly reduced.

Many commercial installations and a couple of years of
production experience has shown that the ASEA robot, together
with the right applications technology, offers possibilities
to automatize a very wide range of deburring operations.

Reprinted from Modern Machine Shop, April 1980

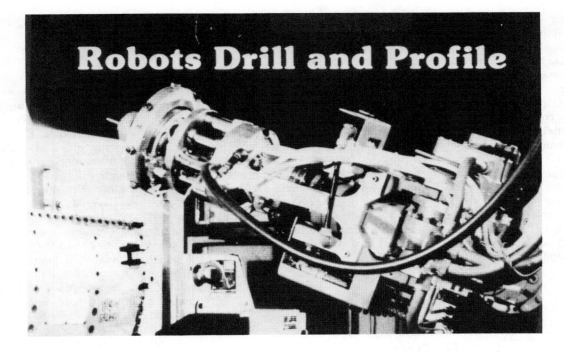

Robots Drill and Profile

The robot has made a quantum leap from a materials handling device to one that directs the cutting tools in the machining processes—part of which are contouring.

By KEN GETTELMAN, Editor

The workpieces appear deceptively simple. They are rectangular plates with rounded corners and roughly one hundred drilled holes, most of which are near the outer edge but some are in a line pattern from top to bottom toward the center.

At first glance, the machining requirements appear to be a simple two-axis hole-drilling routine plus a final profiling to shape. Instead of a basic 2½-axis NC unit, the machining is done by a robot arm that picks up a drilling or routing spindle and guides it through its paces.

There is good reason for this. The workpieces are not flat plates. They are part of an airfoil surface and thus have complex contours that go far beyond simple arcs. If the holes were drilled on a numerically controlled machine tool it would require the full five-axis variety. There would have to be full control over the X, Y and Z rectangular coordinates plus a tilt and swivel of the head to insure that the holes were drilled normal to the workpiece surface. The same would hold true for profiling the edge so that the profiling cutter would main-

tain a proper orientation to the workpiece surface. NC could be a viable approach, but it would require a heavy investment in a highly sophisticated machine tool plus a lot of complex programming. The robot shows good promise of a more economical approach to certain types of aerospace machining requirements. By implication, the same holds true for general manufacturing.

The workpieces now being machined by the robot are removable panels that are integral fuselage components on the F-16 fighter aircraft produced by General Dynamics in Ft. Worth, Texas. There are numerous panels on each aircraft to allow access to the interior for servicing and maintenance. When bolted into place, the panels become a continuum of the sculptured fuselage surface.

While the numerous panels are similar in appearance, each is slightly different and must be drilled and profiled to its own individual set of specifications. The traditional method of manufacturing the panels, made from a 1024 aluminum alloy, was to saw the panels from basic

plate, press them to contour, and mount them on drill jigs. The panels were drilled and then profiled by operators with portable power tools. The method was subject to all the vagaries of any manual operation, but the hand method had been the best manufacturing approach prior to the robot.

About two years ago, the United States Air Force Materials Laboratory at Wright-Patterson Air Force Base in Dayton, Ohio began its seven-year $100 million integrated computer-aided manufacturing program. One of the first efforts in that program was the sheet metal segment (airframes are basically sheet metal fabrication). And one of the first projects was a robotic system for aerospace batch manufacturing that would simplify some of the complex sheet metal machining efforts. The project was divided into three tasks called A, B and C. Task A, undertaken by General Dynamics, had four objectives:
1. Assess the state of the art in robotics.
2. Prepare a Robotics Application Guide. (The preliminary draft has been completed and its 141 pages

cover just about everything in the state of the art today.)

3. Develop a robotic work station for drilling and routing a limited number of aircraft sheet metal panels.

4. Expand the capabilities of the work station with automatic part identification, integrated station control, and material handling, which thus forms a work cell. The cell, shown in Figure 1, is where numerous aircraft sheet metal panels are drilled and routed in a batch manufacturing environment. The $650,000 contract was completed by General Dynamics in late 1979.

Task B, which will tackle the problem of programming the robot cell off-line, is in the hands of McDonnell Douglas Automation of St. Louis. Task C is designed to apply the robot to assembly operations—a project being undertaken by Lockheed Aircraft in Marietta, Georgia.

Task A has been completed and the sheet metal cell is operational in the General Dynamics' Ft. Worth plant. The purpose of the work cell is to provide sheet metal parts at high production rates with improved quality and lower manufacturing costs. That goal has been achieved. Previously, it required 80 to 100 minutes to complete the drilling and profiling of a panel; it now requires 20 minutes with the robot, and a consistently higher quality workpiece is produced.

Cell Makeup

The heart of the cell is a Cincinnati Milacron T³ (The Tomorrow Tool—Cincinnati Milacron's registered trademark) robot arm with six axes of motion as shown in Figure 2. The robot positions the drill and router tool systems at the workpiece for machining. Drilling is essentially a positioning operation. But the router operation is directed by a programmed path of the robot arm involving both a continuous movement and a feed rate to profile the rough-cut panel into a finished shape.

The robot is controlled with a programmed Cincinnati Model 2200B computer numerical control (CNC). It is equipped with a cassette loader for entering the operating system tape and individual workpiece programs. This allows the robot to operate in a stand-alone mode if the integrated controller malfunctions. In the stand-alone mode, there is a limit on the number of parts the controller can process automatically since the data-point-storage capacity is only 685 points. With very careful programming, five or six small workpiece programs can be stored within the CNC. Due to this limitation, the CNC is supplemented with a larger integrated control computer.

It is an Intel System 80/20 microcomputer linked to the CNC with the standard RS232C communications data link and a dedicated electronic computer interface. It is possible to store hundreds of programs in the microcomputer's memory.

Two different types of portable tools are used in the cell. The first is a compliant drill system and the second is a compliant router system. Either tool system may be coupled to the robot arm through a quick-change adapter fitted to the robot arm. The pickup and clamping of either tool system are under full program control.

The drill motor is housed in a highly compliant (able to make a shift in any direction even though the robot arm is rigid) cage fitted with a guiding nosepiece. Used in combination with a fiberglass jig equipped with drill bushings, it forms a hole locating system capable of holding tolerances within ±0.005 inch (0.0127 mm) even though position-

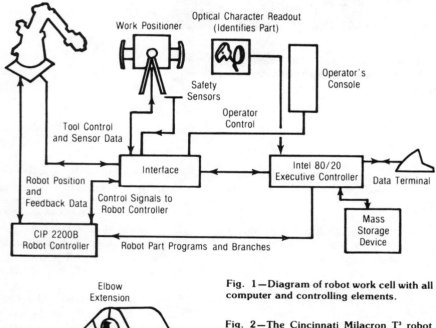

Fig. 1—Diagram of robot work cell with all computer and controlling elements.

Fig. 2—The Cincinnati Milacron T³ robot has six axes of motion and can carry a 100-pound load at full extension.

ing accuracy of the robot arm is no better than ± 0.050 inch (1.27 mm). The compliant drill cage also houses the sensors that detect cage extension and retraction. drill extension and retraction, and part contact. The router housing is similar. but it has skate bearings that ride along the edge of the jig so that the cutter can profile the workpiece to the finished contour.

The part positioner for the work station is an Aronson weld positioner retrofitted with a rotating box made of aluminum plate. The positioner holds the workpiece for the drilling and routing operations. At the same time, the cell operator can unload the previously machined workpiece and then load the next workpiece to be machined. When the machining is completed, the positioner rotates the finished workpiece to the outside position and the mounted work to the machining position.

The workholding method consists of two fiberglass shells with the workpiece sandwiched in between. The workpiece is actually held by clamps through two index holes that are drilled by hand. The outer shell is the drill jig fully equipped with bushings. Spokesmen state the fiberglass tooling is extremely durable and has shown no signs of wear or degradation through prolonged and sustained use. Fiberglass tooling has proved to be equally durable when used for hand drilling and routing operations.

As part of the computer and CNC system there is, of course, an operator's console and a data entry terminal to give complete flexibility to the cell.

One of the unique cell features is an optical character readout. The part number is printed on a panel on the face of the holding fixture. When the workpiece is to be machined, the cell operator merely passes the optical character reader over the part number where it is picked up and fed into the computer. The correct part program is automatically activated and fed through the CNC. Thus, there is no chance of calling up the wrong workpiece program as would be the case with an incorrect manual data entry of part number.

In addition to the operating components, the entire operating cube of the robot arm is surrounded by an optical safety system. If any object enters the cube and interrupts the safety light beam, the system comes to a halt. It's the same optical approach used for press safety. The operator's control panels are outside the light curtain.

How It Works

Assuming a cold start, power to the system is turned on and the robot hydraulic system is allowed to stabilize. The computer numerical control and the storage computer are turned on. The compliant tool cages must be mounted in the proper spaces in their racks. Sharp cutting tools must be in the cage spindles. Once the startup preparations are complete, the operator loads the workpiece/fixture assembly onto the rotatable part positioner. He loads it in the outside position. The part number is then read by the optical character recognition device (or entered by keyboard, but the optical reading is preferred). The computer then searches its system for the part information and retrieves it. The information is then fed to the CNC for actual robot control. At the time of startup, the robot arm is in the wait station. The first workpiece program instruction directs the robot arm to the tool rack where the drilling cage is picked up as shown in Figure 3. The robot then proceeds to drill the one hundred or so holes in the workpiece (see Figure 4). The drill cage includes the motor, spindle, overall housing, probe, and drill in addition to a coolant system. Since the robot has a ± 0.050-inch tolerance on point location, it is not the actual drill that enters the drill bushing; rather, it is the probe. Once the probe reaches the programmed depth, the drill is then advanced through the hollow probe to the workpiece surface. The fluid and chips are carried back through the probe to an opening beyond the surface of the fiberglass fixture.

The Cleco high-speed (2700 rpm) air-driven spindle automatically couples to the shop air supply as the spindle-cage assembly is picked up by the robot arm. The drill has a special 90-degree point with two high-helix flutes.

During all normal operations the robot arm will place the compliant drill cage close enough to the drill

Fig. 3—The robot arm advances (left) and then picks up a compliant drill housing.

bushing mouth that probe penetration, followed by drilling, is no problem. But there are those times when the probe is unable to enter the bushing. If it is unable to enter, the program will call for a retract of the robot arm and a second attempt. If the second attempt fails, a third attempt will be made. If the third attempt fails, a diagnostic printout (see Figure 5) is made on the printer strip, which is an integral part of the CNC unit. The robot simply moves on to the next program point and continues its routine. When the operations are complete, the operator checks the diagnostic printout and manually

drills any holes that have been missed. The average workpiece will have from one to three holes that were unable to be drilled by the robot operation.

When the drilling is completed, the program directs the robot arm to place the drilling cage back in the tool rack and move over and pick up the router cage. The router cage, with its skate-type bearing, rides on the two halves of the fiberglass fixture assembly. The routing cutter then mills the final profile on the panel. This is a continuous path milling operation and involves a travel rate as well as path definition. The pro-

grammed path is tight enough that pressure is always applied to the router cage, forcing the cage to comply with the fixture profile.

When both the drilling and profiling are completed, the robot arm returns the router cage to the tool rack and the arm itself returns to a rest position. The part positioner rotates 180 degrees and the previously mounted workpiece rotates into the work position while the completed workpiece is either removed or finished by the operator after he checks the diagnostic printouts. The cell operator then mounts a new workpiece in the wait position and

Fig. 5—Reviewing a diagnostic printout of the completed workpiece are (left to right) Dean Golden, Program Manager of Robotic Applications, General Dynamics; Dennis Wisnosky, ICAM Program Manager; Rosann Stach, ICAM Project Engineer; and Bob McMahon, Group Supervisor CAM Development, General Dynamics.

the process continues.

Training The Robot

In the present state of the art, the only practical means of programming or training the robot is a "walk-through" process. The operator actually positions the probe in each of the holes in the workpiece and captures the position by depressing a memory button. The same holds true for the profiling. The CNC captures a whole succession of closely sequenced points, which are then related to a programmed feed rate. This method of programming requires four hours or more.

The very logical question to ask is, "Why can't the robot be programmed the same as a machining center?" The answer lies in the type of resolvers and motions. For the most part, the regular NC machine tool is operated by a series of straight motions along the X, Y or Z axis. This may be combined with one or more rotary motions. But anything above the three-axis level gets into a very sophisticated programming effort.

The robot used in this program has six axes of movement and all of them involve some kind of rotary motion. Translating the combined moves of six rotary axes to get to a specific point within rectangular coordinates is beyond the current state of the art. As mentioned previously, task B of the robot project will be the development of an off-line programming capability. Thus, while off-line programming is not now feasible, it is well within the realm of possibility that a processor language for this purpose will be ready in the near future—possibly a year or two.

One objective of task A, along with developing a robot work station, was the preparation of a Robotics Application Guide. The final draft of this 141-page presentation is just now becoming available. Numerous experts in the field contributed to the publication, which contains a summary of the present robotic state of the art. It makes an excellent source document for those wishing to obtain a very detailed insight into robot design and application.

The only restriction upon the distribution of the Guide is a two-year limitation of delivery to only United States' recipients. Any U.S. company within the 50 states, territories, and District of Columbia wishing to receive a copy may obtain one by contacting Rosann M. Stach, AFML/LTC, Wright-Patterson AFB, Ohio 45433. Recipients must agree to keep it within United States environs. Two years after its formal introduction there will be no limitation upon its distribution. **MMS**

Presented at the Robots V Conference, October 1980

Robotic Drilling and Riveting Using Computer Vision

By Richard C. Movich
Lockheed-California Company

An experimental program using computer vision as a sensory feedback method for automated fastening of aircraft structures is described. Computer vision was investigated for training, calibration, recognition, determination of position and orientation, fine positioning, verification, and inspection. A hierarchical microcomputer system was developed using a 16-bit microcomputer for overall control and subordinate microcomputers for dedicated control of a vision module, a robot, and fastening equipment. Computer controlled drilling and riveting equipment with a two-axis servo-controlled table for positioning of small, flat aircraft assemblies was developed. Preliminary work using a five-axis robot for handling and positioning of small assemblies is described.

INTRODUCTION

Drilling and riveting is one of the most tedious and repetitive tasks in aircraft manufacturing. This task is generally highly labor intensive and has overall low productivity. A variety of tools, equipment, and techniques are employed to drill holes and install rivets; a human using visual perception is almost always in the drilling position control loop. Even when automatic drill-riveters under computer control are used, an operator monitors and adjusts the drilling locations using CCTV. With few exceptions, the billions of holes drilled for installing rivets for aircraft assembly in the last half century have been guided by human vision.

Recent developments in microcomputers and robotics have enabled the development of automation technology which is being applied to batch manufacturing. This technology has been

termed "programmable industrial automation" and is characterized
by flexibility and adaptability. These characteristics require
a large amount of information flow which can well be handled by
computer. Application of computer vision as a versatile,
sensory feedback technique promises to enhance the flexibility
and adaptability of programmable automation systems.

The purpose of the experimental work described in this
paper is to utilize programmable automation technology including
microcomputers, robotics, and computer vision to improve the
productivity of the aircraft drilling and riveting task. The
intent has been to investigate and demonstrate generalized
concepts which can then be applied to specific production
requirements. The program has emphasized microcomputer hierarchical control and computer vision using simple X-Y manipulation. These techniques are being evaluated for production
application to small, flat aircraft structural assemblies.
Experimental work has progressed to multi-axis manipulation
using a robot for workpiece handling and positioning. Future
work will involve robot manipulation of tools for drilling
and riveting.

NEED FOR PROGRAMMABLE AUTOMATION

The key characteristics of programmable automation, as
stated previously, are flexibility and adaptability. Flexibility is the ability to manufacture a wide variety of configurations with minimum effort and cost. This implies ease of
training or programming, ease of set-up, and minimum part-
specific tooling. This scenario is the opposite to the normal
aircraft manufacturing situation where tedious programming,
difficult set-ups, and extensive special purpose tooling
prevail. Lot sizes in aircraft fabrication are generally quite
small, ranging from 10 to 50 parts. Lot sizes of 1 to 5 parts
for replacements or spares are not uncommon. The need for
flexible manufacturing processes, equipment, and systems which

can produce parts and assemblies as economically in very small
batches as in larger batches is obvious.

Adaptability is the ability to adjust or conform to
changing environmental conditions. These can be workpiece
dependent such as changes in location or part-to-part configuration variations. Changes can be due to equipment and tooling
wear, malfunction, or improper set-up. Improved capability to
adjust to changes would improve quality, decrease dependence on
human surveilance of the process or equipment, and contribute
to increased productivity.

USE OF COMPUTER VISION

Computer vision has been emphasized in this program to
achieve flexibility and adaptability in aircraft drilling and
riveting operations. A major objective has been to reduce
tooling requirements by using vision in conjunction with
computer-stored configuration data. The following uses of
computer vision have been explored:

1. Training. The use of vision in training can eliminate
   the need for exact positioning of the manipulator at
   the location to be trained. By using a mark or a hole,
   training can be accomplished by gross positioning to
   bring the point within the camera field of view
   followed by vision processing to determine the exact
   location.

2. Calibration. Equipment and workpiece calibration can
   be performed by moving the manipulator to a defined
   position followed by vision processing to determine
   the offset from the defined position.

3. Recognition. Parts can be recognized and differen-
   tiated from other parts by vision.

4. Position and Orientation. Randomly orientated parts
   can be acquired by manipulators using vision to
   determine position and orientation.

5. Fine Positioning. Part or assembly details can be used for fine positioning. In this program, visual edge-tracking was used for final adjustment of the drilling location.

6. Verification. Correct equipment operation can be verified by using vision. For example, if a picture is taken after hole drilling and a hole is not found, it can be assumed that a malfunction occurred.

7. Inspection. In-process inspection is an important application for vision. In this program vision was used for drilled hole measurement, hole-to-edge distance measurement, and rivet upset measurement.

COMPUTER SYSTEM DESCRIPTION

The sensory-feedback computer system developed for this program is shown in Figure 1. The system design criteria were: implementation in a dedicated microcomputer; capability of multichannel analog-to-digital and digital-to-analog conversion; provision for operator interaction with the computer; mass program storage capability; and a high level user programming language. The system consists of four major components: a Digital Equipment LSI-11 microcomputer with 64K-byte RAM memory and an expander chassis containing analog-to-digital and digital-to-analog converters; a Digital Equipment RX-01 dual floppy disc drive; an SRI International Vision Module; and a Soroc IQ-120 terminal.

The vision module was developed as an industrial prototype by SRI International. It consists of three major components: a LSI-11 microcomputer with 64K-byte memory; up to four General Electric TN-2200 solid-state TV cameras; and an interface pre-processor between the TV cameras and the microcomputer. The monitor located to the right of the preprocessor in Figure 1 is a Hewlett-Packard 1340A and is used to display either an analog or binary-thresholded image. A Tektronix 4051 graphics terminal

is used as a control console for independent operation of the
vision module and also to display computer processed information.

The vision module has the ability to recognize patterns
independent of position and orientation; to determine position,
orientation, and size of objects; and to be trained by showing.

Figure 1. Sensory-Feedback Computer System

When a picture is taken, the preprocessor thresholds the camera video signal into binary (either black or white) data. A run-length encoder converts the binary data into run-length code to reduce the image processing time performed by the microcomputer. The image is processed by the LSI-11 microcomputer using a method known as connectivity analysis, a procedure that breaks a binary image into its connected components. The connectivity analysis program builds a description of each blob (a connected component--either an object or a hole) as the image is processed. An array, which is called a "blob descriptor" is created to hold information about the blob and its shape. Finally, a number of shape and size feature-values characterizing the blob are derived

The feature-values derived from an aircraft sheet metal part, superimposed on the computer-processed outline of the part, are shown in Figure 2. Before the picture was taken, the vision module was calibrated so that the feature-values appear in

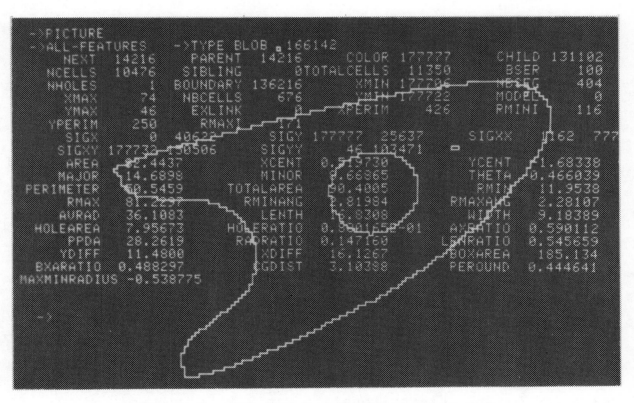

Figure 2. Vision Module Feature-Values

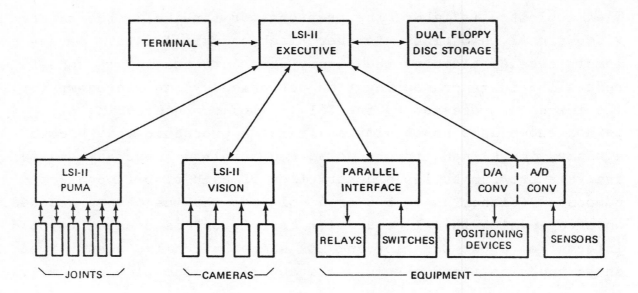

Figure 3. Computer System Organization

inches. Some of the feature-values are useful for recognition and inspection because they are independent of position and orientation. Others can be used to determine position and orientation.

As shown in Figure 3, the sensory-feedback computer system is organized in a hierarchical manner with system control residing in the LSI-11 executive computer. One advantage of this organization is that the system is capable of expansion to include additional computers or other devices controlled by the executive computer. A PUMA robot has recently been incorporated into the system in this fashion as indicated in Figure 3. Another advantage is that subordinate microcomputer hardware and software can be dedicated to specific tasks. Thus, the executive computer is free to control, coordinate, and interface the system and is not concerned with the details of local processing. The executive computer performs the data handling and outside communication functions for the system.

The software system being used at the present time for high-level control is RPS (Robot Programming System) which was developed by SRI International to facilitate writing and

debugging of application programs for factory tasks. RPS
includes a user language called RPL. Application program statements consist of calls to executable library subroutines or to
interpretable user-defined subroutines. The current library
includes subroutines for the following functions: defining and
computing position/orientation transformation matrices; visual
sensing and processing to recognize, locate, and inspect objects;
opening and closing equipment relays; terminal input/output;
arithmetic, trigonometric, and vector functions; and branching.

In order to operate the sensory-feedback computer system
under RPL control, several low level software routines were
written. Driver routines link RPL with the analog-to-digital
converter, the digital-to-analog converter, and the parallel
interface bit register. A communication routine links RPL with
the vision module through a parallel interface. The link
between RPL and the PUMA robot is through a serial interface
at the present time.

EXPERIMENTAL DRILLING AND RIVETING EQUIPMENT

The equipment shown in Figure 4 was developed to provide a
test bed for vision experimentation applied to drilling and
riveting. The equipment was designed for drilling and riveting
flat structural panels in order to simplify the manipulation
task to two-degrees of freedom. A framework supports an X-Y
positioning table and top and bottom tool clusters. The
positioner moves the workpiece in X and Y coordinates under
computer control between the tooling clusters. The tooling
clusters use pneumatically operated drilling and riveting tools;
all functions are under computer control and position sensors
monitor operation of all moveable compnents. A General Electric
TN-2200 solid-state camera is mounted on the bottom tool cluster
for viewing a panel with stringers on the side opposite the
drill. The electronic system for control of the positioner and
tool clusters is located in an enclosure attached to the rear
of the framework. The interface between the electronic system

Figure 4. Experimental Drilling and Riveting Equipment

and the sensory-feedback computer system is through flat cables
to the LSI-11 parallel interface, the analog-to-digital converter, and the digital-to-analog converter. The camera interfaces
through a cable to the vision module preprocessor. All equipment functions can be operated independent of the computer for
check-out or maintenance using a portable control box. The box
incorporates a joystick to position the X-Y table for training
and calibration.

USING VISION FOR TRAINING AND CALIBRATION

The use of vision in conjunction with computer-stored
configuration data has been emphasized in this program for
determination of drilling locations. The goal of this approach
is to eliminate the need for templates, fixtures, pilot holes,
hand layout, etc. for this purpose. Ideally, a drilling data

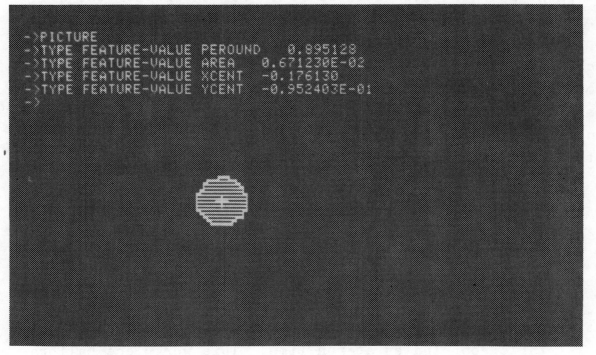

Figure 5. Using Vision to Train Pilot-Hole Location

base would be established by off-line programming based on
design data. In some cases training may be required to establish, correct, or add drilling locations. Figure 5 shows vision
processing for training a pilot-hole location. Prompted by a
training program written in RPL, the operator moves the X-Y
table, using the joystick, so that the pilot-hole can be seen
in the monitor. A picture is taken by the vision module on
command of the RPL program. The PEROUND (a measure of roundness)
and the AREA feature-values of the blob are checked by the RPL
program to verify that the blob qualifies as a pilot-hole. The
feature-values XCENT and YCENT, which are the X and Y coordinates of the centroid of the blob in relationship to the center
of the field of view, are then found. The RPL program calculates the X and Y coordinates of the hole in relationship to a
previously calibrated reference frame and enters this into the
data base which is subsequently stored on floppy disc.

The use of computer vision with computer coordinate trans-
formations has been found to be an effective method for both

equipment and workpiece calibration. For the former the X-Y
table is moved under RPL control to a tooling-hole location.
A picture is taken and the XCENT and YCENT values obtained as
shown in Figure 6. The X-Y table is then moved to a second
tooling-hole location where the same procedure is followed.
The RPL program calculates the table scale factors, offsets,
and camera rotation. A hole is drilled in a scrap piece of
metal and drill-to-camera offsets determined by vision.

For workpiece calibration, an RPL program prompts the
operator to load the workpiece and move the X-Y table so that a
specified tooling hole can be seen in the monitor. A picture
is taken and the XCENT and YCENT values obtained. The X-Y
table is then moved to a second tooling-hole location where the
same procedure is followed. The RPL program establishes a
matrix which will transform the data base workpiece coordinates
to X-Y table coordinates at run time. This workpiece calibration procedure obviates the need for precise workpiece location
using special tooling and measurement means.

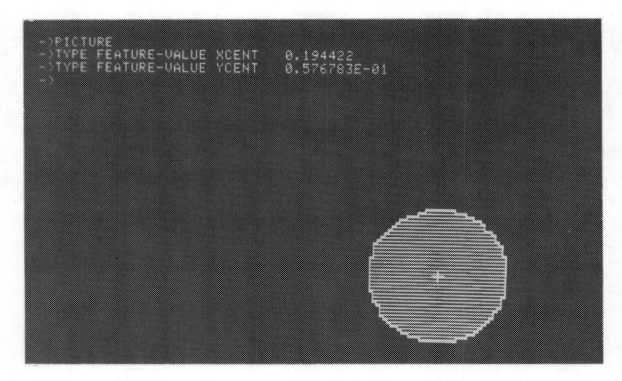

Figure 6. Using Vision for Equipment or Workpiece Calibration

USING VISION FOR SERVO CONTROL AND INSPECTION

The aluminum alloy test panel shown in Figure 7 was fastened using the experimental drilling and riveting equipment, operated by an RPL program, and a special vision application program. The RPL program runs in the executive LSI-11 microcomputer and has four major subroutines as follows: MANCON; DRILL (X,Y); RIVET (X,Y); and FASTEN. MANCON operates all machine functions under computer control by manual input at the terminal keyboard. DRILL (X,Y) drills a hole at a typed-in X-Y location. RIVET (X,Y) installs a rivet at a typed-in X-Y location. FASTEN drills holes and installs rivets under visual control after typing-in the starting X-Y location, the number of rivets desired, the rivet spacing, and the edge distance. Future improvements will provide for accessing program or disc stored data in lieu of typing on the terminal.

Figure 7. Drilled and Riveted Test Panel

A special vision application program which runs in the vision module LSI-11 microcomputer was written for edge tracking and inspection for drilling holes and installing rivets. The significance of edge tracking is that most riveted aircraft structure designs are based on a minimum hole-to-edge distance with a common tolerance of plus or minus 0.030 in (0.76 mm). If the tolerance was greater, wider flanges would be required resulting in a heavier structure.

The vision processing scheme recognizes and determines the size and relative location of round blobs (holes and rivet upsets) and long blobs (edge shadows). Figure 8 shows the edge recognition and location vision processing. The blob is actually a picture of the shadow formed by the light from a small incandescent lamp which is located off-center from the camera axis. The command FEDGE (Find Edge) takes a picture and selects the object in the scene (the edge shadow has been out-lined) which is long and narrow by using the feature-value

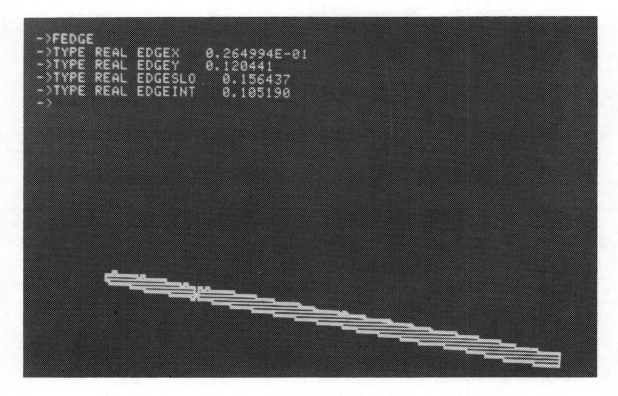

Figure 8. Edge Recognition and Location

LENRATIO (refer to Figure 2). LENRATIO is width divided by length with a value of 0.20 maximum being used. The point on the perimeter of the selected blob closest to the center of the field of view is determined. Additional commands have caused the coordinates of this point to be displayed. A segment of the edge centered on the point is defined and a least-mean-square fit to the edge segment determined. This straight line appears in Figure 8 but is very difficult to see. Additional commands have caused the slope and intercept of the best-fit line to be displayed.

The hole recognition, size, and location vision processing is shown in Figure 9. The command FRND (Find Round) takes a picture and selects the blobs in the scene (the hole has been outlined) which meet a selected "roundness" criteria using the feature-value PEROUND (refer to Figure 2). PEROUND is equal to the quantity four times pi times area divided by the perimeter squared. For a perfectly round object, PEROUND would equal one. The value of PEROUND being used is 0.75 minimum. The round object closest to the center of the field of view is selected and its center identified with a cross. Additional commands have caused the radius of the hole and its X and Y locations to be displayed.

Figure 10 shows the hole-to-edge distance vision processing. The command FEDGE operates as explained previously except the point on the edge blob perimeter closest to the center of the hole is used. The perpendicular distance between the center of the hole and the best-fit line is determined and displayed.

The FRND command is used for rivet upset recognition and size determination. The acceptance criteria defined by the minimum value of PEROUND can be changed if desired to meet specific inspection requirements. Also, acceptance criteria based on the relative center locations of the hole and rivet upset can be used. The radius of the rivet upset must be within specified limits to be acceptable.

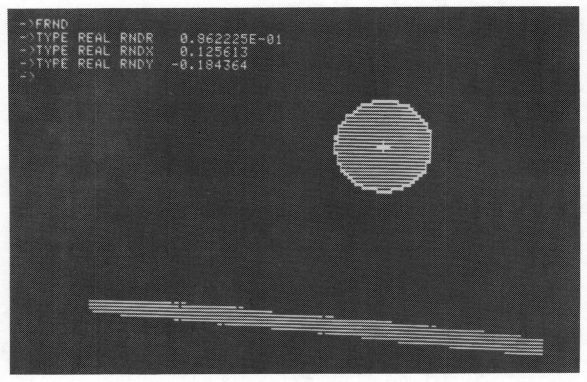

Figure 9. Hole Size and Location

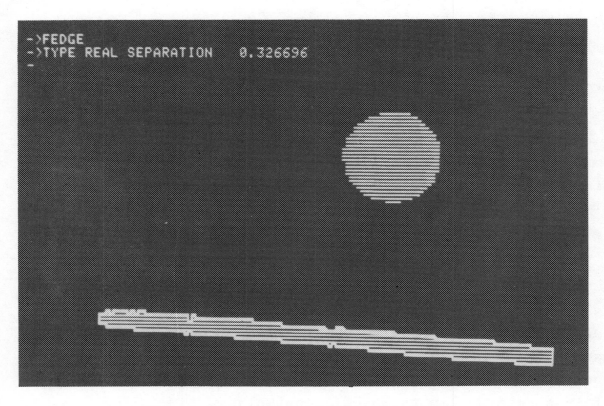

Figure 10. Hole-to-Edge Distance

An inspection record which was printed out for a group of ten rivets installed in the test panel is shown in Figure 11. The printout repeats the design data which was typed into the terminal (X-Y starting position, number of rivets, rivet spacing and edge distance). The inspection data (hole diameter, edge distance, and quality assurance code) is printed out for each hole as the riveting operation proceeds. In this example, the rivet upset diameter was not printed out. As can be seen from the inspection data, the system is capable of visually tracking a straight sheet metal edge and installing rivets within specified edge distance tolerances. Tests of the vision system itself have shown tolerance capability of plus of minus 0.004 in (0.10 mm) for drilled holes and plus or minus 0.008 in (0.20 mm) for edge distance under the lighting conditions and the 1.00 in (25.4 mm) square field of view used in this program.

DESIGN DATA ENTRY
-------- ----- -----

STARTING POSITION: X =   14.0000   Y =   2.50000

NUMBER OF HOLES: 10

HOLE SPACING: 1.000000

EDGE DISTANCE: 0.340000

INSPECTION   DATA
------------- -----

HOLE NO.	DIAMETER	EDGE DISTANCE	QA CODE
1	0.163268	0.353478	A
2	0.164720	0.327326	A
3	0.167883	0.318506	A
4	0.163829	0.320754	A
5	0.165590	0.318555	A
6	0.168493	0.343454	A
7	0.168827	0.331169	A
8	0.169555	0.348026	A
9	0.169822	0.333687	A
10	0.167901	0.336365	A

Figure 11.   Test Panel Inspection Data Printout

USING VISION FOR ROBOT PART ACQUISITION

The test set-up for conducting vision experiments with a 5-axis Unimate PUMA robot is shown in Figure 12. A General Electric TN-2200 camera is overhead with a field of view 2 ft (0.6 m) square. An aircraft sheet metal assembly is below the camera. Work is in progress to interface the PUMA to the LSI-11 executive computer as shown in Figure 3. When this is completed, an RPL program will make calls to the vision system

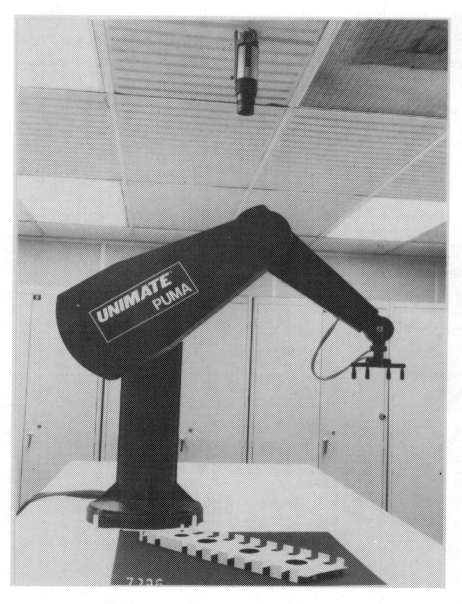

Figure 12. PUMA Robot With Camera Overhead

to determine the position and orientation of an object; the information will be passed back to the executive computer where calculations will be made based on previously trained part pick-up locations; and corrected pick-up location parameters will be passed to the PUMA robot operating program. The robot will then pick up the part and manipulate it for drilling and riveting in lieu of the X-Y table. Visual tracking and inspection will be conducted with the robot in the control loop instead of the X-Y table.

The method to determine part orientation is shown in Figure 13. In this case it is not necessary to view the entire part. However, to insure that enough of the part is in the field of view to accurately determine the orientation, the feature-value NHOLES is called. Three holes must appear for the operation to continue. THETA, the angle of the axis of least moment of inertia, is then obtained. This value is used

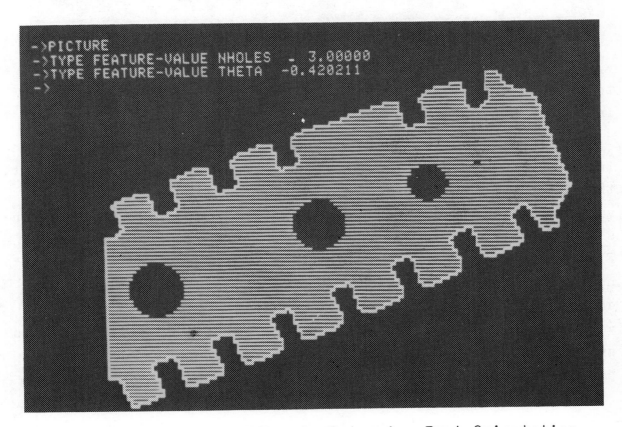

Figure 13. Using Vision to Determine Part Orientation

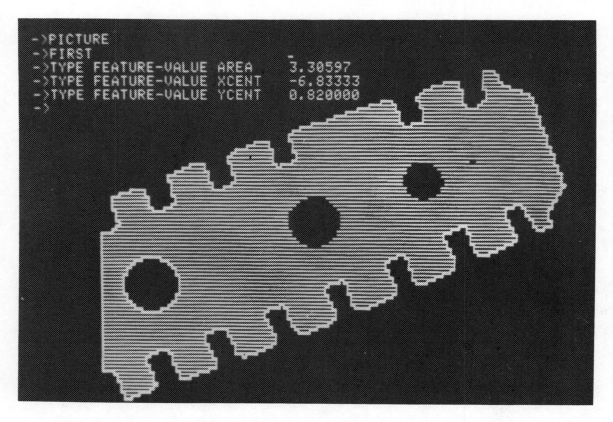

```
->PICTURE
->FIRST
->TYPE FEATURE-VALUE AREA      3.30597
->TYPE FEATURE-VALUE XCENT    -6.83333
->TYPE FEATURE-VALUE YCENT     0.820000
->
```

Figure 14. Using Vision to Determine Part Location

to calculate the PUMA gripper rotation angle which is entered
into the PUMA program using the terminal.

Figure 14 shows the method used to determine part location
in robot X and Y world coordinates. The Z location is constant,
the part height being fixed. Since the entire part is not in
view, a hole in the part which has a known location in relationship to the periphery of the part is used. The vision command
FIRST causes the hole on the left to be circled and causes that
hole to be the blob of interest. The AREA of the hole is
checked and the values of XCENT and YCENT obtained. These
values are used to calculate the PUMA gripper location which is
entered into the PUMA program using the terminal. The robot
can then pick up the part as shown in Figure 15.

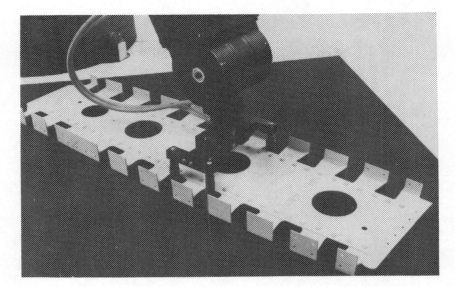

Figure 15.   Vision Data Used to Position Robot

CONCLUSIONS

Results of experimental work to date have shown that computer-controlled, visually-guided drilling and riveting is feasible for fastening aircraft structures. The feasibility of using computer-vision for robot acquisition of aircraft parts subsequent to fastening or other manufacturing processes has been demonstrated. Robotic drilling and riveting using computer vision has the potential to reduce training and programming cost, reduce tooling cost, reduce set-up cost, provide in-process inspection, and improve the overall productivity of aircraft assembly operations.

ACKNOWLEDGEMENT

The author would like to thank the following members and former members of the Programmable Automation Laboratory of the Lockheed-California Company and the Industrial Automation Group of SRI International for their contributions, assistance, and advice in carrying out the objectives of this program: Carl Linden, Jim McRoberts, Gary Shubert, and Mark Rosenblatt of Lockheed; Charles Rosen, David Nitzan, Gerald Gleason, John Hill, Bill Park, Dennis McGhie, Tony Sword, and Alfred Brain of SRI.

CHAPTER 11

FINISHING

Commentary

Finishing operations with robots involve the spray application of paint, stain, plastic powder, plastic sealer, sound deadener or similar materials. Application methods may utilize air-atomizing, electrostatic or airless spray equipment.

The robots involved in finishing operations are servo-controlled, continuous-path units, with six or seven degrees of freedom. They are hydraulically driven and intrinsically safe for use in a volatile atmosphere.

Spray finishing with robots provides a number of advantages to the user. These include: labor replacement, consistent quality, material savings, energy savings, reduced booth maintenance and removal of workers from exposure to harmful or toxic substances.

Presented at the Finishing '77 Conference and Exposition, October 1977

Robots—The Answer To Painting In The 1980's

By Jack Shaneyfelt
Steelcase, Inc.

This paper is a chronicle of the experience of a leading office furniture manufacturer, Steelcase Inc. of Grand Rapids, Michigan, in their successful application of robotics in finishing of metal office furniture.

It all started in 1975 with our search for cost reduction ideas during a business downturn. Frankly, we were searching for anything that had the potential to produce cost savings.

Each cost reduction idea was evaluated and assigned a priority. Robotics fell under what we called sophisticated machine ideas. It was considered a pretty far out idea but we felt it was interesting enough to warrant investigation.

First we wanted to see robots in action and determine which one was best suited for our purposes. During that period of time we were working on another project with a leading spray equipment supplier and learned of a forthcoming robot show being held in Chicago where they would be exhibiting along with other robot manufacturers.

I attended the show and came away rather disappointed since all the exhibitors were either doing material handling or welding operations. None of them were painting. After asking numerous questions I was told that there were only a few painting robots in the U.S. at that time, although they were being used extensively in Europe with a great deal of success.

After observing all robots at the show it was concluded that the DeVilbiss-Trallfa painting robot was the smoothest operating and seemed to fit our requirements. They suggested that I visit their lab and observe some of the research they had done in robotic painting. After the trip I came away quite apprehensive about a robot's ability to ever paint our products. Later a test was scheduled to evaluate the robot in a simulated production run. This test was a failure and it proved nothing.

Finally, in order to perform a reasonable evaluation, Steelcase and DeVilbiss entered into an agreement to ship a robot to the Grand Rapids plant so it could be evaluated under production conditions. Time limits were to be of no importance. We only sought an answer to the question "could it be made to do the job?"

I had only one restriction: current production was not to be tampered with in any way. The robot had to perform as well as, or better than, the existing method. My excitement over a new exotic machine could not be put ahead of our current operating schedule, and this became quite a problem during the next few months.

At that time we were running our conveyor at 22 FPM with hangers on 2' centers. Two men were spraying approximately 30 different styles and parts on this line in a normal production day. The robot had to have the ability to do them all and replace both painters without sacrificing quality.

When the machine arrived, worker acceptance was not of the optimistic nature. Some of the questions from plant personnel were: "Is this where all the profits go?". "Is it a relative of yours?" and many more. Oddly enough, the question I was expecting never came up and that was: "What is going to happen to my job if it works?"

As in the case of most new and revolutionary equipment, problems seemed to develop everywhere at once. Teach time on the tape control memory was too slow and could not keep up with the close part centers at 22 FPM. Programming was almost a joke. Everyone had a try. We just did not have the expertise to do the job.

The 30 different styles of parts could not all be programmed because the memory bank only held one program at a time. It was necessary to change the tape cassettes for different runs. Another problem was fluid flow; we could not get enough paint on the part in the time we had available in front of the robot.

About this time DeVilbiss-Trallfa introduced a new computer with a floppy disc memory. This turned out to be the salvation of the project, at least as far as I was concerned. All of a sudden everyone was optimistic, and the project took on an almost "can't miss" look overnight.

With the new computer in place, work started again in earnest. We now had the capacity to store 64 programs with almost instant search and replay capabilities. The computer could also copy and edit programs.

The next requirement was to increase the delivery of paint. The answer to this was an airless electrostatic gun, but once again we met with adversity; the right nozzle and restrictors had to be manufactured from scratch. This took almost three weeks, but again the supplier rose to the occasion to put the project back on stream.

Finally the time arrived for our company to make a decision: was the robot a worthwhile investment or not? Could it maintain production? Would it meet our quality standards? Could we keep it running and, of course, could we show a savings and cut costs? All these questions had to be answered and cost justification had to be shown. The studies proved positive and we had a "go" project.

The line where the robot was installed originally had a five-man crew; this was cut by two people. The better control of overspray due to the robot resulted in a forty percent reduction of paint usage. After running for a few months we also noticed a drastic reduction in the amount of booth maintenance required. The reject rates decreased from seven to two percent. The final plus factor was the elimination of absenteeism problems.

Presently the robot is doing a very good job. It will get even better as we gain experience in its use. At the present time we are also experimenting with high solids paint in conjunction with the robot. This looks to be very promising.

In summary, as environmental and health and safety people become more involved in painting, I'm sure that robotics will look even better and better. We are just getting started; their future is unlimited. As experience and demands go, so will robots. They will almost assuredly become the hand sprayer of the 1980's. In my opinion, painting robots are a "can't miss" proposition and definitely are here to stay.

Presented at the Westec '79 Conference, March 1979

The Increasing Demand for
Robots in Manufacturing Applications

By Clell D. Routson
Nordson Company

This paper is concerned with the factors moving companies to examine Robotics as an answer to many economic, governmental, and social problems within the corporate boundaries. This alternative is based upon the flexibility of the system to perform a wide variety of current manual operations, especially in hazardous locations such as painting.

INTRODUCTION

Economic factors, government legislation and job enrichment programs are causing all branches of industry to consider Robots for an increasing number of operations. Robots are being evaluated, not only to do simple repetitive tasks, but also to perform more complex jobs in an ever-widening variety of sophisticated applications. Application of sprayed materials is the specific area of my involvement with Robots. Thus, this paper deals with factors involved in product finishing and Robots.

Since many of you may not be familiar with the various types of application methods, a couple of definitions before we start might be helpful.

1) Air Spray — the conventional method of applying paint which was first developed before the 20th Century. This process uses air to atomize the paint particles.

2) Airless Spray — this process, developed in the early 1950's, utilizes hydraulic pressure and fixed orifice to create atomized spray.

3) Heated System — by adding heat to a closed material flow of paint, the viscosity changes, and lower spraying pressures can be obtained. Thus, less solvent used and less overspray of the paint. Heat is used with both air and airless systems.

4) Electrostatics — used with both air and airless, an electrostatic charge is induced into the paint flow. The part being sprayed must be grounded so the net effect is like a magnet.

Economic Factors Leading to Robots

Much has been said and written concerning justification procedures on capital purchases. We have developed models to project cost figures over a period of time (usually ten years). In this manner, our customers can look at investment, comparing long-term cost to overall efficiency, and make a decision based upon current corporate position and capital posture. We utilize two types of cost analysis to determine a recommended course of action to a customer. The first is cost avoidance. For example: Figure #1 is a comparative chart of hand air spray system, semi-automatic hot airless system (requires touch-up) and Robot hot airless electrostatic system. The simple hand air spray system requires only a small capital investment, but over the life of the equipment, due to labor and system inefficiency, becomes very expensive. The semi-automatic system, even though a considerable capital expense is required, the overall cost of the system is substantially less. Finally, the fully automatic Robot system, which contains the largest comparative capital outlay, costs the least over the ten year period of time. The same philosophy, in more detailed form, will set up a cost avoidance comparison for any systems analysis.

Figure #2 compares a typical, semi-automatic hot airless electrostatic system to a Robot hot airless electrostatic system. By examining the data, we can see that even though the initial capital purchase is approximately four times greater, the total cost is less than one half. When the savings differential ($714,000) is spread over the ten years, our payback is really only two years.

The other type of cost analysis we look at is cost savings. Under normal conditions, our customers have some historical record of their finishing costs. These figures are used as a base from which we can compare the savings of more automated equipment. As you can see from the example in Figure #3 showing historical cost figures, we have gone back four years to get a true picture of their growth in expenses. XYZ company has been using two painters on each of two shifts for several years and has maintained good cost records. They are using manual air spray guns with pretty good efficiency. This type of operation is typical in industry today. Next, we have compared his current costs to the costs we expect with the proposed equipment. This data would normally be based upon tests our representative would have performed in the customer's plant and would be conservatively stated. The savings generated by the new equipment would provide a savings which would be divided into the new equipment cost to determine a payback period.

In addition to economic analysis, companies need to examine proposed production requirements. Since many thousands of dollars are spent to increase production by building new plants or adding new assembly lines, it makes sense to increase productivity in the existing plants before new ones are built. One heavy equipment manufacturer has committed to a new spray booth and spray operation. His costs for the booth alone are about two million dollars. By automating with Robots, and improving productivity, he believes he can delay adding additional booths by several years. And, at the same time, he will be experiencing the kinds of savings we just discussed.

Government Regulations

Most paint operations are very much concerned with EPA, OSHA, NFPA, other federal and state governmental bodies and insurance agencies. The overriding concerns of these groups are solvent emissions, effects on human health and safety and fire hazards caused by the material. Technology might never solve the problem of providing a paint material that people can work in and breathe safely, but industry can eliminate a great deal of their problems by taking the man out of the spray operations. In many plants the only way this can be done is by Robots. Conventional automatic systems work very well when there are repetitive parts or when parts are hung on a "Christmas Tree" rack, but not when there are wide differences in size, complexity and shapes. These types of jobs can only be done with a man or a Robot. Because many of these jobs require a man to climb into a pit or become exposed to severe solvent fumes, government agencies are putting ever-increasing pressure on companies to get the man out. In many cases, the ultimate automation is the answer: a Robot.

European labor has for years looked upon some jobs as demeaning and they have, therefore, found ways to automate. We are gradually coming to that position in North America. Painting is one of the jobs people basically don't enjoy. It's dirty, dangerous and damaging to one's health. So, why should the American worker do it? Well, a good many of them will not. The turnover rate in many plants is faster in the paint area than in any other single area of the plant. In many cases, the painted appearance of the product has a great deal to do with the quality associated with the product, so companies need a stable work force in this area. So, because of the nature of the work, they're not getting the consistent, high quality finish they need. Several companies have solved this problem by upgrading the painter to a technical operator and given him the responsibility to not only get the product painted, but maintain the required automatic equipment. In many cases, the transition must be made by other people. But, because of the schooling supplied by most of the vendors, the training is relatively inexpensive to develop personnel from within the plant. Now, when a company automates, they upgrade a man's worth and self-esteem, solve a turnover problem within their company, and have a finish on their product of which everyone is proud.

Robot Technology: Increasing Opportunity

The first Robots were developed to move from point A to point B and back; and, those are still in very high demand today. But we've advanced a great deal to this date. Robots are now capable of sensing a variety of parts and even making decisions, depending, of course, upon the complexity of the ancillary equipment that is linked to the Robot. For instance, we currently have Robots in operation that will recognize a variety of parts; move on a track forward and backward to perform different tasks; turn conveyors off and on -- to call parts up; change colors; and ring bells and whistles when something isn't right.

In the very near future, I suspect there will be Robots that will be able to sense a part out of position and make adjustments within their own program. As it stands right now, part positioning is very important. If Robots were able to adjust our program, automation would be considerably easier. The industry is moving in that direction with cameras and other types of sensing devices, but reality is still some time off.

What all of this really is leading to is total automation where entire plants are manned by a few technicians who maintain the Robots and their machines. The Japanese expect to have a plant of this type in operation in the mid 1980's.

Economic pressures, the four-day work week, governmental regulations all are pushing corporations to look further at Robots. At the same time, Robot manufacturers are developing new technologies to advance the state-of-the-art and Robot opportunities around the world.

FIGURE #1

TEN YEAR COST OF PAINT FINISHING SYSTEMS

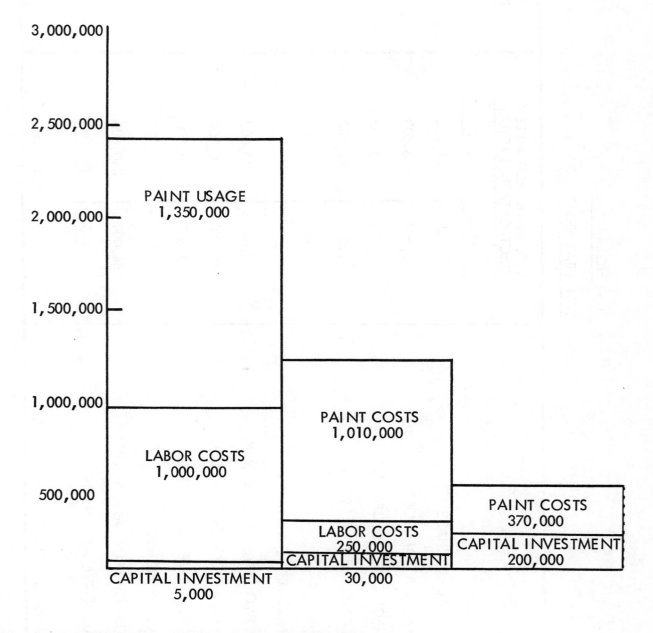

MANUAL AIR SPRAY

AUTOMATIC HOT
AIRLESS SPRAY

ROBOT HOT AIRLESS
ELECTROSTATIC SPRAY

PAINT USAGE
1,350,000

LABOR COSTS
1,000,000

CAPITAL INVESTMENT
5,000

PAINT COSTS
1,010,000

LABOR COSTS
250,000

CAPITAL INVESTMENT
30,000

PAINT COSTS
370,000

CAPITAL INVESTMENT
200,000

(FIGURES ARE IN APPROXIMATE US DOLLARS)

FIGURE # 2

TEN YEAR COST OF

	AUTOMATIC HOT AIRLESS ELECTROSTATIC SYSTEM		ROBOT HOT AIRLESS ELECTROSTATIC SYSTEM	
	1st YEAR	TOTAL	1st YEAR	TOTAL
INITIAL PURCHASE	45,000	45,000	200,000	200,000
INSTALLATION COST	3,500	3,500	6,500	6,500
LABOR COST TO OPERATE	25,000	250,000		
MAINTENANCE COST	1,000	12,500	2,000	13,500
SPARE PARTS	5,000	17,000	6,500	18,500
ANCILLARY EQUIPMENT				
TWO SPRAY BOOTHS (WW)	15,000	15,000	15,000	15,000
AUTO. COLOR CHANGER	10,000	10,000	10,000	10,000
PROGRAM CONTROLLER	7,500	7,500	7,500	7,500
PAINT EFFICIENCY	75%		85%	
COST OF PAINT AND SOLVENT	84,000	840,000	36,960	369,600
TAX DEPRECIATION ON CAPITAL EQUIPMENT	(7,750)	(77,500)	(23,250)	(232,500)
TOTAL (FIGURES IN US DOLLARS)		1,123,000		408,100

392

FIGURE # 3

PROPOSED NORDSON HOT AIRLESS ELECTROSTATIC SYSTEM

	1979	1980
INITIAL PURCHASE	200,000	---
INSTALLATION COST	6,500	---
LABOR COST TO OPERATE	25,000	25,000
MAINTENANCE COST	2,000	2,000
SPARE PARTS	6,500	1,000
PAINT EFFICIENCY 85%		
COST OF PAINT AND SOLVENT	60,400	60,400
TAX DEPRECIATION ON CAPITAL EQUIPMENT	(20,000)	(20,000)
TOTAL ROBOT SYSTEMS COSTS	280,400	68,400

SYSTEM PAY BACK

1979 AND 1980 AIR SPRAY COSTS	545,000
1979 AND 1980 ROBOT SYSTEM COSTS	348,800

SAVINGS

$ 196,200 OR 8,175 PER MONTH

PAY BACK ON SYSTEM INVESTMENT IS 24.5 MONTHS

(FIGURES IN US DOLLARS)

FIGURE #3

SUMMARY OF XYZ COMPANY'S FINISHING COSTS

	HISTORY				PROJECTED COSTS	
	1975	1976	1977	1978	1979	1980
INITIAL PURCHASE	5,000	---	---	---	---	
INSTALLATION COST	500	---	---	---	---	
LABOR COST TO OPERATE	66,000	70,000	78,500	85,000	100,000	112,000
MAINTENANCE COSTS	600	800	600	1,000	1,000	1,000
SPARE PARTS	500	100	300	400	500	500
ANCILLARY EQUIPMENT TWO WATER WASH BOOTHS	10,000	---	---	---	---	
PAINT EFFICIENCY 30%						
COST OF PAINT SOLVENT	86,400 (80 gal. per day at 4.50)	113,400 (90 gal. per day at 5.25)	109,140 (85 gal. per day at 5.35)	132,000 (100 gal. per day at 5.50)	171,000 (125 gal. per day at 5.70)	171,000 (Same as 1978)
			ACTUAL 1978 COSTS:	218,400		
			PROJECTED COSTS:		272,500	272,500

(FIGURES IN US DOLLARS)

Presented at the Robot III Conference, November 1978

Robot Painting of Microwave Oven Cavities and Other Job Shop Sheet Metal

By Ronald G. Jacob
Watertown Metal Products, Division of Western Industries, Inc.

Painting the inside of sheet metal boxes is not an easy task, especially if that box becomes a microwave oven on the housewife's kitchen counter. Watertown Metal Products sprays a liquid baking enamel on the inside of microwave oven cavities and other sheet metal products. The learning curve is becoming a reality.

HISTORY

Watertown Metal Products (WMP) is a sheet metal job shop located in Watertown, Wisconsin since 1966. We are a division of Western Industries of Milwaukee and are one of five sheet metal job shops that comprises Western Industries. Fabrication of light gage sheet metal parts of cold rolled steel, stainless steel, and aluminum on a contract basis is our specialty.

One contract that was started in 1969 was a consumer model of a microwave oven cavity of stainless steel. We were also fabricating a commercial cavity of stainless steel and still make this model. Due to the cost of stainless steel and the poor acceptance in the consumer market of stainless steel appliances because of their "institutional" appearance, our customer wanted to produce a painted cold rolled steel cavity.

Since finishing a sheet metal box which must have an excellent aesthetic appearance and some corrosion protection, an "end-result" finishing specification was written. Several paint companies were involved and several different coating processes were considered including spray, dip, flowcoat, and electrocoating.

One coat electrocoat was first considered to be the best system to provide the necessary requirements. After samples were coated and evaluated it was apparent that a single coat of electrocoat (Epoxy) would not meet all of the specifications. Aesthetics, uniformity of color and stain resistance were questionable. Therefore, the electrocoat was to be used as a primer with an acrylic top coat sprayed on the visible areas.

An electrocoat system was added to an existing spray finishing line (Fig. #1). We started to paint cavities in 1972.

FIRST ROBOT

With the requirement of a uniform coating of 0.9-1.2 mil. of top coat over the inside of an electrocoated cavity, a robot was a good consideration. At that time the Trallfa robot was distributed in the United States by Unimation. The application equipment was supplied by DeVilbiss and eventually DeVilbiss became the representative for the finishing type robots in the U.S. for Trallfa.

A five function, single cassette control unit was purchased with an automatic spray gun with an extension and a 360° fan pattern. This system produced all motions of a human painter except the wrist rotation. The single cassette control meant that a different cassette was required for each cavity, i.e. each program. This was not a problem because we needed to batch load, by customer, because of color and packaging after paint.

We experienced an extremely rapid growth in our painted microwave oven cavity production and the robot did help in accomplishing those production goals. Paint was originally done with 3 painters at 8-10 feet per minute (FPM). Eventually painting was done with one robot and 2 painters at a line speed of 12-14 FPM. As we introduced more models of cavities, new programs were developed.

The full potential of the robot was not realized when used on that paint line for several reasons. Part presentation was not uniform because of the hangers that were used. With the single cassette control system all programming had to be done at the actual production line speed which was difficult with the first generation of hydraulic seals.

Our major problem from a usage standpoint was maintenance of the unit which required a cleaner atmosphere and preventive maintenance program that our shop was not accustomed to. Since the robot did not stop in the middle of the program if the conveyor stopped, parts were occasionally ruined and/or the spray gun damaged.

Acceptance by the leadmen (programmer's) and the painters was slow. Once they realized the robot was reducing the work load and giving more consistant results the robot (nicknamed "Wally") was accepted.

NEED FOR EXPANSION

The system was intended to produce 150 cavities per hour; a sizeable increase from 50 units per day when the stainless steel model was first started. That was in 1972. By October 1975, ten months after I started with WMP, we were forecasting production requirements of 2,000-3,000 cavities per day for several microwave oven manufacturers.

I was asked how we could get more units out of the existing system. Since we were running a washer that was designed for 10 FPM and an electrocoat system that was designed for 8 FPM both at 12-14 FPM, I suggested a complete new line.

DESCRIPTION OF NEW LINE (FIG. #2)

A line to run at 24 FPM was considered but was changed to 12 FPM by double hanging the cavities. The economics of designing a system higher rather than twice as long is quite obvious. Because we are a job shop, I did not want to restrict our package size to just cavities. Therefore, a package size of 36" wide, 72" high, and 30" with the conveyor on 24" centers was designed. We are actually hanging cavities double with dimensions of 18" wide, 57" high, and 20" with conveyor on 48" centers.

At 12 FPM this system has a theoretical capacity of 5760 cavities per two-shift day. Since our forecasts are for 3,500-4,000 cavities per day in 1978, there is still capacity without altering any equipment. As you can see our growth from 50 cavities per day in the late 60's to 360 cavities per hour is significant.

An eight foot sanding booth (dry baffle with filters) was installed to allow for collection of any dust from minor touch-up sanding between electrocoat and top coat. This booth is an approved paint spray booth if so needed in the future.

After sanding, the parts enter three 16 ft. water-wash spray booths with 12 ft. flash-off vestibules between the booths. With the double hanging method it was necessary to put 2 ft. deep pits in each booth to put the painters at the proper spray height. Again, flexibility for a job shop type line was designed into the booth sizing.

DESIGNED FOR ROBOTS

The use of robots was designed into the booth construction. Vestibules were used to connect the booths so the robot controls could be located in a non-hazardous area. Access to the controls through doors in the vestibules and windows in the booth walls allow an operator to view the parts and robot action from the control area.

We learned our lesson about hangers from our older line and a lot of time was spent in designing a hanger that would accommodate nine different microwave oven cavities for six different customers. With the double hanging concept, hanger uniformity was even more critical. We have invested in portable storage racks to store the hangers when not in use. We also have a complex fixture that is used to repair and check all critical points of the hanger if they are damaged.

DECISION TO PURCHASE THE SECOND ROBOT

"Wally" was relocated on the new paint line shortly after start up. The limitations of the control system were evident even after we accomplished two passes per cavity at a line speed of 12 FPM. Test cavities were sent to DeVilbiss where different spray methods, guns, and line speeds were tested. A dual disc-type mini-computer control with the ability to use line synchronization was also observed and filmed.

The decision to buy a second robot with the mini-computor control, line synchronization, work area limiters and retrofit the existing robot with a mini-computor was made in January 1978. The new robot, also a DeVilbiss Trallfa, has the new "Flexiarm" manipulator (included wrist rotation) which is easier to program and "cleaner" than the 5-function servo manipulator, "Wally". Of course the new robot needed a name and "Sally" was christened in March.

DESCRIPTION OF 2 ROBOT SYSTEM

Included in our package from DeVilbiss was a 40 hour school held at our plant by a DeVilbiss robot engineer. Foremen, maintenance personnel, and our technical support people attended the sessions. I would strongly recommend this type of instruction for any robot installation.

The acceptance and learning curve was much faster with the second robot. The robots are now located in the first and second spray booths with "Sally" putting two passes on the front panel of each cavity and one pass on the inside back panel. "Wally" is making 3 passes inside each cavity for a total of 6 passes and washes his "nose" after each two cavities. This is a significant improvement from the 2 passes in one cavity at the same line speed on the old line.

A desire by the supervisors to make the robots do the job has been the biggest factor in our production improvement with the use of the robots. We can paint manually with 4-5 painters or 2 robots and 2-3 painters. We expect to get to 2 robots and 1-2 painters in the future.

From an equipment standpoint, the line synchronization, easy-program hydraulic seals, and mini-computor controls have proved to be very helpful. Line synchronization also allows for programming at a significantly lower line speed, and play or repeat at our normal 12 FPM. We have not used this provision to its fullest extent yet, but having the robots start and stop, during a program, with the conveyor has been a big advantage. Enineering is being done on a special tip for "Wally" to produce a slightly forward tipped cone instead of the flat 360° fan. This should help to reach the corners of the back panel which is the hardest part of the cavity to paint now.

SECOND PLANT WITH ROBOTS

As stated earlier, Western Industries is comprised of five sheet metal job shops. Northern Metal Specialties is another of our companies and is located in Osceola, Wisconsin. They have an identical paint system to our new line and are in the process of installing their second DeVilbiss Trallfa robot.

They also have the first production all-powder porcelain enamel installations in this country. They are coating range cavities and cook tops.
Robots for this spray application are being considered but the severe
abrasive characteristics of powdered porcelain are a concern.

FUTURE PLANS

Our plans for future robots include a possible third robot on the new
line to eliminate all painters. Trying other application techniques such
as airless electrostatic and/or heating the paint are possibilities.

A robot for the old line is also being considered. That line is now used
for our job shop type work which includes various line speeds, a tremendous product mix, and a variety of materials. The cost of hangers for
consistant product presentation for the robot is the hardest cost to
justify. Products on this line include various cabinets, phone booth
parts, light fixtures, and microfilm reader parts.

Robots for press loading/unloading and welding operations could also be
an advantage in a sheet metal job shop. The flexibility of using programmable robots for a job shop type application has advantages over fixed
automatic equipment.

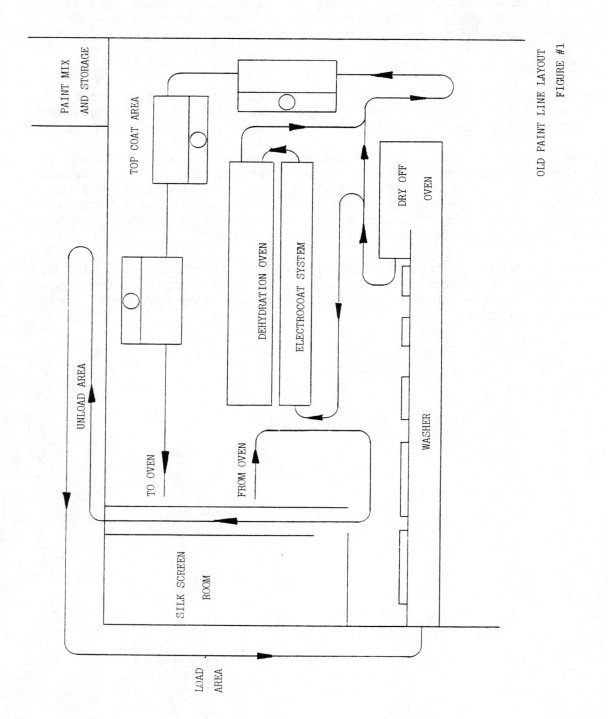

OLD PAINT LINE LAYOUT
FIGURE #1

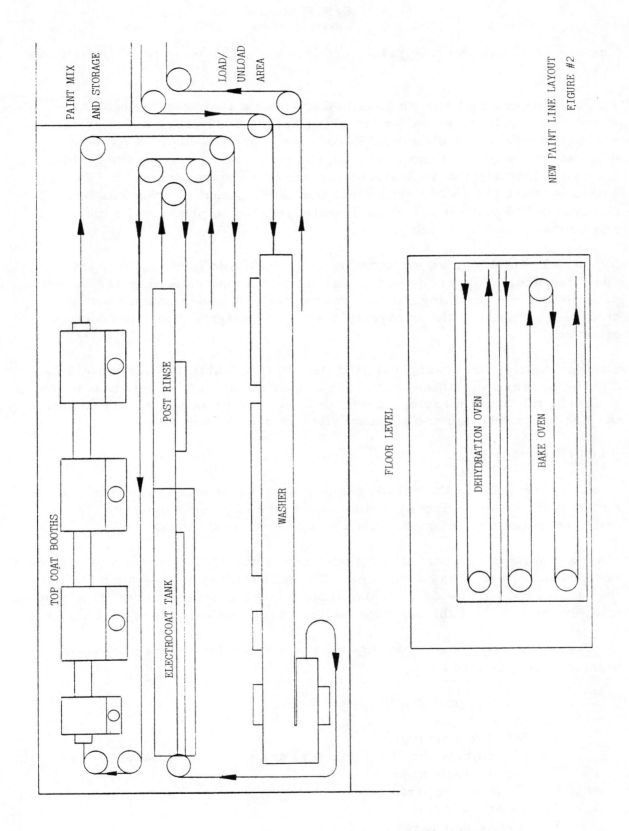

PAINT MIX AND STORAGE

LOAD/ UNLOAD AREA

TOP COAT BOOTHS

POST RINSE

ELECTROCOAT TANK

WASHER

FLOOR LEVEL

DEHYDRATION OVEN

BAKE OVEN

NEW PAINT LINE LAYOUT

FIGURE #2

Presented at the Robot II Conference, October/November 1977

Robotic Painting—The Automotive Potential

By R. R. Mosher
Chrysler Corporation

The use of robots to apply paint certainly qualifies as an idea whose time has come.

To get to where we are, we have had to come a long way. We find the Egyptians developed an expertise using colors 8000 years ago. There is evidence that the inhabitants of Europe were using a form of paint to decorate the walls of their caves 50,000 years ago. During the middle ages painting skills were lost and rediscovered many times. It is only within the last 100 to 150 years that the development of paint has been documented to provide the basic knowledge used to produce the materials and processes we use today.

Industrial Finishing, as we know it, is typically defined as "half art" and "half science." We are proud of our technical understanding and accomplishments, yet the things about process we don't understand or can't control we assign to the sprayman's skill and judgment in a production situation.

Fundamentally, successful painting depends on stablizing our materials and the process variables in the work environment. The potential to use robots in the paint spray booth will depend greatly on our ability to define the process variables and our expertise in controlling them.

BACKGROUND

The starting point is the existing process. The process is people oriented, working in an environment that, measured by contemporary standards, is considered hostile. Let us explain what we mean.

Most automotive companies paint with conventional horizontal and vertical reciprocater type spray machines. These machines are designed to paint the large areas of the car. When using these machines properly, approximately 70 to 80% of the exterior surface area can be covered. (Exhibit I)

Sprayers are used to support this type of automation and perform such painting assignments as:

1. Lower grill and headlight area.
2. Wheel wells.
3. Window openings.
4. Cut in between the horizontal and vertical surface.
5. Lower deck area.
6. Engine compartment.
7. Door facings.
8. Trunk compartment.

9.   Final dress up - particularly on metallics.

Spray booth design is based on codes which have been developed to insure the safety of the operator. Requirements such as ventilation are based on providing fresh, clean and comfortable air for the spraymen. Control velocities and air volumes are designed to protect him against particulate matter and paint fumes. It is significant to note that spray booth design would be quite different if based on the requirements of process. Exhibit II illustrates the differences relative to exhaust air requirements.

Conventional air dispersion spray equipment is universally accepted as the most viable, high production, quality producing, top coating application method available for automobile painting. However, we have a noise problem with most of the air caps on this type of spray gun. This is based on the current OSHA Standard for minimum exposure for an eight-hour period. (See Exhibit III) There is no known technology that will allow us to attenuate to a lower acceptable level. As a result, operators are required to use protective ear plugs.

Certain pigments are currently suspected as having carcinogenic tendencies. OSHA has promised new and more comprehensive standards for carcinogenic materials in the next few months. We expect this standard to affect the materials we will be using. This, in turn, can influence the process.

The quantities of fresh air for spray booths require very large amounts of energy. In this climate region for the 200 ft. booth we mentioned earlier, this translates into a very significant heating system and power requirement. This is illustrated by Exhibit IV.

In certain painting applications such as the painting of van interiors, there is great concern because of particulate matter produced by the process with no good way to control overspray in order to protect the sprayman, other than a protective mask.

The solvents that are used in paint are designed to provide the fluid properties necessary for spraying. Once the paint is deposited on the part and the desired mil thickness, the solvent evaporation rate becomes important since this controls the smoothness of the finish and the time to "skin-over."

A quality finish must be free from orange peel. There should be no evidence of solvent haze, trapping or popping. The color should be "matched"; and, for metallics, there should be no evidence of mottle. Imperfections such as dirt or "sagging" are not acceptable. The finish should exhibit a high gloss.

Spray painting with solvent based materials has been a process way-of-life for the automotive industry. The advent of new EPA Standards and guidelines, which the States are expected to enact into law, will have far-reaching impact on existing processing methods. The proposed 1980, 1982 and 1985 standards for Hydro-carbon emissions from a manufacturing facility will accelerate changes such as the use of waterbase and urethane materials, both of which have serious shortcomings in terms of their processing ability, environmental effect and the need to conserve energy. (See Exhibit V)

The dilemma facing the manufacturing engineer is placed in perspective when we attempt to resolve the many individual painting problems we face today. The dilemma is further complicated by government programs to reduce energy consumption.

Based on energy consumed per unit manufactured, we required approximately 25 million BTU's to build a car in 1972. The energy reduction proposals we are looking at request a 20% reduction by 1980 and 20% reduction again in 1985 and 1990. Simultaneously, we expect energy costs to double by 1980 and double again by 1990.

Exhibit VI illustrates what this can mean in terms of incentives to develop an energy efficient process. It becomes quite obvious that paint shop facilities will undergo significant changes within the next decade.

THE ROBOTIC POTENTIAL

Comparing the current paint shop needs and the robot's potential to paint automobiles, we come to the conclussion that fully automatic painting will become a reality in the near future. Once the system technology has been demonstrated, we conclude that OSHA requirements and the need to conserve energy will become the impetus to resolving the spray booth's problems simply by removing all the people from the process.

In all probability, from the government's point of view, the question will become, How long will it take to convert all U.S. assembly plants to using robots?"

Robotic development in recent years allows the user to think seriously about the ultimate automatic painting system. Noteworthy developments have been:

1. Micro-processor control of the robot.
2. Expanded disc memory.
3. Development of the flex-arm.
4. Improved teaching methods.

Items 1, 2 and 3 collectively represent the kind of breakthrough needed to catch our attention.

From the user's point of view, the robot's potential can result in the benefits outlined in Exhibit VII. Before you decide to run down to the corner hardware store and pick up a couple of dozen robots, let's take a good look at the process problems that will have to be met to realize the first viable system.

First and foremost, the user is interested in using the painting robot in his existing spray booth. The average U.S. spray booth has a width varying between 15 and 17 feet. Utilization of existing packages results in compromising the robot's working envelope.

The alternative to rework the spray booth to accommodate the existing robot design is very costly and prohibitive in terms of plant down-time. The correct solution is, design the robot to work in existing booths.

One concept of automatic painting is based on using all proven technologies assembled into a viable operating system controlled by a master or host computer. This system would be designed to meet all of the processing needs automatically with the exception, perhaps, of spot priming. The concept is represented by Exhibit VIII.

The concept is based on utilizing the following proven technologies:

1. Automatic body recognition.
2. Automatic body color (based on schedule.)
3. Existing horizontal and vertical reciprocators with micro-processor control.
4. Automatic color change.
5. Automatic purge and pre-purge.
6. The robot system with micro-processor control.
7. The flex-arm assembly.
8. Expanded robot memory.
9. Improved teaching methods for the robot.
10. Proven interrupts and interlocks.

Development work that is considered prerequisite to a viable system must include:

1. Automatic viscosity and specific gravity control.
2. Automatic fluid regulation at the color changer, controlled from the host computer.

Development work that is considered important to the ultimate system might include:

1. Complete system diagnostics.
2. Automatic calculation of paint usage by color.
3. Hard copy print out.

    (a) Diagnostics.
    (b) Material usage.
    (c) Tabulation of interrupts.
    (d) Periodic tabulation of environmental parameters, such as temperature, humidity, barometric pressure, air balance, control velocities, etc.

SUMMARY

This presentation has tried to identify the painting robot's potential related to the needs of a total body spray painting system. The potential is based on the robot's ability to work in a hostile environment and allow the facility to operate using significantly less energy. The viable automatic system requires further definition and development in terms of material control as well as the configuration of a master or host computer control system.

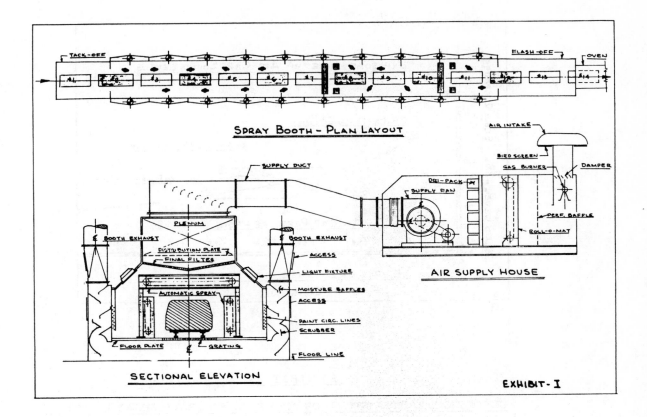

SPRAY BOOTH – PLAN LAYOUT

SECTIONAL ELEVATION

AIR SUPPLY HOUSE

EXHIBIT - I

EXHIBIT II

SPRAY BOOTH EXHAUST REQUIREMENTS RELATIVE TO VARIOUS DESIGN CONDITIONS

Design Condition	Booth Description & Solvent Load	Control Velocity	Req'd. Exhaust
1. Explosion/ Fire	200 Ft. Lg. X 17 Ft. Wd. X 10 Ft. High Ins. Down Draft Design – Wet Scrubbers 65 Gals. Solvent/Hr.	4 (L. E. L.)*	11,440 SCFM
2. Overspray Control	(Same as Item 1.)	100 FPM	340,000 SCFM
3. Operator Safety	(Same as Item 1.)	150 FPM**	510,000 SCFM

*Assumes Solvent Is Acetone (L. E. L. =2650 Ft^3/Gal.)
**Table G10 Page 90 Gen. Ind. Std. Rev. Jan. 1976 OSHA 29 CFR 1910

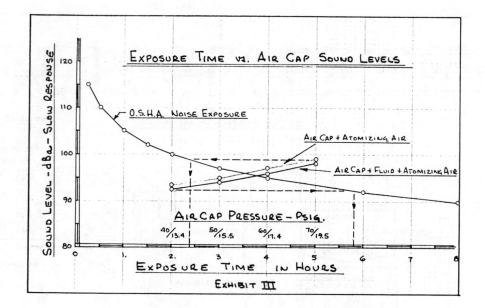

EXHIBIT IV

ENERGY CHARACTERISTICS FOR 200 FT. DOWN DRAFT SPRAY BOOTH

(MANUAL VS. AUTOMATIC)

Description	Conventional Design Manual Operation	Conventional Design Automatic Operation
Length	200 Ft.	200 Ft.
Width	17 Ft. (Ins.)	17 Ft. (Ins.)
Height	10 Ft. (Ins.)	10 Ft. (Ins.)
Heating System Size	55,000,000 BTU/HR.	36,720,000 BTU/HR.
Power Demand (Sequential Start)	1050 KVA	780 KVA
Connected H.P.	1400 H.P.	1050 H.P.
Electrical Energy Used/Yr.	4,323,820 KWH	3,242,860 KWH
Natural Gas Used/Yr.	910,800 THERMS	608,085 THERMS
Total Equiv. Energy Used/Yr.	1,058,675 THERMS	719,330 THERMS
Current Estimated Energy Cost*	$253,825.00	$178,365.00

*Nat. Gas @ $1.60/1000 Ft.^{3}; Electrical Energy .025¢/KWH

EXHIBIT V

E.P.A. PROPOSED STANDARD
FOR PAINT HYDROCARBONS

YEAR	PRIME-lbs/GAL.	TOP COAT-lbs/GAL.
1980	2.8	4.4
1982	1.9	2.8
1985	1.3	2.2

POUNDS OF SOLVENT PER GALLON OF PAINT-
VARIOUS MATERIALS

Lacquer	6.6	
Clear Coat	5.2	
Enamel	4.8	
Spray Primer	4.8	
Water Dip Primer	3.2	
Electro-Coat		
35% Surfacer	2.8	(Present Mat'l.)
62% "	1.9	(Future Mat'l.)
75% "	1.3	(" ")
Water Base Top Coat	2.8	
70% Solids	2.2	

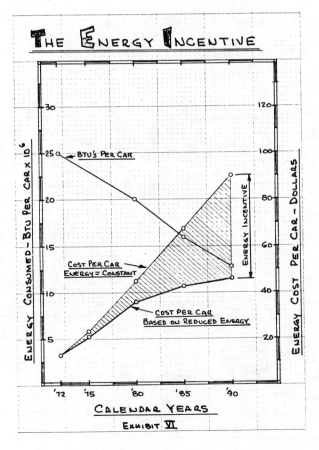

THE ENERGY INCENTIVE

EXHIBIT VI

EXHIBIT VII

THE ROBOTIC POTENTIAL AND BENEFITS

I Robots Will Allow Us To Deal With A Hostile Environment

    (a) Noise

    (b) Carcinogenic Materials

    (c) Particulate Matter

II Robots Will Allow Us To Process With Less Energy.

    (a) Reduced Fresh Air Requirements

    (b) Reduced Exhaust

    (c) Reduced Energy Cost

III Robots Will Allow Us To Improve Paint Quality.

    (a) Less Dirt

    (b) Uniform Build

    (c) Consistent Quality Level

    (d) Cope With Specialized Spray Techniques

IV Quality Improvements Result In

    (a) Reduced Warranty

    (b) Reduced In-House Repairs

V Reduced Material Costs.

VI Reduced Direct Labor.

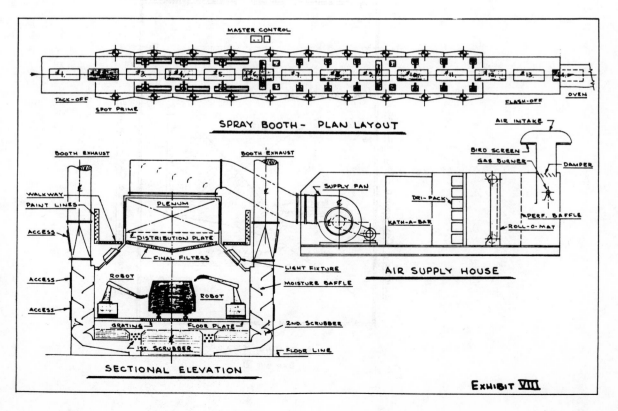

SPRAY BOOTH - PLAN LAYOUT

AIR SUPPLY HOUSE

SECTIONAL ELEVATION

EXHIBIT VIII

Reprinted from Robotics Today, Spring 1980

Robot Spray Painting at GM's Guide Division

Sixteen computer-controlled

robots provide

cost savings and

improved finish quality

in painting

automotive bumpers

Primer and finish color spray painting operations on urethane plastic bumpers have been automated with 16 servo controlled robots at the General Motors Guide Division in Anderson, IN. The new installation consists of four painting lines, with four robots in each line. Initial results indicate the robots are living up to expectations in terms of cost savings and improved finish quality.

The DeVilbiss Co., Toledo, OH,

supplied the versatile DeVilbiss-Trallfa robots used in this installation. They provide the continuous path motion necessary for precise duplication of the human sprayer's motions and techniques. Each of the four painting lines includes one robot for priming and three robots for applying color paint.

Figure 1 shows the three color-spraying robots in one of the lines. The double-hung monorail also car-

1. A bumper with dark prime coat approaches the first of three robots in one of the four new painting lines installed at GM's Guide Division.

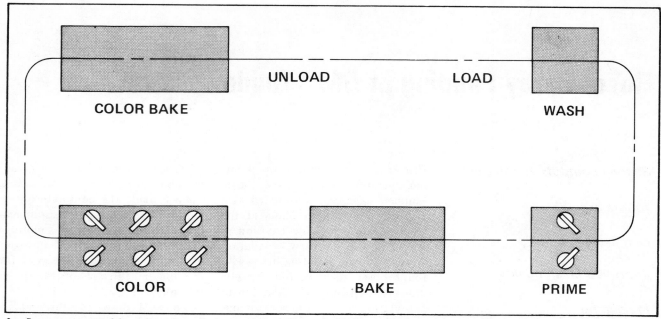

2. *Bumpers are carried through two lines of paint spraying robots on a double-hung monorail. The baking cycle in both ovens is 30 minutes at 250° F.*

ries bumpers past another group of paint spraying robots located on the opposite side of the screen at the right in the photo. The complete monorail loop, *Figure* 2, carries the bumpers through a washer and two bake ovens. Two such installations, as depicted in the drawing, comprise the four automated painting lines.

Paint savings on the order of 10% were originally anticipated, but actual results have exceeded that figure. This advantage comes, in part, from the robot's ability to precisely follow part contours and to trigger the spray gun with better-than-human consistency and precision. The result is a more uniform and thicker coating, providing improved durability. Application of three color coats on each bumper provides the required paint thickness—from 0.0015 to 0.0025" (0.038 to 0.063 mm).

Energy savings are another important benefit. A major factor here is the reduced requirement for ventilation compared to that where human operators are involved. The Guide Division has not completed its evaluation of these savings, but DeVilbiss notes that in some systems of this type exhaust volume can be reduced as much as 60% while still conforming to governmental regulations and local codes. With a reduction in

spray booth exhaust, air replacement is proportionately less.

In each of the new lines, an operator is stationed beyond the last robot to apply a final light coat of paint to

assure a good color match on each part. The robots are programmed— that is, taught the various spraying patterns—by another employee who also monitors operation of the robot

3. *Robot control cabinets are located away from the spray booths in a separate control room. Spraying programs are stored on floppy discs like the one shown loaded here.*

spray painting lines.

The DeVilbiss/Trallfa robot incorporates arm and wrist actions that provide six axes of motion. Its hydraulically powered manipulator features a unique, patented Flexi-arm device which essentially duplicates the pitch, yaw, and rotation motions of the human wrist. The horizontal sweep of the arm spans 135° —an optional feature selected by Guide over the standard 93° arm swing. The counterbalanced arm is also hydraulically depressurized while the robot is being programmed, making the arm easy to move manually.

The microprocessor-based control for each robot consists basically of a microprocessor, a dual floppy disc memory system, and an LED display. Up to 64 spraying programs can be stored in memory. Recall time for any one program is just one-half second. Other control features include random program selection, extended memory times, copy and edit capability, and synchronization with monorail movement. Solid state circuits mounted on plug-in boards facilitate maintenance and trouble shooting.

At Guide Division, all of the robot control cabinets are located remotely in a separate control room, *Figure* 3. This isolates the computer equipment from plant dust and dirt, paint overspray, and temperature variations. Interface panels located in this room also simplify trouble shooting by bringing all signal wires to readily accessible terminal strips.

Synchronizing robot movements with the two monorails is an important requirement. Guide and DeVilbiss engineers worked out a system in which a pulse generator on each monorail feeds appropriate signals to the robots. Monorail starts and stops automatically start and stop the spray guns.

Part sensing is also a prerequisite for determining part presence and the style of bumper hanging on each rack. Limit switches are considered preferable to photocells for this purpose based on cost and maintenance.

An interesting feature of the sensing system in this installation is the

4. A lead-through teaching system is used to program each robot. All moves of the gun, including triggering, are stored in memory for subsequent automatic spraying.

use of a Modicon programmable controller to simplify signal requirements. The controller, in effect, ties the three color paint robots together in each line. Thus, only one incoming signal is required. The programmable controller steps the signal down from the first robot, to the second, and then to the third. The signal can not only flag a change in paint color or bumper style, but it can also be used to stop the spraying action in the event of an empty or missing rack.

Programming the robots is a simple matter. The operator attaches a teaching handle to the manipulator arm, *Figure* 4, and plugs it in electrically to a receptacle on the robot base. He then leads the arm through the desired program sequence to define the path and relative velocity of the arm and spray gun. When programming is complete, the operator removes the handle and the electric plug, switches the control from "programming" to "repeat" and puts the robot into automatic mode.

During the lead-through teaching process, position transducers measure and generate position information. This analog information is then converted to digital form and stored in the control's memory. During subsequent automatic operation, the data is compared to the arm position, producing an error signal. This signal actuates a servo valve to position the arm and spray gun in a smooth, continuous path at the programmed relative velocity. ■

CHAPTER 12

INSPECTION

Commentary

Inspection is a relatively new applications area, where the robot may have a passive or active role.

In the passive mode, the robot is used primarily as a handling device. It may load a gage or inspection machine and then segregate or sort the parts, based upon the signals from the gaging system.

In the active mode, the robot handles the measuring device itself, positioning the gage in programmed locations relative to the part being inspected. A data processing unit is often incorporated to record the gage readings for reference. Gaging devices may be either contact probes or non-contact devices, such as lasers.

These robot systems can significantly increase productivity of inspection operations and assure high reliability of data, as well.

Using cameras to visually inspect parts, a robot for
tedious work-sorting and a programmable
controller to tie everything together, signals a new
era in testing and inspection

"Smart" Robot with Visual Feedback Boosts Production 400%

Auto-Place Inc., Troy, MI, has added a new dimension to automatic parts testing and inspection. It's a system which uses four solid-state image cameras to visually determine if valve covers for V-6 engines are properly assembled, and generates data that tells a robot when to accept parts or reject them according to the nature of the defect.

Jerry Kirsch, president of Auto-Place, says: "To the best of my knowledge, this is the first time vision coupled with a robot has been used as an inspection tool; it's the first instance where a camera vision system makes the 'accept/reject' decision. Camera-gathered data directs the robot via a programmable controller. However, the system could be structured so visual data would directly trigger robotic functions."

The Auto-Place system installed and operating at Chevrolet Motor Div. of General Motors Corp., Flint, MI, consists of a dial index machine with six dual-fixture stations on a 12-ft diameter aluminum table. Its operation is best described by tracing the testing and inspection sequence.

Twelve hundred valve covers for V-6 engines are pressure tested for weld soundness, visually inspected by computerized cameras and automatically sorted and unloaded by a robot. This reportedly is a four-fold increase in testing and inspection rates

To initiate the automatic testing and inspection routine an operator manually loads valve covers on fixtures at the right side of station No. 6 and the left side of station No. 1 (see schematic). Two styles of valve covers are inspected. The right-hand valve cover has one hole on top, a clinch nut, five metal brackets and a baffle. The left-hand style has two holes on top and one small bracket. The parts are placed on cast aluminum fixtures which fill the valve covers to within .125 in. of the wall. Parts are seated on a urethane gasket to support and position them. (The operator can randomly load either style valve cover on any fixture.) At this point, the operator pushes palm buttons to advance the parts to station No. 2 for 100% leak testing for weld soundness.

Leak Testing. Station No. 2 consists of a pair of Uson Corp. (Houston) electronic memory pressure decay units interfaced with a Texas Instruments (Attleboro, MA) programmable controller (PC). Each fixture at station No. 2 has magnetic sensors to indicate the presence of a part. If a part has not been loaded on one of the fixtures, the Uson tester is instructed to test only the loaded fixture. When the presence of a part on a fixture is sensed, a vertically descending ram seals the holes in the top of the valve cover and clamps the bottom flange of the part to the urethane gasket to form a seal. The test cavity then is charged with approximately 5 psi air pressure through a solenoid valve controlled by the tester. Next, the

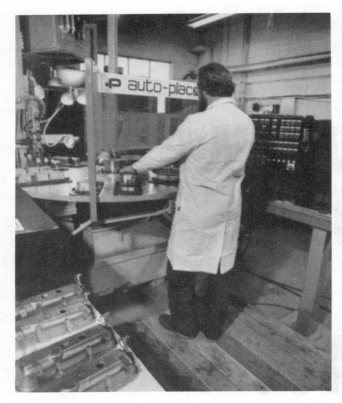

The Opto Sense/Robotic testing and inspection system can be mounted on a 9-ft sq platform. The grey boxes (top left) house four GE cameras

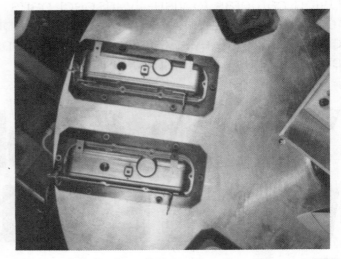

This is what the Opto Sense system "sees" as parts are visually inspected for the presence of holes, baffles, brackets, etc. at station No. 4

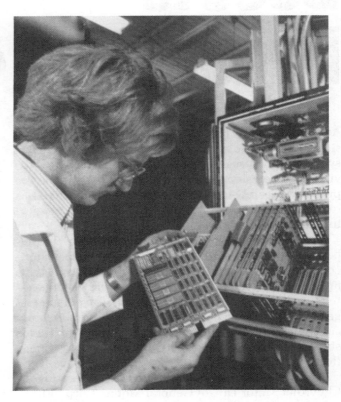

The Auto-Place robot is directed by a Texas Instruments programmable controller which stores "accept/reject" signals made by the Uson leak tester and Opto Sense system. It is easily reprogrammed to handle other parts

solenoid valve is closed, and an electronic pressure transducer continuously measures pressure trapped in the part. Pressure readings are individually compared to pre-programmed limits and an "accept" or "reject" decision is made for each valve cover. The Uson tester transmits this decision signal to the Texas Instruments PC where it is stored in memory for later use.

After leak testing, the valve covers index to station No. 4, the visual inspection station. Station No. 3 is an "idle" station. It is there because Chevrolet had available a six-station Ferguson indexing unit which it elected to use rather than wait four months for delivery of a four-station unit. This also is the reason parts are loaded at stations No. 6 and No. 1. So, in essence, there are two unused positions on the index table.

Visual Inspection. The visual inspection station is served by four General Electric Co. solid-state imagers with matrix arrays positioned to view the parts. They are mounted overhead (two above each fixture) in protective enclosures with Lexan windows. The cameras' lenses are focused on the valve covers, not the windows, so dirt or dust on the windows will have little or no effect on the vision system's operation.

The GE cameras are tied to Auto-Place's Opto Sense camera control and comprise what Auto-Place calls its Opto Sense system. The cameras have a solid-state matrix array with approximately 60,000 picture elements. Each element is sensitive to black, white and

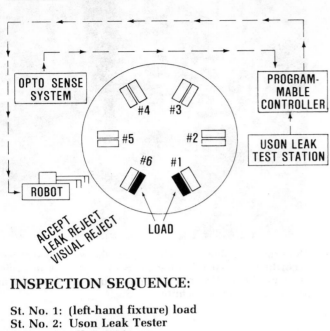

INSPECTION SEQUENCE:

St. No. 1: (left-hand fixture) load
St. No. 2: Uson Leak Tester
St. No. 3: Idle position
St. No. 4: Opto Sense System
St. No. 5: Auto-Place Robot sorts
and unloads valve covers
St. No. 6: (right-hand fixture) load

Data Transmission Path: (Red)

The "accept" or "reject" signals made by the Uson Tester and the Opto Sense System are relayed to the programmable controller where they are held in memory until the inspected parts reach station No. 5. At this point, the "accept/reject" decision signal is transmitted to the robot which sorts and unloads the parts according to the nature of the defect.

shades of grey. Reflected light and shadows permit the camera system to discern the absense or presence of detail. If no light is reflected the camera sees black, indicating the hole or holes in the part had not been made. The baffle inside the valve cover and beneath the hole also would appear white if properly installed.

The pairs of cameras work as a team with each viewing one half of a part with a slight overlap in their fields of vision. If one pair of cameras is down, the line can still inspect at 50% of capacity because everything is programmed in parallel. The first task performed by the cameras is to establish part identity—is a left-hand valve cover or a right-hand valve cover being inspected? (It is because of this capability that an operator can randomly load either style valve cover at stations No. 6 and No. 1.) The second task is to "look" for the presence or absence of all necessary characteristics: clinch nuts, metal brackets, baffles, one or two holes, etc. This step also includes determining whether any extra parts have been added. Information gathered by the GE cameras is analyzed by the Opto Sense camera control, and an "accept" or "reject" decision is made. The decision signal made by the Opto Sense system then is transmitted to the Texas Instruments PC where it is stored for later use.

The final step in the testing and inspection sequence is indexing the parts to station No. 5 where they are sorted and unloaded by an Auto-Place Series 50 pneumatically operated robot. This unit has two independently controlled, slide mounted arms which can extend 18 in., rotate 90 degrees and retract 24 in. All motions are directed by the Texas Instruments PC.

Automatic Parts Sorting. The PC combines the "accept" or "reject" signal from the Uson leak tester (stored in its memory) with the signal from the Opto Sense system to determine whether a part will be accepted or rejected. If either signal indicates a "reject," the part will be sorted according to the nature of the defect—leak test failure or a missing (or extra) component—and rejected.

The robot's "hands" are positioned above the unload station. When parts come into position, the arms descend, grip the parts (one in each hand) and lift them off the fixture. The robot then rotates to its "dispose" position which is near three chutes leading to part bins. If only one part is fixtured, both hands descend but only one will grasp the part.

As the robot rotates, the first position at which it can release the part is above the "accept" or "ok" chute. If the PC memory signals the part is a leaker, the most probable cause for rejection, the robot arm extends approximately 18 in. and releases the part into the leak rejection chute. In the case of a PC-signalled visual reject, the robot rotates, reaches and slides horizontally to release the part into the visual reject chute.

Increased Production. Once valve covers have been manually loaded at stations No. 6 and No. 1, leak testing, visual inspection, and parts sorting and unloading is automatic. According to Duane Carlington, general manager of Auto-Place, the system is designed to operate at 600 indexes per hour. However, because of dual fixturing, actual production is 1200 valve covers per hour.

Carlington says the Opto Sense/Robotic system replaces two testing and inspection lines which were operated on a two-shift basis. Installation was triggered by opportunities to improve inspection efficiency and productivity. Carlington isolated these advantages.

Space Utilization: The complete Auto-Place system is mounted on a 9-ft sq unitized base.

Increased Production: Inspection rates have been increased to 1200 valve covers per hour versus 300 parts

Auto-Place President, Jerry Kirsch (right) and Duane Carlington, general manager, expect to sell 200 Opto Sense
systems within the next two years

The programmable controller directs the robot to sort parts
according to inspection decisions (accept, leak test reject
or visual inspection reject) then unloads them to the appropriate take away chute

per hour by two lines operating two shifts.

Direct Labor: The Auto-Place system is loaded and
operated by one person whereas the previous inspection method required eight people, two persons per
line per shift.

Inspection Assurance: The Auto-Place system assures 100% inspection of all parts; it is not subject to
fatigue and is unlikely to pass defective parts.

Assembly Savings: By eliminating the likelihood of
defective valve covers moving to subsequent assembly stations, the time and expense of replacing defective valve covers after final assembly is eliminated.

Ease of Changeover: Should there be a need to make a
part change affecting the visual inspection procedures, it only is necessary to make software changes
in the camera computer system. Were an entirely different part scheduled, it also would be necessary to
change fixtures.

According to Jerry Kirsch, Auto-Place has sold approximately 1000 industrial robots. The robot at
Chevrolet Flint is the only one to be tied to an Opto
Sense system. Four Opto Sense systems not associated with robotics have been sold. One performs a
measuring task. Three others are located at the end of
assembly lines to visually determine whether sub-
assemblies have been properly put together.

Kirsch is convinced that a great demand exists for
systems combining an industrial robot with visual
testing and inspection techniques. Such an approach
is practical, he says, anywhere a robot is not on a job
where people are manually handling and inspecting
parts at the same time.

Certainly an attractive feature of the Auto-Place
Opto Sense/Robotic systems is that it is priced within
the budgets of a broad range of companies. Consider a
typical installation. The robot (which can handle
parts weighing up to 30 lb) cost approximately
$16,000. The four cameras cost about $42,000. The
two Uson test stations cost $21,000. The Texas In-
struments equipment cost about $2,000. The basic
machine (hydraulics, dial table, etc.) cost about
$49,000. And fixtures cost approximately $10,000.
For approximately $140,000, a company can pur-
chase a complete testing and inspection system
which is fully automated and provides 100% inspection assurance.

Market Outlook. Auto-Place believes it is the first
manufacturer to marry visual inspection techniques
to robotics. It also believes it has the inside track on
marketing opportunities. General Motors reportedly
is evaluating other applications. One division is be-
lieved to be looking at a system to visually inspect
parts coming from a punch press. Another automaker
and a major aerospace company also are working with
Auto-Place.

"Within two years," says Kirsch, "I see a market for
more than 100 Auto-Place robots to be used with or
without the Opto Sense system. I see a market for
many other Opto Sense systems which may or may
not incorporate robot handling."   □

Brian D. Wakefield

Reprinted from Robotics Today, Summer 1980

Robots Combine Speed and Accuracy In Dimensional Checks of Automotive Bodies

A dimensional inspection system developed by Ford features four Cincinnati Milacron robots that bring speed, accuracy, and repeatable results to a time-consuming task

GENNARO C. MACRI
Supervisor of Automation
and
CHARLES S. CALENGOR
*Facility and Equipment Specialist
Automotive Assembly Div.
Ford Motor Co.*

In a unique application of robots, the Automation Section of the Automotive Assembly Div., Ford Motor Co., conceived, designed, built, and installed an Automatic Body Checking (ABC) System at its Wixom, MI, assembly plant. The system is used to check Lincoln Continental and Mark VI luxury automobiles at a rate of five or six bodies per hour, depending on the mix of 2-door and 4-door styles. This compares to one or two bodies per shift using conventional manual methods.

The importance of proper dimensional control of the completed body cannot be overstated. The body is the foundation of the total vehicle. Good door, hood, deck, glass, and molding fits are all predicated on the dimensional stability of the body.

Quality Control checks the body shell at points located in and around the windshield, door, and back window openings. In essence, the body

1. The Automatic Body Checking System at Ford can check five or six bodies per hour and provide extensive quality control data.

2. *Selected bodies are routed into the body checking station with a conveyor subsystem that picks bodies from the main line.*

shop of an assembly plant does not build just bodies, but "body openings" which must be of the right size, shape, and relationship to one another for the proper fit of doors, glass, and moldings.

There has always been a need for a fast and accurate means to check automotive bodies and at the same time provide data which can be analyzed easily by the tool engineer. Ford's new system incorporating robots meets this need.

The ABC System, *Figure* 1, provides the assembly plant with a greater ability to improve the response time for making corrections to the assembly process and/or tooling by identifying body dimensional trends and errors promptly. This improvement in response time reduces defects, repairs, and warranty costs by raising the quality of the bodies to a higher and more consistent level. In this connection, the new system has a total system proven accuracy of ±0.020″ (±0.51 mm).

The complete system consists of six subsystems, all of which must operate properly and at the right time for the total system to perform as intended. Following are details on each subsystem.

Conveyor Subsystem

The ABC System has a bypass since it can only check five or six bodies per hour while the main body construction system produces units at a rate considerably above that figure. The conveyor subsystem, *Figure* 2, removes a body from the main construction line and moves it into the ABC System. It also removes the "checked" body and returns it to the main conveyor into the space or line gap generated by the removal of the succeeding body.

The subsystem has its own sequence controls and is only activated on command of the operator. The ABC System could have been made completely automatic, but it was necessary to have an operator on hand to watch over the system, input the body serial numbers, and make the body selections.

Automatic Body Locating Fixture

This fixture receives the body on its construction pallet and locates the body and pallet to a predetermined location. At this point the pallet is clamped to an elevator which lowers the pallet down and away from the body by 4″ (101.6 mm). This leaves the body suspended on four air-operated counterbalance details, while four other details, *Figure* 3, pin the body through its four master locating holes in a set sequence.

All assembly tooling is designed and performs its function in relationship to the master locating holes. The holes are located at each front door hinge pillar (right and left side) and at the rear of each quarter panel in the tail lamp area (right and left side). With the body suspended and fixed by only these four holes, it is ready for checking.

Robots

Four Cincinnati Milacron TC-3

robots are used in the system. They were modified for this application by changing the servovalves to a maximum 5 gpm (18.95 L/min) on the major axis and 1 gpm (3.79 L/min) on the wrist axis. These changes were made primarily to obtain greater accuracy.

ance of the servo-driver boards are necessary before the robot can be put back into service. In addition, the hydraulic oil and complete system must be maintained at a temperature of 100° F (37.8° C). Since the TC-3 robot is designed basically for a different class of applications, the modi-

robot's control. Approximately 150 points are programmed for checking on a 4-door body. In operation, all 150 points are checked in a total cycle time of 12 minutes.

A self-checking feature is incorporated in the program for each robot. Before moving to the car, the probe

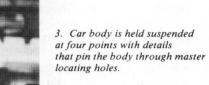

3. Car body is held suspended at four points with details that pin the body through master locating holes.

The four robots have a repeatability of ±0.005″ (±0.13 mm). As for accuracy, the TC-3's specification of ±0.050″ (±1.27 mm) was improved to ±0.010″ (±0.25 mm) through several changes involving programming and robot operation. This figure can probably be further improved by 50% by using more accurate resolvers on each axis and more accurate servovalves.

Initial development work revealed that under certain conditions the TC-3 would repeat within a range of ±0.015″ (±0.38 mm). Tests were conducted with a sensitive probe mounted on the robot's tool faceplate. It was learned that at each step of the robot program where a probe reading was to be made, a DELAY function of five seconds had to be programmed to allow the robot's hydraulic system to settle out. This resulted in readings in a range of ±0.010″ (±0.25 mm).

It was also learned that the robot controls must be left on 24 hours a day. If a shutdown is necessary, a warmup of 8 to 9 hours and a rebal-

4. A master gage block at each corner of the car is used for checking each probe before and after the actual body checking routine.

fications required for its successful use in this dimensional checking system are not unexpected or unreasonable.

The robots were programmed using Milacron's standard hand-held entry terminal. This terminal is connected by a flexible umbilical cord to each

goes to a checking block, *Figure* 4. Probe positioning (relative to the body's longitudinal centerline) is checked fore and aft, up and down, and in and out. This check is made again after completion of the inspection program. If a probe setting is found to be off by more than ±0.020″

(±0.51 mm), the data is discarded.

The system also includes a built-in safety feature. If the operator hits the "emergency stop" button, all four robots go into a "hold" condition. To resume operation, the robots must be released individually and a master release button on the control panel activated.

LVDT Probes and Schaevitz Interface

Each robot is equipped with three LVDT probes mounted in different positions so that they can reach each programmed body check point with a minimum of program steps. The probe toolholder was developed by experimentation in a cooperative effort with Cincinnati Milacron. Configuration of the holder facilitates

5. *Selector buttons on control panel permit selection of checking routines on each car body.*

making all of the physical moves required for checking.

Schaevitz Model GCA-121-500 probes are used. The Schaevitz interface system consists of: an LVDT signal conditioning subsystem employing 12 PC boards; an analog-to-digital subsystem consisting of 8 PC boards and one 3½ DVM; an 18-column digital printer; and the interface to the data control system.

The interface system, a Cincinnati Milacron Maximizer programmable controller, and the operator control buttons are all mounted in one air conditioned enclosure. If desired, the operator can use certain buttons,

Figure 5, to select specific points on the body for checking and exclude the remainder. This provides the ability to obtain data quickly relative to any particular dimensional trouble spot and to monitor the effect of corrective measures taken further upstream in the assembly process.

The 18-column digital printer outputs all the raw data that is inputted to the data control system. In the event the data control system is inoperative for any reason, it is anticipated that bodies could still be checked. However, the raw data would have to be manually charted to be understandable.

Sequence Controller

The operational sequence of the body location fixture functions, robot program selection, and cycle start of the ABC System are controlled by the Maximizer PC. This controller is equipped as follows: a-c inputs from all sources; a-c outputs to the body location fixture; d-c outputs to the robots; TTL outputs to the Schaevitz probe system; and a-c output to the computer.

The PC is one of the key elements in the entire system. Its performance has met all expectations for a unit of its size and cost.

Data Control System

Although this system contributes

nothing to the mechanical operation of all the other parts of the ABC System, it is most important in terms of making the total system worthwhile. The data control system is essentially a minicomputer with graphics capability. A Hewlett-Packard 9845 computer is used.

The system has the capacity to store information on 3000 checked bodies. Data stored on discs for each body is made up of the following:

- Body type
- Six-digit body serial number
- Date and time of body check
- Master gage block probe readings
- The 150 identified check point readings
- Body framing fixture number
- Body side assembly fixture number, right and left hand

On demand, the following information can be obtained from the data control system:

- Data on any checked body by a serial number search
- Data of the last body checked (all or any particular body opening)
- Data of the last 20 bodies checked (all or any particular body opening)

The above information can be obtained in any of the following formats:

- Raw data for 1 to 20 bodies (with the requested body opening illustrated and check points labeled)
- Chart histograms of each check point of a specific type body opening covering up to 20 bodies
- A trend chart of any specific point covering the 20 bodies

The data control system biases the reading of each body check point to correct any probe or accuracy error (within limits) that might occur. The system bases its correction (if required) on the reading taken before and after each body check for the three X, Y, Z readings made originally at the four master gage blocks. ∎

—Adapted from "Automatic Dimensional Inspection Using Robots." Presented at Robots IV Conference and Exposition, October 30-November 1, 1979, Detroit, sponsored by SME and RIA.

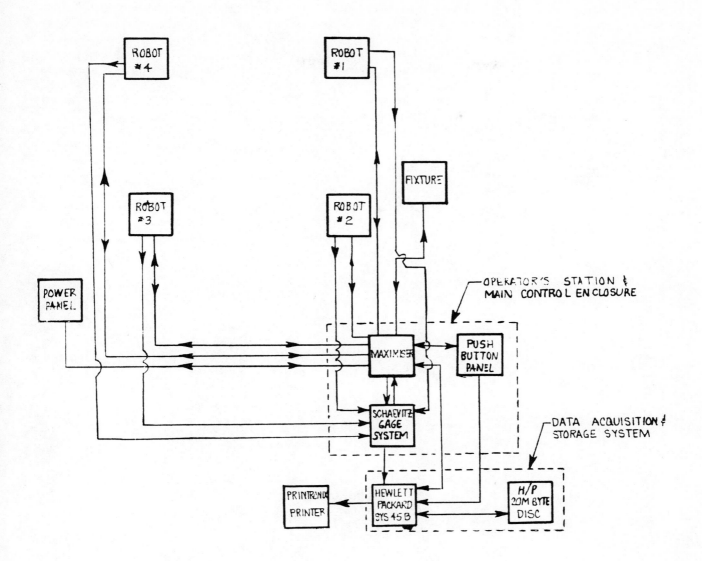

Block Diagram
Automatic Body Inspection System

CHAPTER 13

ASSEMBLY

Commentary

Programmable assembly systems utilizing robots can effectively automate low- to medium-volume and batch assembly operations. Assembly operations are a major use of labor in manufacturing and thus represent a robot applications area of significant economic potential.

A number of new, smaller, servo-controlled robots have recently been developed for use primarily in assembly operations. Tactile, force and vision sensing capabilities are being developed for this new generation of robots. In addition to robots and sensors, programmable parts orienter/feeders, passive compliance devices, system control hardware and software and interchangeable robot hand tooling are under development.

Currently, few programmable assembly systems are in operation in production. However, research and systems development is underway at a number of major facilities. The potential economic benefits from labor replacement, improved quality and increased productivity will encourage continued research and development in the assembly area.

Presented at the Robot I Conference, October 1976

Cost-Effective Programmable Assembly Systems

By Richard G. Abraham and James F. Beres
Westinghouse R & D Center

Manufacturing normally contributes approximately 30% of the gross national product of modern industrialized countries. In spite of this amount, manufacturing, although normally thought of as a highly productive and efficient activity, is not generally so efficient. This misconception exists because most people think of manufacturing in terms of mass production of products such as automobiles. In actuality, approximately 75% of our total national outlay for parts manufacturing is accounted for by batch-production methods where the individual batch size is 50 or fewer. Products made in this fashion may cost anywhere from 10 to 30 times more than they would if they were mass produced.[1]

Fortunately, revolutionary developments in manufacturing technology are opening the door to a new kind of factory, a factory where, in comparison to present day factories, the work environment is greatly improved, where labor requirements are vastly reduced, where raw materials are more effectively utilized and where, as a result, goods can be manufactured at substantially lower cost. The developments responsible for these advances involve an intensified application of computers to the control and management of manufacturing machines and systems.

Such systems are termed "programmable" because they can be easily changed to accommodate different products or styles or, in fact, can even be used on a completely different application (e.g., a robot originally used for part handling could be used for spot welding). This versatility is achieved through the use of a computer or memory in the automated system. By changing the contents of memory or using different computer software, the "programmable" automated machine can be adapted to the inspection, machining, handling or assembly of a wide variety of parts. Thus, programmable systems are ideal for batch manufacturing.

Significant progress already has been made in implementation of N/C, CNC, and DNC machining systems, programmable inspection and testing equipment, and robot parts-handling systems. Assembly operations, on the other hand, are very labor intensive and in need of considerable productivity improvement. Recognizing this, the National Science Foundation has been supporting research work on programmable assembly systems at the Stanford Research Institute (SRI),[2] Charles Stark Draper Laboratories,[3] and other institutions.[4,5]

Westinghouse has been an industrial participant in the SRI and Draper Labs programs for the past two years. During this time, we have analyzed over 20 product lines in 12 different batch-manufacturing plants and configured programmable assembly systems for the most promising product line. The purpose of this paper is to present the results of this work and our conclusions on how to accelerate the transfer of programmable assembly technology so that productivity improvements in batch assembly operations can be realized.

PRODUCT LINE ANALYSIS

In order to focus attention on only those product lines which are technically and economically suited to programmable automatic assembly, we had to develop some simple guidelines for plant manufacturing personnel to use in selecting candidates for flexible automatic assembly. We purposely considered only labor productivity improvement in our initial guidelines since the other potential economic benefits, such as product quality improvement, reduced manufacturing cycle time and lower in-process inventory, although significant, are difficult to quantify. An evaluation of these benefits must await operational experiences.

Our product line selection method[6] is an iterative procedure employing two simple graphic aids. By first establishing the actual man-hours currently required to assemble one product line and the total annual volume of that product line, we are able to determine whether this product line can be economically assembled automatically simply by using a graph such as the one in Figure 1 and determining into which region the application falls. For the assumed initial investment and labor cost, the boundary line depicts the combination of assembly time and annual volume that produces a desired return-on-investment (ROI). (In the example graph, a 25% ROI is the cut-off point; however, any value of ROI can be used.) Equivalently, the graph shows that if the present assembly man-hours per year are greater than a particular value (in the example, 10,000 man-hours), then the application should be considered further using the second step of the methodology, which is the application of a plot depicting robot labor hours per unit as a function of annual volume. The example graph shown in Figure 2 assumes the plant is operating three shifts. Similar graphs are available for one- and two-shift operations. For the annual volume required, we use such a graph to determine the available robot hours to assemble each product line. If the particular programmable assembly system configuration being considered can achieve this time value, the product line is a potential candidate for flexible automation.

A more detailed analysis is then conducted taking into consideration other significant variables, such as: number of styles, annual volume per style, batch size, set-up time, set-up cost and downtime.[6] Following this, a more detailed layout of the programmable assembly system is developed, and a new assembly task sequence is generated. Appropriate revisions to the analysis are made. If the application still is feasible technically and economically, our standard computer program for calculating return on investment is applied to the data. Time value of money, anticipated volume increases, other potential savings, etc. are incorporated in the computer program.

Our aim in developing a family of graphic aids was to make the guidelines simple and the methodology iterative. This enables plant personnel to utilize a minimum of data on the initial pass. We proceed with a more detailed analysis only if the preceding pass reveals that the product is still a viable candidate for programmable assembly.

The product lines analyzed by Westinghouse include CO relays, compressors, DPM contactors, fuses, lightning arrestors, low voltage bushings, motor rotors, tap changers, Type-400 resistors and W-2 switches. The first product, a CO relay, is an induction overcurrent relay used to disconnect circuits or apparatus when current exceeds a threshold value. The unit requires over 300 assembly steps and is very labor-intensive. Although the annual volume is relatively low, the large assembly time makes the CO relay economically attractive for automatic assembly, since the estimated ROI is 26%.

The heat pump compressor product line consists of compressors with one to five cylinders which require slightly more assembly time than the CO relay. However, since the annual volume is also larger, the compressor line lies further into the automatic assembly region than the relay.

The DPM contactor is a new device scheduled for production in 1976. It is one of the most labor-intensive assemblies considered, but the anticipated low-annual volume puts it well below the automatic assembly line.

The fuse assembly product line has 83 different styles based on the current rating and physical size of the device. Nevertheless, the number of parts in each fuse and the assembly sequence remain nearly the same. Since the estimated ROI is 23%, which is quite close to the desired 25% value, the fuse assembly was subjected to further analysis using the second graphic aid.

As the name implies, lightning arrestors are used to protect devices from the large high-voltage surges encountered when power lines are struck by lightning. Final assembly alone of this candidate is inappropriate for automatic assembly, since the product of annual volume and manual assembly time is not greater than 10,000 man-hours per year.

With only five parts, the low-voltage bushing is one of the simplest assemblies analyzed. The basic assembly procedures are similar for all 56 styles, which exist mainly because of dimensional differences in parts. Despite its relatively high annual volume, automatic assembly of the bushing product line alone proved not to be attractive because the relatively simple assembly tasks required little time to complete.

The rotor assembly for small motors (Figure 3) is an excellent candidate (75% ROI) for automatic assembly for two reasons: it has a very high annual volume and reasonably long manual assembly time. Furthermore, the assembly and handling tasks are performed in an unpleasant environment.

Tap changers are large rotary switches designed for very limited usage. When considered alone, automatic assembly of the tap changers is uneconomical, because the assembly man-hours per year is too small. However, since tap changers, fuses, and low-voltage bushings are assembled at the same location, assembly of all three product lines using the same programmable assembly system may be feasible.

The Type-400 resistors are large wire-wound resistors with various ratings. They are uneconomical to automatically assemble for the same reason and should not be considered alone.

W-2 switches are low-voltage switches made up of 35 parts (Figure 4). The assembly procedures are very complex. Each switch consists of stackable assemblies such as contact housings, rotating contacts, spacer plates, as well as washers and stops. A rotor shaft and through-bolts complete the final assembly. Lengths of completed switches range from 3" to 30" depending on the number of decks. Over 100 styles are possible. Preliminary analysis shows the final assembly to be uneconomic for automatic assembly, but if subassembly tasks are considered, the situation changes.

The ROI's for each of these 10 candidates are summarized in Table 1. The ROI calculations assumed an initial investment of $125,000 for all of the automatic assembly equipment and tooling. Only four of the 10 candidates crossed the border into the flexible automatic assembly region. As alluded to during the product descriptions, it may be possible to assemble different product lines with the same programmable assembly equipment if they are, or can be, assembled at the same location. Furthermore, when the final assembly tasks alone appear unfavorable, consideration can be given to using the same equipment for performing subassembly, testing, and/or packaging tasks as well. Table 2 summarizes the results of this expanded analysis. With this approach, four more of the original 10 candidates pass the first economic test and are worthy of being considered in the second step of the analysis; the DPM contactor and Type-400 resistor are still uneconomic and will not be considered further.

The next step in the analysis incorporates a consideration of the maximum available robot labor hours/assembly and the maximum number of assembly steps at 2.4 seconds per step. The latter is based on the fact that for a two-robot system, a combined retrieve and assemble sequence takes about 2.4 seconds on the average. This number was obtained from Draper Labs and confirmed by a potential robot supplier. These data are summarized in Table 3 which also reveals the investment that will yield an estimated 25% return on investment for each candidate or combination thereof.

By first examining the lightning arrestor data, we note that in all three cases, the estimated number of assembly steps is greater than the corresponding maximum number of robot steps available. This implies that without a faster robot assembly system, we must perform some assembly steps using one of the following: (a) another robot, (b) a piece of fixed automation equipment dedicated to one or more tasks, or (c) a human to augment the robot labor. As a result, our original investment cost estimate of $125,000 in all probability will be inadequate. In addition, the original fixturing cost estimate of $25,000 will increase since sub-assembly, packaging and testing tasks for the lightning arrestor are being considered as well as its final assembly. A more detailed layout of a complete programmable system is needed in order to make a judgment on whether or not the job can be done within the economical constraint for maximum total investment shown in Table 3.

The compressor appears to be an economic application with a relatively large number of assembly steps available for the programmable system. The next step is to analyze the current assembly operations to see if the 250 steps are, in fact, reasonable for a robot configuration.

Another economical candidate, the CO relay, also has many assembly steps available. However, this product possesses many technical problems, such as flexible leads, delicate springs, and intricate assemblies, all of which are too difficult to perform with current state of the art.

The W-2 switch final assembly was economical when considered along with the stationary contact sub-assembly; final assembly along with stationary contact and star-wheel sub-assemblies were also economic when considered together. Based on the number of assembly steps involved, there appears to be sufficient time to perform these operations with the speed of existing robots. But again, fixturing costs for changing from final assembly to sub-assembly may be significant. Technically, the automatic assembly appears to be difficult, but possible. Fortunately, this product will be undergoing a design change in the near future thereby providing us with an opportunity to make design changes which will make the product more favorable for automatic assembly.

Even though the fuse assembly does not meet the established economic requirement, we note that an initial investment of just $10,000 less than the assumed $125,000 permits it to meet the 25% ROI criteria. Thus, if the robot configuration were to drop in price by 8%, this assembly application crosses the manual/automatic assembly border. From the standpoint of assembly steps, current robot speeds are sufficient, making this product a good economic and technical candidate.

By using the same programmable system to assemble the fuse and low-voltage bushing, the application becomes more economical but the time in which the robots must perform the assembly tasks gets smaller. To make the automatic assembly of two different product lines technically feasible, a mix of people, special purpose equipment and robots must be included in the system configuration and then an economic re-evaluation must be performed. Likewise, the combined assembly of tap changers and fuses or tap changers, fuses and bushings using one system is economical but further configuration work must be done to compensate for inadequate robot speeds.

The rotor assembly appears economical, but, here again, the available robot time value demands that additional assembly aids be incorporated in the system for performing some of the assembly tasks. Even so, the application holds much more potential than the other candidates because the total investment can be $375,000 and remain economically attractive.

The 10 products already analyzed clearly illustrate the importance of part recognition, acquisition and inspection in assembly operations. Although inspection tasks are not specifically delineated in the documented manual assembly instructions, an operator performs the inspection when he picks up a part. Eight of the products examined contain threaded parts. Typical defects encountered include flat spots on bolts and screws, an excess of zinc in the threads of galvanized bolts, unslotted heads on screws, and chipped ceramic parts. Products assembled automatically using faulty threaded or ceramic components would ultimately be rejected, thereby increasing rework or scrap costs, or might fail in the field increasing warranty expense. It is therefore important to incorporate inspection capability in programmable assembly systems.

One product, the W-2 switch, emphasized another problem closely related to inspection. Each of the molded parts is symmetrically designed with respect to its center axis except for some small raised numbers that are molded onto each piece. These numbers, in actuality, determine the correct orientation of the part during the assembly process. However, the same recognition performed manually can only be performed automatically by using time-consuming and expensive visual pattern recognition routines. Simply adding a notch, dimple, paint spot or other quickly recognized feature to the component in order to make its orientation less ambiguous would probably be much more cost-effective. Thus, in many cases, small design changes can make automatic assembly easier.

This point was also emphasized by researchers at Draper Labs in their peg-in-the-hole analysis. They found that addition of a small chamfer can greatly improve the speed at which parts can be automatically assembled.

We also learned that assembly operations requiring press fits, stamping, or simply "knocking" or hammering could utilize the robot arm and/or hand forces directly instead of special purpose equipment. To our knowledge, robot manufacturers have not specified this capability.

Furthermore, for those assembly operations that require frequent style changeovers, the computer should be utilized to automatically change both the robots' tasks and programmable fixtures, so that set-up time is minimized. The ability to reprogram robots quickly for different tasks has always been emphasized, but if fixtures are manually changed for different styles, the advantage of quick robot changeover may be lost.

For nearly every case that we have examined, the cost for fixturing and parts presentation equipment was almost as much as the cost of commercially available robots. One particular firm has addressed this problem by incorporating an overhead camera, light table, and general parts presenter in their system. Special vibratory feeders have wide tracks that can handle a wide variety of parts. These feeders are then used to feed parts one at a time, independent of orientation, to the parts presenter. The general parts presenter flips parts one at a time onto the light table where the camera, in conjunction with the minicomputer and vision software, identifies the part, performs gross inspection, and sends location and orientation information to the robot. Such a system becomes overly complicated when the number of parts is large since each part demands its own vibratory feeder.

SRI, on the other hand, has developed a system that can identify, classify, inspect, and determine orientation and location of parts moving on a conveyor belt. This system is therefore not limited to a small backlighted table station and can handle a large variety of parts. However, each part must still be fed one at a time onto the conveyor belt, thus the basic part feeding problem remains.

Two basic solutions become apparent after close examination of the ten candidate products and their assembly operations. First, many of the batch assembly applications use tote bins for parts storage. Robot part selection from tote bins could thus make the system more flexible and easier to implement than complicated mechanical feed type systems.

The second solution to the part selection problem is the use of egg crate-type packaging. Several of the manual assembly operations currently receive parts in packaging; thus, programmable robots could take advantage of the gross orientation provided by the carton. After grasping the part, the robot could more precisely orient the object with the aid of vision or tactile sensors in a special hand.

PROGRAMMABLE ASSEMBLY SYSTEM CONFIGURATIONS

Since the motor rotor appeared to be the best candidate for programmable automatic assembly, we proceeded to develop alternative equipment layouts or configurations. To do so, we first determined the following operational requirements and constraints:

Environmental - Identify and quantify special atmospheric conditions (e.g., temperature, humidity, cleanliness, radioactivity), noise level, temperature of parts, etc.

Fixturing/Tooling - Analyze current fixturing and tooling, but interpret in terms of basic product line assembly requirements so that we do not constrain ourselves to the use of existing fixtures and tools.

Inspection - Identify the inspection operations required prior to and during actual assembly of parts and quantify the accept/reject criteria.

Parts Presentation - Analyze current methods to determine if parts are already oriented roughly in cartons, if they are in bins or if feeders are used; identify special part orientation or recognition requirements.

Special Equipment - Determine if presses, staking machines, welders, etc. are essential for successful assembly and document basic requirements.

Adaptability/Programmability - Determine the degree of programmability required for different styles and product lines and the degree of adaptability required to accommodate manufactured part tolerances and variations.

Close examination of the assembly tasks for the motor rotor (Table 4) reveals that there are many different parts feeding requirements, including those for cores, shafts, switches, sleeves, rubber washers, and nylon washers. All of these parts vary in size, weight and geometry. There are no special orientation requirements for sleeves and washers, i.e., it does not matter which side is up when they are placed over the shaft. Core and shaft orientation must be known; the side of the core with the hole must be identified and the longitudinal axis of the shaft must be known. Unlike these parts, which are symmetrical about one axis, the centrifugal switch is an irregular shape with more stable states. Here again, though, the hole is the key to determining orientation. Whereas, the sleeves and washers may be handled using conventional small part feeders, the cores, shafts and switches could require special handling equipment.

Analysis of the tasks also reveals that actual assembly operations are primarily of the insertion or peg-in-the-hole class. There is, however, a choice as to whether the shaft is inserted into the holes in a part or the part is placed over the shaft. Fixtures and tooling (grippers) could differ for each case. Tolerances, part variability and the expected clearance between parts to be mated define the degree of adaptability required.

The core heating, quench cooling, shaft staking and subassembly testing operations point out the need for special purpose equipment or fixtures and tooling. There could be many alternative ways of implementing these operations, including the use of existing equipment. The core heating requirement defines an environmental constraint, a high temperature work area, and a special handling requirement, hot cores.

Inspection requirements may not be apparent from an examination of the sequential tasks alone. Many component parts for the product family must be examined to determine how frequently defects occur and the nature of the defects. Part size variations are more of a problem for the motor rotor than defects, although inspection for gross defects is necessary. Furthermore, there may be a need to inspect and classify parts according to size variations in order to accomplish successful assembly. However, the alternatives to inspection and classification could be the use of special presses to assist in successful insertion or the use of force sensors to detect those cases when a press-assist is needed.

Since the motor rotor is used in a family of motors, there are many style variations. Major differences are in the position of the core on the shaft and in shaft length. These differences dictate the need for programmable robots and fixtures that can be easily and quickly changed for different batch requirements. The motor rotor is built in relatively small batches with up to seven changes per shift. Some changes, such as the use of different cores, occur less frequently than others and limit the degree of programmability for core style differences.

The basic assembly task sequence for the motor rotor, presented earlier in this section, can be accomplished in many different ways. Tables 5, 6 and 7 illustrate how these tasks can be divided between people, robots and special equipment with programmable fixtures. The division of tasks was made considering cycle time requirements (speed), task complexity (positioning accuracy, degrees of freedom, etc.) and cost. The alternate configurations for these three different task sequences are illustrated in Figures 5, 6 and 7.

The first configuration for motor rotors, Figure 5 and Table 5, has five computer-controlled robots (two 3-axis, two 4-axis and one 5-axis), a 10-position index table with programmable fixtures, and special equipment such as an oven, press and staking machine. Cores are picked up two at a time by Robot #1 from a gravity fed conveyor and placed over a post in the existing oven (cores are stacked two high). Robot #1 then retrieves a single hot core and places it over a shaft held by Robot #2. A seating press applies extra push-down force on the core whenever Robot #1 detects out-of-tolerance parts (force-sensing wrist). Robot #2 picks up shafts one at a time from a gravity fed conveyor. Grip points for different styles of shafts are accounted for by the programmable control of Robot #2. After grasping a shaft, Robot #2 holds the shaft in an automatic staking machine that puts notches in the shaft. It then holds the shaft while Robot #1 places a core over it (coordinated assembly using two robots). Once the core is seated on the shaft, Robot #2 places the core/shaft assembly on a gravity fed, in-process buffer conveyor.

Robot #3 takes a core/shaft assembly and places it on the index table. Fixtures on the index table are programmable to account for different shaft and core styles. Once the subassembly is dynamically tested, it is indexed to the next station where Robot #4 picks up a centrifugal switch from a vision table and places it on the shaft. These switches are fed one at a time by a general parts presenter onto a conveyor watched by a TV camera. Position and orientation are automatically determined by the computer and communicated to Robot #4 so that it can access the part. Next, Robot #4 takes a top sleeve from a bin and places it on the shaft. Whenever defective rotor assemblies are found at the test station, Robot #4 can also take the assembly off the index table and place it on a reject conveyor.

A dedicated press pushes the top sleeve and switch on the shaft. Likewise, dedicated assembly equipment puts the bottom sleeve and washers on the shaft. Robot #5 takes the rotor assembly and places it on a gravity fed output conveyor. Then, two dedicated assembly units complete the assembly by placing the nylon washers on the shaft.

The second configuration, Figure 6 and Table 6, uses two computer-controlled robots, one pick-and-place mechanism, one manual assembly station, the existing oven, and an existing press. Again, just as in the 5-robot configuration, Robot #1 picks up two cores from a conveyor and places them in the oven. It also takes one hot core at a time and places it on a shaft. This configuration, however, uses an 8-position index table with programmable fixtures to hold the shafts. A dedicated seating press pushes every core over the shafts.

A quench station on the index table cools the assembly. Each shaft is staked at the next station with a dedicated staking machine. Core/shaft subassemblies are then transferred to an in-process buffer by a dedicated pick-and-place unit. Robot #2 takes shafts from the gravity-fed shaft feeder and loads the index table. It also takes centrifugal switches from a vision table and places them on rotor shafts at an assembly station. The two sleeves are put on shafts at the same station by dedicated units. An operator takes the subassembly from this assembly station, places it in an existing press to push on the sleeves and switch, then manually assembles the washers. Final units are manually loaded into boxes.

The third configuration, Figure 7 and Table 7, uses one computer-controlled robot, one pick-and-place unit, and dedicated assembly equipment. The major difference between this configuration and the other two is the use of a new vertical in-line oven to heat rotor cores. For this system, cores are conveyed through the oven and exit in a position ready for shaft insertion. This insertion is handled by a dedicated unit that pushes a shaft through a core until the shaft strikes a programmable stop. After being conveyed through a quench station, core/shaft subassemblies are transferred to a vertical in-line indexing system. Robot #1 attaches the centrifugal switch, fed onto a vision table, and top sleeve as in the 2-robot configuration. All other assembly operations are performed by dedicated assembly equipment (bottom sleeve, pressing, and washer assemblies). A pick-and-place unit takes subassemblies from the indexing conveyor and places them on a gravity fed conveyor.

A comparison of all three configurations shows that all systems use a robot with vision to assemble centrifugal switches. These switches have presented problems for conventional parts feeding and are suitable for more general parts feeding systems. The 5-robot configuration uses two cooperative arms to assemble cores over shafts. The 2-robot system combines robot core/shaft assembly with programmable fixtures to hold shafts. The 1-robot configuration uses an adjustable stop and a dedicated shaft insertion.

CONCLUSIONS

The product line analysis and configuration work performed thus far has led to the following conclusions:

1. The number of programmable automatic assembly applications will remain relatively small until lower cost robots with speeds faster than humans and simple, low-cost parts presentation methods become commercially available.

2. Annual volume and assembly task complexity (and therefore time) must be considered together when determining feasibility of product lines for programmable automatic assembly.

3. The assembly of two or more product lines (which are made in the same manufacturing plant) on the same programmable assembly system is technically feasible and may be more cost-effective than the automatic assembly of either one alone.

4. Inclusion of subassembly, test and packaging tasks as well as final assembly can make a product line more attractive for programmable automatic assembly.

5. Product redesign to facilitate automatic assembly may be more cost-effective than configuring a system for an existing product design.

6. Programmable fixtures are essential to minimize changeover for different styles.

7. A mix of man, programmable and fixed automation equipment can improve the system cost effectiveness; furthermore, cycle time requirements may dictate the use of people or special purpose equipment.

8. Complex assembly tasks are more appropriate for humans and, in fact, are not technically feasible with existing programmable assembly robots.

9. Quantification of economic benefits other than labor productivity improvement, such as reduced in-process inventory and improved product quality (and therefore lower rework and warranty costs), is essential if the number of programmable automatic assembly applications is to increase.

We feel that technical and economic feasibility must be proven by conducting experiments during which we attempt to assemble real products using programmable assembly systems. SRI and Draper Labs are working on research that could lead to more versatile, cost-effective, second-generation systems that automatically assemble many more products. Our active participation in these programs coupled with our independent work led to the conclusion that the transfer to industry will be quite slow and may never occur unless a user firm like Westinghouse selects those research results from SRI and Draper Labs with the most promise and develops, installs, operates and evaluates a complete second-generation programmable assembly system.

The specific questions which we hope to answer during such a program include:

1. Do the SRI and Draper Lab developments to date meet real assembly application requirements? What modifications are required to make them practical, economic and acceptable to industry?

2. What assembly tasks can actually be accomplished with the existing or near-term programmable assembly technology?

   a. For what tasks are force sensing and feedback control desirable?

   b. How does bin-picking compare to other parts presentation alternatives?

   c. How valuable (from a practical, economic and job performance viewpoint) are teach-by-doing, teach-by-showing and teach-by-speaking during set-up?

   d. Can the SRI parts inspection and recognition features be used in a factory environment?

   e. What trade-offs exist between the use of mechanical fixturing or vision for determining parts orientation?

3. Can a product line with many style variations be completely assembled using programmable automation? Can it be done economically?

   a. Can programmable fixturing be developed so that changeover time for the different styles is minimized?

b. What effect does product redesign have on simplifying assembly and improving the economics of programmable assembly?

c. Does programmable assembly reduce manufacturing cycle time and improve product quality?

4. How does programmable automatic assembly compare economically with fixed automatic and manual assembly? What is the effect of mixing these three assembly alternatives in one assembly system configuration?

5. How well do the human operators accept the system? Is their productivity improved? What degree of job retraining or reassignment is required? What is the desirability of job upgrading?

By successfully answering all of these questions and disseminating the results to researchers, equipment suppliers, and potential users of programmable assembly systems, the transfer to the batch manufacturing industry will be accelerated. The data gathered during the experimental program on actual improvements in productivity, product quality and manufacturing cycle time will enable manufacturing managers to conduct sound economic analyses on their particular batch assembly operations. The actual demonstration that real products can be successfully assembled with programmable automation will be invaluable in satisfying the "show me" attitude so prevalent in manufacturing personnel. It will also lower the risk factor for programmable automated assembly and greatly enhance the probability of acceptance by business managers. Furthermore, by letting unions know of our intentions from the start and working closely with union personnel to assure human resource issues are properly addressed prior to introduction into manufacturing plants, we will greatly increase the probability of acceptance.

REFERENCES

1. H. Cook, "Computer-Managed Parts Manufacture," Scientific American, 232: 22-29 (February 1975).

2. C. A. Rosen and D. Nitzan, "Developments in Programmable Automation," Manufacturing Engineering, 26-30 (September 1975).

3. J. L. Nevins and D. E. Whitney, "Adaptable-Programmable Assembly Systems: An Information and Control Problem," IEEE Intercon 75, (April 1975).

4. Proceedings of Second RANN Grantees Conference on Production Research and Industrial Automation, University of Rochester (January 1975).

5. Proceedings of Third RANN Grantees Conference on Production Research and Industrial Automation, CASE Western Reserve University (October 1975).

6. R. G. Abraham, J. F. Beres and N. Yaroshuk, "Requirements Analysis and Justification of Intelligent Robots," Proceedings of the 5th International Symposium on Industrial Robots (September 1975).

Table 1
Estimated ROI's for Ten Candidate Products

Candidate Assembly	ROI for an investment of $125,000, %
CO Relay	26.3
Compressor	28.3
DPM Contactor	12.5
Fuse	22.9
Lightning Arrestor	19.0
Low-Voltage Bushing	7.6
Rotor	75.0
Tap Changer	6.2
Type-400 Resistor	3.1
W-2 Switch (Final Assembly)	15.0

Table 2
Estimated ROI's for Candidate Product Combinations

Candidate Assembly	ROI @ I = $125,000 (Percent)
Lightning Arrestor (Sub-Assembly)	8.9
(Final + Sub)	25.9
(Test/Package)	12.0
(Final + Test/Package)	29.0
(Final + Sub + Test/Package)	37.9
Type 400 Resistor + DPM Contactor	15.6
W-2 Switch (Starwheel Assembly)	2.19
(Final + Starwheel)	17.3
(Stationary Contact)	11.0
(Final + Sta. Con.)	26.2
(Final + Star. + Sta. Con.)	28.4
Low Voltage Bushing + Fuse	30.5
Low Voltage Bushing + Tap Changer	13.8
Tap Changer + Fuse	29.1
Low Voltage Bushing + Fuse + Tap Changer	36.7

Table 3

Most Favorable Candidate Products

Candidate Assembly	Max. Available Robot Labor Hours/Assembly	Max. Number of Assembly Steps @ 2.4 sec/steps	Estimated Average Assembly Steps	Investment @ 25% ROI
Lightning Arrestor (Final + Sub)	.016	23	24	129.4
(Final + Test/Package)	.016	23	26	145.3
(Final + Sub + Test/Package)	.010	16	21	189.6
Compressor	.167	251		141.3
CO Relay	.171	256	300	131.3
W-2 Switch (Final + Sta. Con.)	.073	110	45	130.8
(Final + Star. + Sta. Con.)	.060	90	39	141.8
Fuse	.012	18	16	114.5
Low Voltage Bushing + Fuse	.005	8	10	152.6
Tap Changer + Fuse	.011	16	17	145.5
L.V.B. + Fuse + Tap Changer	.005	7	11	183.7
Rotor	.003	5	18	375.0

Table 4:   Motor Rotor Assembly Tasks

1.   Retrieve cores
2.   Place cores in oven
3.   Retrieve shaft
4.   Stake shaft
5.   Remove hot core from oven
6.   Place hot core over shaft or insert shaft in core
7.   Quench cool subassembly
8.   Test subassembly
9.   Retrieve switch (governor)
10.   Place switch on shaft
11.   Retrieve top and bottom sleeves
12.   Place top and bottom sleeves on shaft
13.   Press sleeves and switch
14.   Assemble two rubber washers
15.   Assemble two nylon washers
16.   Transfer subassembly to conveyor

Table 5:   Tasks for 5-Robot Configuration

Sequential Tasks	Task Implementation Method
1. Retrieve core 2. Place core in oven 3. Remove hot core from oven 4. Place hot core over shaft	Computer-controlled robot #1 (4-axis force sensing) (Utilizes existing oven)
5. Retrieve shaft 6. Stake shaft 7. Hold shaft in programmable fixture 8. Move subassembly to quench station	Computer-controlled robot #2 (4-axis, compliant wrist)
9. Transfer subassembly to index table	Robot #3 (3-axis conventional)
10. Test subassembly	Automatic test equipment
11. (Optical - remove reject subassembly) 12. Retrieve switch from conveyor 13. Place switch on shaft 14. Retrieve top sleeve from bin 15. Place top sleeve on shaft	Computer-controlled robot #4 (5-axis force sensing) Vision system
16. Press top sleeve and switch 17. Assemble and press bottom sleeve 18. Assemble two rubber washers	Dedicated assembly units
19. Transfer subassembly to conveyor	Robot #5 (3-axis conventional)
20. Assemble two nylon washers	Dedicated assembly unit

Table 6

Tasks for Two-Robot Configuration

Sequential Tasks	Task Implementation Method
1. Retrieve core 2. Place core in oven 3. Remove hot core from oven 4. Place hot core over shaft	Computer-controlled robot #1
5. Retrieve shaft 6. Place shaft in index table fixture	Computer-controlled robot #2
7. Seat hot core on shaft 8. Quench cool 9. Stake shaft	Dedicated assembly units
10. Transfer subassembly to buffer conveyor	Pick and place device
11. Retrieve switch from vision table 12. Place switch on shaft	Computer-controlled robot #2
13. Place two sleeves on shaft	Dedicated assembly unit
14. Load subassembly into press 15. Place washers on shaft 16. Place finished unit on tray	Manual

Table 7

Tasks for One-Robot Configuration

Sequential Tasks	Task Implementation Methods
1. Heat core in oven	New vertical in-line oven
2. Place shaft in hot core	Dedicated assembly unit
3. Quench cool	Water spray
4. Transfer subassembly to in-line conveyor	Pick and place device #1
5. Stake shaft	Automatic stake machine
6. Test subassembly	Automatic test device
7. (Optional - remove reject sub-assembly)	Computer-controlled robot #1
8. Retrieve switch from vision table	
9. Place switch on shaft	
10. Retrieve top sleeve	
11. Place top sleeve on shaft	
12. Press top sleeve and switch	Dedicated assembly units
13. Assemble bottom sleeve and press	
14. Assemble rubber washers	
15. Transfer subassembly to conveyor	Pick and place device #2
16. Assemble nylon washers	Dedicated assembly unit

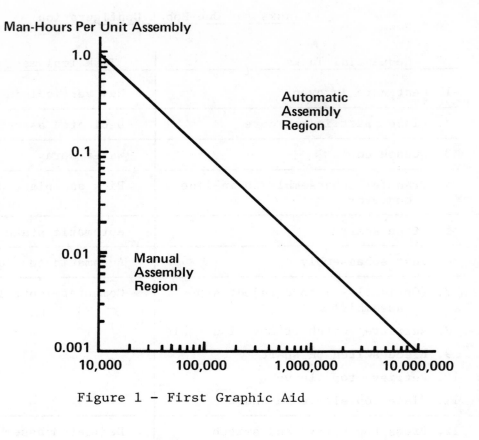

Figure 1 – First Graphic Aid

MAXIMUM AVAILABLE ROBOT HOURS/UNIT ASSEMBLY

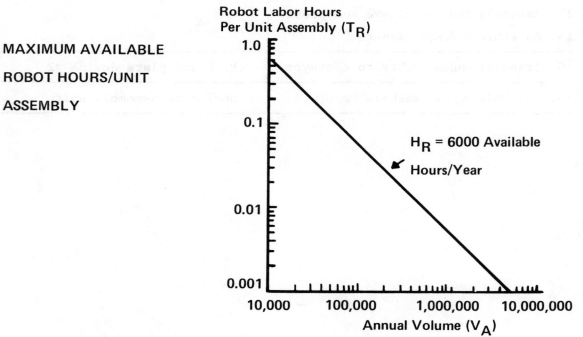

Figure 2 – Second Graphic Aid

Figure 3 — Motor Rotor Product Line

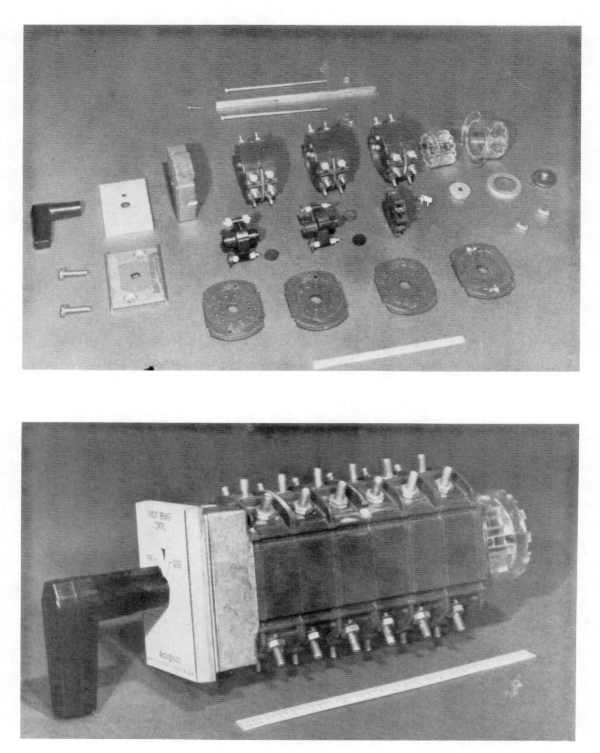

Figure 4 — W-2 Switch Product Line

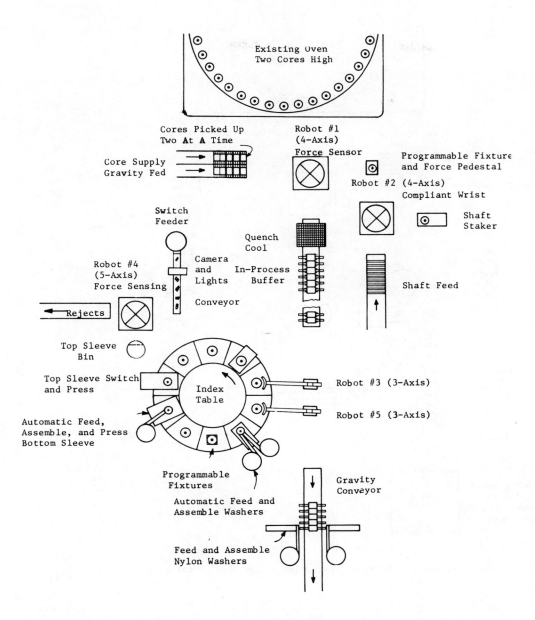

Figure 5 - Five-Robot Configuration for Motor Rotor

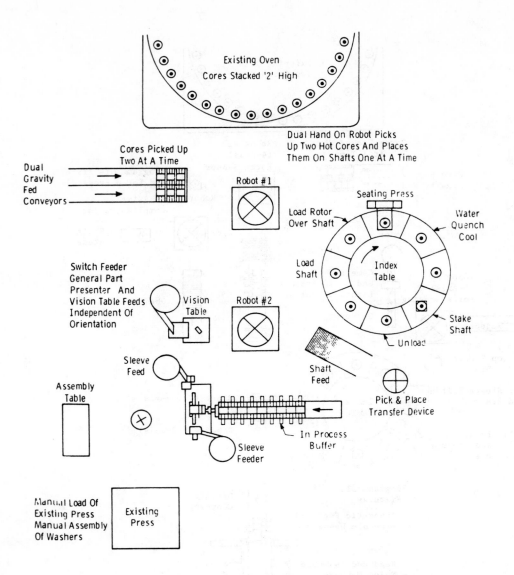

Figure 6 – Two-Robot Configuration for Motor Rotor Assembly

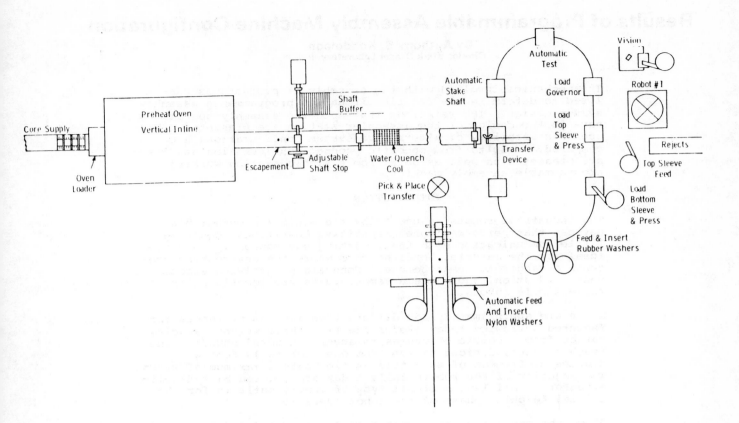

Figure 7 – One-Robot Configuration for Motor Rotor Assembly

Presented at the Robot II Conference, October/November 1977

Results of Programmable Assembly Machine Configuration

By Anthony S. Kondoleon
Charles Stark Draper Laboratory, Inc.

The mechanics involved with the assembly of products is analyzed to determine the relation it has to programmable assembly system design. The relations that specific assembly operations have to the design of the assembler station are presented. The economic viability of alternative system configurations are compared. Results from specific product configuration studies are presented as well as data from a single station working programmable assembly machine.

INTRODUCTION

The industrial robots in use today are minimal programmable devices that perform simple, repetitive operations requiring little communication with their working environment. They are ideal for the material handling jobs where the environment presents a danger to human beings. Conditions involving excess dust, radiation, and extreme temperatures are locations for these simple robots.

There are approximately 250 different variations of robots manufactured worldwide today (Reference I). Their weight capacity ranges from a couple of ounces to several hundred pounds. The reach of these devices varies from one inch to I0 feet with degrees of freedom of as little as two up to a maximum of seven. The majority of the robots built today are powered by hydraulic actuators. All have a simple type of memory built in for the initial teaching phase of the robot operation.

Since the passage of the Occupational Health and Safety Act of I970, employers have been trying to replace men by machines in hazardous assembly jobs. At the present time, only very simple material handling is performed by industrial robots. For assembly work, a new type of robot has to be developed. This device is called a programmable assembler. It will have the ability to be programmed to perform major assembly operations similar to what people do now. It will have sensors that enable it to communicate with its environment. For example, a sensor in a system that measures force could be used to automatically readjust the position of the assembler during assembly and also report the force encountered. With a record of the contact forces, the operators of the assembly system can determine if there is a trend of part tolerance as production time increases. This can help stop major breakdown or recall of assemblies due to bad piece parts.

Products are presently assemblied by two modes of production. The products are either produced in batches of hundreds or thousands (such as clock radios) or have millions of identical items produced continuously (such as with simple ball point pens). The method of batch production is used when products contain models similar in nature but vary in some specific property. Products of this type are grouped together and called a product family.

Fixed automation is well suited for the production of the products in the millions. This method of production cannot handle the variation contained in product families (especially size and number of part variations) easily to make it an economical method for assembly. The intelligent automatic assembly system (or programmable assembly system) would be suited for the assembly of such products. Figure 1 shows a prototype single station programmable assembly system. This station was designed to assemble an automobile alternator.

Figure 1. Prototype Programmable Assembly Station

PRODUCT ASSEMBLY STUDIES

When a product is selected for automatic assembly, a method must be developed for obtaining relevant data necessary for the sizing of the system. The sizing of a system includes determining the degrees of freedom, the speed, and the reach an assembler needs to accomplish assembly of products. Also involved in the initial sizing of a system is the amount of tools needed for production of the assemblies under study. Certain initial facts must be known for this sizing; such as the number of parts to be handled, the type of operations involved, and the number and location of directions of approach needed for assembly. These facts help in the sizing of a system.

The necessary data for product statistics analysis for the initial system sizing is accomplished by use of an assembly flow sequence diagram (Reference 2). This diagram is similar to a flow chart in computer programming in that it maps flow of information through various operations until a final assembly emerges at the end.

The procedure for the construction of this diagram is:
1. Identify each of the piece parts that are to be assembled.
2. Identify and number the assembly axes of the product.
3. Identify and number all the directions which the assembly axes lie along.
4. Construct a sequence flow diagram by the assembly of each part onto the next until the final assembly is reached.

Any new axes and/or directions encountered in the sequence flow diagram are recorded for the product statistics study. When subassemblies are encountered in this step, they are identified by capital letters inside a triangle. Different sequences of assembly for the same product are analyzed by separate flow diagrams for each sequence.

Once the flow diagram is constructed for a product, the data is analyzed to size the assembler system to produce a product. The number of axes per direction is analyzed first since there can be several axes in one direction. The number of occurrences of each operation is next totaled for each direction on a particular axis, since there can be several directions on each axis, and related to the total number of occurrences for that axis. This information can be used to determine the minimum number of degrees of freedom an assembler must have for each approach direction.

For example, one axis per direction allows one to consider a minimum design of a one degree of freedom device. With two axes per direction, the minimum consideration is a two degree of freedom device. With three or more axes per direction, a minimum configuration could be a three degrees of freedom device. When there is more than one direction associated with each axis, this leads to either an assembler with at least three degrees of freedom or a combination of two or more lesser degrees of freedom assembler. The question as to which configuration is better economically is dealt with later. Figure 2 is the assembly flow sequence diagram for the alternator being assembled by the prototype programmable assembler.

The data acquired by examing the distribution of axes per direction gives a first step in sizing an assembly system. The types of operations encountered in the assembly of a product also influences the sizing of the system. Table I lists the operations encountered in the analysis of ten assemblies (Reference 2) and the effect these operations have on the sizing of the assembler system.

FIGURE 2

ALTERNATOR ASSEMBLY SEQUENCE

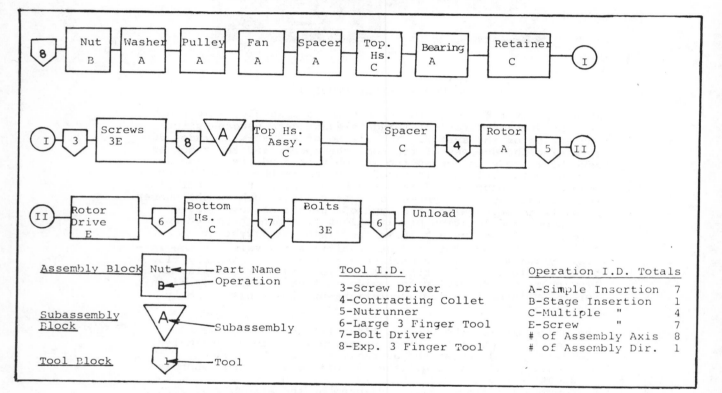

The normalized data from these products as for the distribution of parts, operations and axes per direction can be found in Figures 3 and 4. Figure 3 shows that for the products analyzed approximately two thirds of all the operations, parts, and axes are in one direction (D1). This is three times the percentage of the next direction (D2). Figure 4 indicates that approximately one third of all operations are simple insertions and 80% of them are in one direction (D1). One fourth of the operations are screw insertions with 50% in one direction. Of the remaining 42% of the operations, four operations account for 95% of them. These operations are the stage insertions, the multiple insertions, the perpendicular insertions and the interference insertion.

With the examination of the the data presented there is strong indication that a generalized group of three or four degree of freedom machines will be able to assemble the products analyzed if at appropriate times a flip operation is used. This statement assumes that the piece parts are presented to the assembler preoriented. This is a valid assumption since it was shown in Reference 3 that a majority of simple piece parts could be oriented by very simple means.

Pieces jamming during the actual operation of trying to mate two parts is a major cause of automatic assembly failure. The relationship between circular peg clearance and hole diameter is called the clearance ratio (C). This is written as:

$$C = \frac{D-d}{D} \qquad\qquad 1$$

where D is the hole diameter and d is the peg diameter. The wobble angle (θ) is defined as the angular offset between the axes of a hole and a peg. The relationship between clearance ratio (C), hole diameter (D), length of insertion (ℓ) and wobble angle (θ) is shown in Figure 5a and by the equation (Reference 4):

$$\frac{\ell}{D} = \frac{C}{\theta} \qquad\qquad 2$$

During the process of insertion of mating parts, a jammed condition can result. This jammed condition occurs when(Reference 4):

    1. Two point contact occurs, and;

    2. A system of forces is applied to a peg which falls in Region II of Figure 5b.

Wedging is a type of jamming that happens when two parts are pushed together in a particular way which results in them becoming fused. An excess amount of force, greater than the parts' own weight, is needed to dismantle them. The wedge condition occurs when:

TABLE I

OPERATIONS ENCOUNTERED IN THE ASSEMBLY OF SAMPLE PRODUCTS & THEIR EFFECT ON CONFIGURATION SIZING

Operation I.D.	Operation Description	Example Occurrence	Effecting D.O.F. of System	Effecting Special Tooling Reqmnts.	Non Assy Operation
a	Simple Insertion	Part onto fixture	*		
b	Stage Insertion	Gear meshing	*		
c	Multiple Ins.	Cylinder head gasket	*	*	
d	Perpendicular Ins.	Spring loaded trigger	*		
e	Screw Insertion	Screw into nuts		*	
f	Interference Ins.	Press fits	*		
g	Part Removal	Locating pins	*		*
h	Test Operation	Compression Check		*	*
i	Flip Operation	180° Flip	*		*
j	Part Placement & Hold	Very loose parts		*	
k	Crimp Operation	Sheet Metal Parts		*	
l	Release Hold	Very loose parts		*	*
m	Weld on Solder	Wires onto motors		*	

    1. The two conditions for jamming occur, and;

    2. $\dfrac{1}{D\mu} \leq 1 \qquad\qquad 3$

where μ is the coefficient of friction

By combining Equations 2 and 3, a critical wobble angle (θ_{crit}) can be determined for wedging to be avoided. This equation is:

$$\theta_{crit} \leq \frac{C}{\mu}$$

In both the wedged and the jammed conditions, the clearance ratio has a strong role in the success of insertion operations. These relationships can be used to determine maximum angular error allowable in assembly.

The process of inserting screws can also be described in the terms of a wobble angle. From simple geometry θ, the cross thread angle, is related to the basic major screw diameter(D) and pitch (P) by the relation:

$$\theta = \tan^{-1}(P/D)$$

This cross thread angle (θ) is also the misalignment angle the screw has at the start of cross threading. The cross thread angle θ is plotted versus diameter for a number of screw series (Figure 6) to determine the trend of θ with diameter. The coarse series screws also are the screws with the largest θ for a given diameter because of the larger relative pitch, and hence are the hardest to cross thread.

FIGURE 3

NORMALIZATION OF 10 PRODUCTS ASSEMBLY STATISTICS

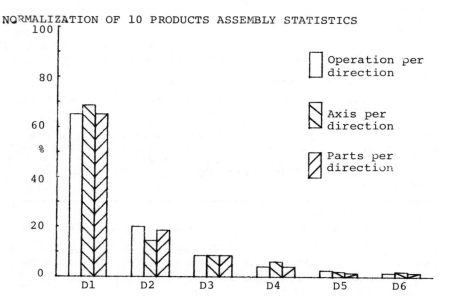

FIGURE 4

NORMALIZATION DATA OF SPECIFIC OPERATIONS FOR 10 PRODUCTS

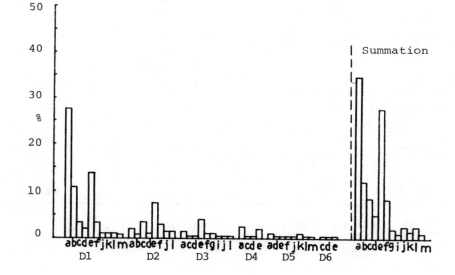

The clearances to which parts are designed have two constraints on them. Functional constraints; the first as in the case of hydrostatic bearings, are determined by the physical laws of science which govern their performance. Manufacturing constraints are the second type of constraints that govern the design of the assembly. Different manufacturing processes working on the same size part will yield different tolerances by nature of the process. The tolerance inherent to a particular manufacturing process will govern the overall clearance involved in any assembly.

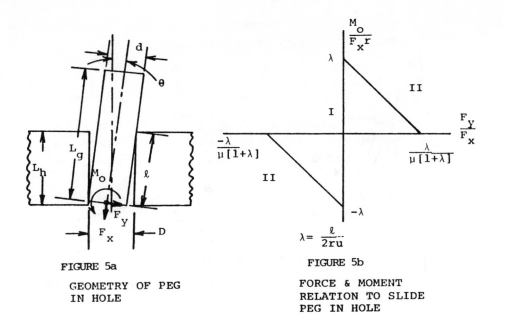

FIGURE 5a

GEOMETRY OF PEG
IN HOLE

FIGURE 5b

FORCE & MOMENT
RELATION TO SLIDE
PEG IN HOLE

To determine what the common range of clearance ratio is, various part assemblies were examined. The ball bearing fit was chosen to represent a lower bound, smallest clearance, while the washer fits were chosen to represent an upper bound on clearance. Assemblies chosen to represent an upper bound on clearance values were the bronze busing bearing and the automobile valve and camshaft fits. Figure 7 has these regions shown as well regions described by Conway (Reference 5) with respect to various types of running class bearing clearances. For a one-inch diameter hole, the clearance ratio can vary from a value of 0.00025 for a ball bearing assembly to a value of 0.054 for a washer assembly.

This range of clearance ratio will determine the amount of angular and lateral accuracy needed in the automatic assembly system proposed to produce the members of a product family by the relationship of the part parameters in Equation (2). See Reference 2 for a more detailed explanation of this process.

ECONOMIC MODEL COMPARISON

It was shown in References 2 and 6 that for a simple broad level model, the comparison index between the various assembly alternatives (manual, fixed automation and programmable assembly) is the unit assembly price of the product.

The total programmable cost per unit (PCPUI) is the sum of the programmable equipment cost per unit (PECPU) plus the programmable labor cost per unit (PLCPU) or:

$$PCPUI = PLCPU + PECPU$$

The PLCPU can be written as:

$$PLCPU = \frac{APLCST}{VOL}$$

where APLCST is annual labor cost for workers ($/year) for programmable systems and VOL is the annual production volume of the system.

The PECPU is defined as:

$$PECPU = \frac{TPSCST}{VOL * PAYPER}$$

where PAYPER is the required payback period in years and TPSCST is the total cost of the programmable system. Payback period is a simple and rough way of indicating the rate of return needed on an investment. Two years or less in payback is the usual goal. Payback period is the answer arrived at after accountants analyzed in depth the proposal of a new system. Here,

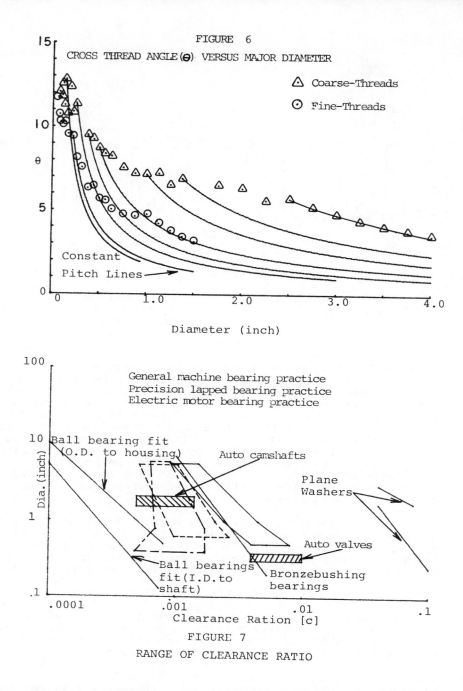

FIGURE 6

CROSS THREAD ANGLE (Θ) VERSUS MAJOR DIAMETER

△ Coarse-Threads

⊙ Fine-Threads

Constant Pitch Lines

Diameter (inch)

FIGURE 7

RANGE OF CLEARANCE RATIO

payback period of two years is used to study the economic possibilities of programmable systems compared with other alternatives of production. The cost of transfer machines was modeled, in Reference 6, as proportional to the number of parts it assembles (NPARTA). This is written as:

TSCST = NPARTA * TMCPP

where TSCST is the system cost for transfer machine and TMCPP is the transfer machine cost per part ($/part).

The transfer machine cost per unit (TCPU) is:

$$TCPU \ = \ \frac{NPARTA \ * \ TMCPP}{PAYPER \ * \ VOL}$$

A manual line cost per unit (MALCU) can be expressed as:

$$MALCU \ = \ \frac{MASCT \ * \ NSTM}{VOL \ * \ PAYPER}$$

where:

NSTM is the number of manual stations
MASCT is the manual station price.

The manual labor cost per unit (MLPU) is modeled as:

$$MLPU = \frac{NWPS * LBCST * NSH}{VOL}$$

where:

NWPS = number of workers per shift

LBSCT = labor cost per worker ($/year)

NSH = number of shifts worked (per day)

The total manual assembly cost (MCPU) per year is then:

$$MCPU = MALCU + MLPU$$

From the economic models, the system with the lowest assembly cost per unit is the system to consider for production of a product, assuming there are no other considerations such as desire for flexibility or desire to use old and traditional methods.

SYSTEM CONFIGURATION COMPARISON

Reference 2 gives an explanation comparison technique. Basically two possibilities for the production facility exist, either to construct several identical systems of low capacity or a few systems with higher production capacity.

The high volume system contains a number of multidegree of freedom assemblers in a series arrangement. In this system, each assembler does a portion, usually two or three operations, of the entire assembly. Such a system could be designed to produce the total annual volume of the product. These systems, with several assemblers arranged in a series configuration, are prone to total system stoppage due to parts jamming in one station than lower capacity systems, unless buffer storage is provided between work stations.

Production facilities made up of low volume systems use several duplicate (parallel) systems to meet the annual production volume. A low volume system would contain one or two assemblers, each doing 50% to 100% of the operations required for assembly. Low capacity systems usually lose their economic competitive advantage when copied several times over.

Figure 8 is a series line approach to the assembly of the automobile alternator. The yearly production volume of this system was estimated to be 625,000 units a year working on a two shift basis. To meet the desired annual production volume of 1.2 million, 2 systems would be needed. The PECPU was estimated to be .30$/unit while the PLCPU was estimated at .06$/unit. This yielded a PEPUI of .36$/unit.

Figure 1 is a prototype of a single arm parallel line approach to the alternator assembly. This configuration has assembled alternators in approximately 160 seconds. This yields a yearly production volume of 71,800 units. To meet the annual production volume, 17 duplicate systems would be needed. The PEPUI of this system was calcuated to be .69$/unit with .56$/unit going to the PECPU and .13$/unit going to the PLCPU.

The present method of production is a semi-automatic line. The total cost per unit of the present system was estimated to be .44$/unit with .07$/unit going to TCPU and .37$/unit going to MCPU.

CONCLUSIONS

By comparing the high and low volume approach to production with the present method, the PEPUI of the high volume approach was lower than the present method and the lower volume system approach. This can be traced to the required number of duplicate systems the low volume approach needed to meet the annual production.

Reducing the number of duplicate system required without increasing significantly the cost of a single system would be a way of lowering the PEPUI. Reducing the percentage the assembler is doing non assembly operations (tool changes) will

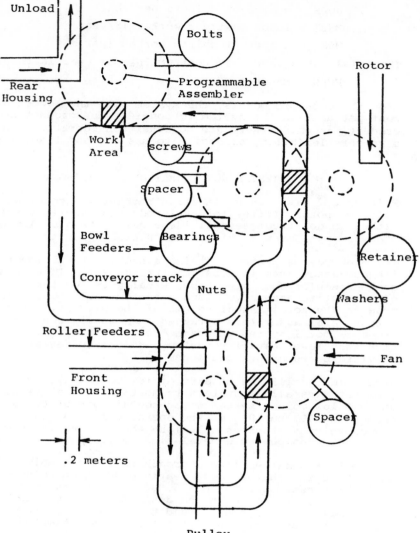

FIGURE 8
SERIES LINE PRODUCTION
FOR ALTERNATOR ASSEMBLY

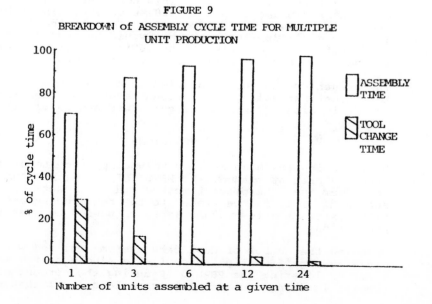

FIGURE 9
BREAKDOWN of ASSEMBLY CYCLE TIME FOR MULTIPLE
UNIT PRODUCTION

increase the annual production of any system. One method of reducing this percentage is by distributing the tool change time over many units. Figure 9 shows the results of this procedure. The unit assembly time of the product is reduced from 160 to 112 when 12 units are built at once by this method or only 12 duplicate systems are needed to meet the annual volume instead of 17. The PEPUI for this procedure of assembly is .50$/unit with .40$/unit going to PECPU and .10$/unit going to PLCPU. This shows that programmable assembly does have a future in being compatible with the present method of assembly.

ACKNOWLEDGEMENTS

The author would like to thank his fellow members of Group 10D for their advice and help in preparation of this paper: Daniel Whitney, Don Seltzer, Mike Lynch, Sam Drake, Sergio Simunovic, Dan Killoran, Don Wang, Jim Nevins and especially to Linda Martinez for her expert typing. This research was sponsored by NSF Contract Number APR74-18173-A03.

REFERENCES

1. Abraham, R.G., and N. Yaroshuk, "Advanced Robotics", Mechanical Engineering, Volume 97, Number 12, December 1975.

2. Kondoleon, A.S., "Application of a Technological-Economic Model of Assembly Techniques to Programmable Assembly Machine Configuration", M.S. Thesis, M.E., M.I.T., May 1976.

3. Boothroyd, G., and A.H. Redford, Mechanized Assembly, McGraw-Hill, London, 1968.

4. Simunovic, S., "Force Information in Assembly Processes," Fifth International Symposium on Industrial Robots, September 1975.

5. Conway, H.G., Engineering Tolerances, Pitman and Sons Ltd, London, 1966.

6. Lynch, P.M., "Economic Design Criteria for Programmable Assembly," Mechanical Engineering PhD Thesis, M.I.T., Cambridge, Mass, August 1976.

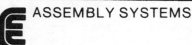
HOUSTON Jan. 28, 1979—In Japan a new era of unmanned, fully automated factories, so flexible that they can readily be reprogrammed to turn out different products, is in the offing, far in advance of similar developments in other countries. *New York Times News Service*

This grim news for the U.S. manufacturing community came from the 1979 annual meeting of the American Association for the Advancement of Science. Various speakers at this meeting pointed out some other disturbing Japanese developments:

• They now have more "flexible manufacturing systems" than all the rest of the world combined according to Harvey E. Buffum, director of operations technology for Boeing Commercial Airplane Co.

• "Japan is the only country close to perfecting such a system," according to Eugene Merchant of Cincinnati Milacron, Inc.

• The Japanese have more robots "on the job" than any other nation, according to Frederick L. Haynes, assistant director of the U.S. General Accounting Office.

From the preceding statements, a challenge to U.S. industry clearly exists—either change or lose whatever remains of our competitive position in the world marketplace. Batch manufacturing currently accounts for over 75% of all manufacturing in the U.S., and flexible automation, i.e. robotics, could help increase productivity in this area by a 3:1 factor—certainly worth investigating by any standards. Fortunately, two major U.S. corporations, Westinghouse Electric Corp. and General Motors Corp., recently described major robotic systems that may both begin to meet this challenge and stimulate U.S. batch manufacturing efforts to expand usage of robotic assembly systems.

GM: Now

General Motors Corp., at the GM Technical Center in Detroit, MI, recently demonstrated some new generations of sophisticated computer-controlled robots, both with and without vision, that they expect to significantly increase manufacturing efficiency. Alex C. Mair, vice president of GM and group executive in charge of Technical Staffs Group, described some motivating factors behind the research and development to increase productivity. "Keeping America in the forefront of industrial technology is one of the best ways of assuring high employment and this country's economic well-being in years to come.

"With other highly industrialized countries—notably Japan and West Germany—aggressively pursuing and applying new technology, U.S. industrial productivity must increase if we are to maintain or improve our competitive position in world markets," he said.

Noting that it now is technically possible to equip robots with a number of sensory capabilities besides vision, Mair explained that GM is concentrating on vision first because robots with visual guidance have the greatest potential for automotive application.

"It's now an accepted fact that the use of robots to perform certain manufacturing and assembly operations will reduce costs. It's also a fact that robots can handle many types of jobs which aren't particularly enjoyed by people—because the task may be repetitive, boresome or fatiguing. Operations which involve extremes of temperature, noise, chemicals or handling of hot or dirty material are prime candidates for robot application.

"We are carefully reviewing our manufacturing and assembly operations to see where these new robots may be introduced to work alongside human beings or where they can take over other jobs, guided, programmed and maintained by human beings." However, he added that he did not foresee any sudden, large-scale introduction of robots into GM plants although they currently have about 150 robots on the job worldwide.

"I don't believe anyone can say precisely what effects this advanced technology will have on our work force in the future. But we do know from past experience that technological advances have enabled the automobile industry to increase employment and remain competitive."

PUMA is an acronym for two related products. Programmable Universal *Machine* for Assembly (PUMA) is the GM designation for its new complete assembly system, including robot arms. The arms were developed jointly with Unimation, Inc., Danbury CT, which calls them Programmable Universal *Manipulator* for Assembly (PUMA™) and offers them as standard products.

The PUMA system was demonstrated by Walter Cwycyshyn, senior

Assembly Robot Update

GM recently demonstrated some innovative robotic assembly systems and the Westinghouse Adaptable Programmable Assembly System is entering Phase II.

This PUMA (Programmable Universal Machine for Assembly) robot was conceived by GM Manufacturing Development engineers in response to the need for a relatively small and inexpensive unit capable of handling simple assembly of small, lightweight automotive components (about 90% of the parts in a GM car weigh less than 5 pounds). In this laboratory demonstration, the PUMA is installing 10 to 15 light bulbs in the backs of instrument panel cluster cases. Although not presently equipped with vision capability, the PUMA is adaptable to visual and other sensing signals. The PUMA robot is seen as part of an assembly system that would include parts feeders, conveyors and people. GM plans its first production installation of a PUMA robot late this spring at the Delco Products Division, Rochester, N.Y.

staff project engineer of the Assembly Processing Department. The PUMA system is a combination of PUMA™ robots, transfer devices and parts feeders' which work alongside employees in an asembly operation. He said that "The latest and probably the most advanced robot on the world scene today is the PUMA robot designed to operate with [these] conveying systems, parts feeders, and people in an integrated system for small component assembly. This robot is designed to handle parts roughly smaller than a bread box and weighing less than 2.3 kg. This package represents about 90% of all components on a GM passenger car.

"Although this robot is adaptable to control by visual and other sensing signals, the first series of installations will depend on dead reckoning and accurate fixturing. Once the reliability and durability of the robot have been proved, further efforts will be directed toward refinement of the system and implementation of true adaptive control by means of visual, acoustic, and other proximity feedback devices."

Ten PUMA systems will be phased into GM component and vehicle assembly operations this year and they are sophisticated enough to accept computer vision guidance when needed for future work applications.

Vision was discussed by Frank Daley, director of Manufacturing Development. He said, "We think the idea of equipping robots with cameras and computers to give them vision is going to open many new avenues to increased productivity. [GM's first industrial application for a computer vision system was SIGHT-I at the Delco Electronics Div., Kokomo, IN.[1]] We expect that our next application of computer vision will be at a GM Central Foundry Division location.

"This pilot application will be followed by many similar applications of machine perception in assembly and material handling. In assembly, vision can recognize the location and orientation of parts prior to their assembly by automatic mechanisms, then assure that the operation has been properly performed, and that details are in place." Daley is convinced that the technology base is here now for programmable and flexible assembly.

Donald E. Hart, head of the Computer Science Department, GM Research Laboratories, described the vision system that permits a robot, directed by a computer using a camera for "eyes," to pick up unoriented parts from a moving conveyor belt and place them where needed.

"A special lighting system was invented to take advantage of the fact that the parts are on a moving conveyor belt," Hart said. "A camera is focused on a line of light projected on the conveyor belt. Where there is no part, it sees light. Where a part blocks the light, it sees dark. With the camera frequently scanning along this line as the part moves past, the computer is able to form a silhouette of the part."

The control computer is programmed so that all six of the robot joints are controlled simultaneously allowing the robot's hand to smoothly follow the proper path and intercept the moving part.

Lothar Rossol, staff research engineer of the Computer Science Department, described research on

advanced vision systems which are capable of solving more difficult problems. He said, "Tomorrow's robots will have two senses: vision and touch. In a limited way, these second generation robots will be able to *sense* and to *react* to their environment. That is, their path will change on each cycle, as the position of parts changes or as the requirements of the job change. Adaptive robots will be used for part handling, for inspection, and for assembly. They'll transfer parts form one machine to another and will not require exact part positioning. They'll automatically inspect parts. And they'll assemble parts and adapt to changing conditions by looking at the parts and touching them and measuring forces. Adaptive robots, however, require new advanced vision systems."

Rossol continued, "CONSIGHT is a [Cincinnati Milacron] robot with a vision system that represents the state of the art in GM today. It can pick isolated parts off a moving conveyor belt. It cannot yet recognize parts that overlap or touch each other. One of the major needs of industry, however, is for systems that can pick a part out of a jumbled heap on a belt or from a bin of parts. No one knows how to do that yet."

CONSIGHT starts without knowledge of a particular part. Computer models are formed by showing it a part so that it can take an instantaneous view and form a digitized picture to record in memory. This picture is used to find edges, straight lines, and curves. The computer is asked to associate a part name with a given image. Then it can locate a similar part in a nonoverlapping group. Overlapping parts can be located with a variation of this method. This is a much more complicated arrangement since the overlapping parts give erroneous edges for the program to compare. The model is called up and the search begins. As each part is located, the program looks at the model again and searches for another part. When all parts are found and no others can be recognized, the program concludes and identifies all five parts and their positions.

As impressive as this system is—and it represents some of the most advanced work in the world today—the vision system still has two major shortcomings and cannot yet be used as an eye for a robot. According to Rossol,

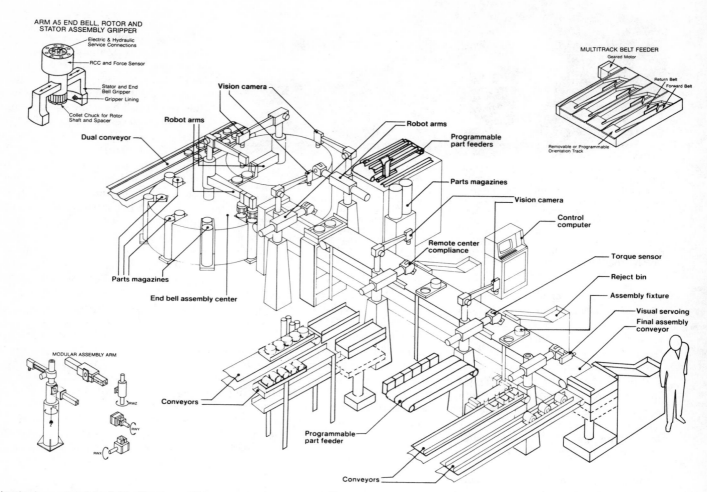

ARM A5 END BELL, ROTOR AND
STATOR ASSEMBLY GRIPPER
Electric & Hydraulic
Service Connections
RCC and Force Sensor
Stator and End
Bell Gripper
Gripper Lining
Collet Chuck for Rotor
Shaft and Spacer

Vision camera

Robot arms

Dual conveyor

Parts magazines

End bell assembly center

MODULAR ASSEMBLY ARM

RWZ
RWY
RWX

Conveyors

Robot arms

Programmable
part feeders

Parts magazines

Remote center
compliance

Programmable
part feeder

Conveyors

MULTITRACK BELT FEEDER
Geared Motor

Return Belt
Forward Belt

Removable or Programmable
Orientation Track

Vision camera

Control
computer

Torque sensor

Reject bin

Assembly fixture

Visual servoing
Final assembly
conveyor

Westinghouse's Adaptable-Programmable Assembly System (APAS) concept is shown. Insets show the configurations of the modular assembly arm, a universal gripper for one station and the multitrack belt feeder. The modular arm concept minimizes costs by using only those modules required for a specific task and also simplifies spare parts and maintenance considerations. APAS components also include force/tactile sensing systems, parts presentation/fixturing equipment and computer hardware/software. Robot arms in the system range from high-speed, 3 axis, fixed sequence versions to 5 axis servo-controlled arms. Vision systems are used extensively for incoming inspection, parts orientation and inspection for correct assembly. Programmable parts feeders, magazines and belt conveyors and remote center compliance (RCC) devices for insertion with visual serving are key technological innovations in the system.

"First: it has no concept of depth—it does not, for example, know which part is on top and which one is on the bottom. Second: it takes too long—even on a large and expensive computer—because the vast amount of data generated by TV cameras simply overwhelms most computers. The solution to these two problems will make the vision system practical."

Westinghouse: Soon

Westinghouse Electric Corp. announced that it received a second award from the National Science Foundation (NSF) to proceed with Phase II of a man/machine program that could increase productivity in low-volume or batch manufacturing by 3:1. Batch manufacturing today accounts for 75%

of all manufacturing in the U.S., which in turn accounts for 30% of GNP.

Richard Abraham, manager of programmable automation at the Westinghouse Research and Development Center, said that the experiment is meant to evaluate the techniques and economics of an automated assembly line to manufacture small motors, a product line selected as representative of batch manufacturing and also as offering significant potential economic benefits. Thus far, no one—in this country or abroad—has worked out a complete adaptable programmable assembly system (APAS) for batch manufacturing. "Programmable" means that the system can change over and assemble different product line styles. "Adaptable" means that it can

adjust to manufacturing and part dimensional variations.

Phase I determined that APAS research results of related NSF-funded programs, such as those at SRI International, the University of Massachusetts and Charles Stark Draper Laboratory could be used to increase the productivity of batch assembly operations. In addition, the research revealed that it was possible to develop an experiment that could provide industry with the necessary data to accelerate technology transfer. Phase II is that experiment. It is scheduled to be completed, with attendant hardware and sofware, in late 1981.

The results of Phase II will be disseminated widely, through film, reports and demonstrations of the pilot system in operation, to hasten U.S. applications of this technological advance.

"For years," according to Abraham, "the National Science Foundation has supported production research and technology programs aimed at productivity improvements in batch manufacturing, with particular emphasis on labor-intensive assembly operations.

"During this program, Westinghouse analyzed 60 different product lines and selected small electric motors as the one most representative of batch manufacturing.[2] The small motor product line has 450 different styles (represented by the eight classes) and over 30 different parts, each of which has numerous variations. The average batch size is 600 and, on the average, 13 changeovers per day are required. The assembly cycle time requirement is 20 seconds.

"Since fixed automation techniques cannot be applied to product lines made in small batches due to the difficult and costly changeover requirements, an extensive, worldwide state-of-the-art review of automated programmable assembly system technology was conducted.[3] The review included all parts of a programmable assembly system: robot arms, parts presentation equipment, fixtures and tools, transfer conveyors, vision and other sensory system hardware, computer system hardware and software, special equipment and people. It was concluded that no one in the world has a complete automatic programmable assembly system (APAS) for batch manufacturing. Furthermore, no quantitative economic data on productivity improvement, product quality improvement, work-in-process inventory reduction, etc. due to utilization of APAS exists. Laboratory experiments on parts of an APAS system (e.g., vision at SRI International and insertion at Draper Labs) have not been adequate for wide-spread acceptance and use of APAS in industry. Vision system experiments have not been concerned with speed or lighting requirements of production; different algorithms may be required. Successful insertion has been accomplished at Draper Labs using compliance devices on a limited number of parts.[4] Programmable parts presenters, fixtures and tooling, and software for managing changeover of a complete programmable assembly system have not been developed. An integration of all subsystems into a complete operational programmable assembly system is necessary, as is a practical proof-of-concept experiment using large numbers of parts for a product line with many styles."

The planned system consists of many computer-controlled work stations that make up a complete assembly line. The system includes modular robot arms, programmable equipment for presenting parts to the assembly area, fixtures and tools, transfer conveyors, vision and other sensory hardware, computer system hardware and software, special equipment and people. Compliance devices such as those from Draper Labs will enable the robots to complete insertion operations without sophisticated and expensive closed-loop force feedback systems. Westinghouse vision research has accomplished a 5:1 reduction in software execution time over the SRI results.

"When we consider," Abraham said, "that 75% of our manufacturing total is accounted for by less efficient, batch manufacturing methods involving batches of 50 or less parts, and that products made in this fashion cost up to 30 times more than if they were mass produced, the economic potential of an APAS system appears significant. It the experimental data from this project support the economical feasibility of an APAS, manufacturers could apply the techniques and results obtained to increase volume and produce lower cost products that would compete successfully with the Japanese and West Germans."

According to Abraham, "It has been shown that the technology base exists for successful design of a complete integrated adaptable programmable assembly system. Novel modular manipulator arms, microcomputer-based binary image processing vision systems, programmable parts presenters and fixtures, and end effectors that are universal for the set of operations at individual stations have been conceived. We are in a position to proceed with pilot system construction and software development and then commence a proof-of-concept experimental program to objectively evaluate technical and economic performance of APAS. A successful cost-effective demonstration that the system can accomplish timely changeover for a variety of part and product styles and can handle parts which contain all the variability in dimensions and tolerances associated with a manufacturing plant environment will accelerate APAS technology transfer to industry, thereby improving productivity of batch assembly operations."

When Abraham was asked to compare APAS sophistication with that of the Japanese modular factory concept, he commented that the Japanese system is much more comprehensive. It includes the entire manufacturing operation from production planning through machining, fabricating and assembly plus testing at all levels of assembly/manufacture. As mentioned earlier, the Westinghouse APAS project has a NSF government grant of $1,386,288 plus $450,000 private funding from Westinghouse over a 39-month period. The Japanese government is 100% funding their modular factory with $50,000,000 over a 5 year period. Japan's target completion date is in the mid-1980's. Despite the vast differences in funding, Westinghouse feels that by concentrating on the labor intensive assembly areas, significant results will be obtained. They are concentrating on the high risk technology first and will supplement it with additional research as needed.

Although the initial project should produce technology transfer benefits for higher volume batch operations, the goal is to produce results that are both applicable and affordable for small firms with lower volumes.

Conclusion

Although the GM and Westinghouse programs differ considerably, both are in agreement that (1) vision is the most important sensory ability to incorporate in an adaptive robotic system, (2) the technology base is here for visual and tactile sensing, (3) their approach is to work closely with outside suppliers of robotic equipment during their research and development stages and (4) robots offer a practical and cost effective means of competing in the world market. Statistics show that technological innovation is the largest identifiable cause for productivity improvement and increased employment. Robots are here and useful. What is your company waiting for? —*TT*

References:

1 T. Thompson, "'I see,' said The Robot," *Assembly Engineering,* April, 1978.
2 R.G. Abraham, "Programmable Automation of Batch Assembly Operations," *The Industrial Robot,* Vol. 4, No. 3, (pp. 119-131), International Fluidics Services, Ltd., England, September, 1977.
3 R.G. Abraham, R.J.S. Stewart, L.Y. Shum, *State-of-The-Art in Adaptable-Programmable Assembly Systems,* Society of Manufacturing Engineers Report No. MSR 77-16, Dearborn, Michigan, 1977.
4 J.L. Nevins, D.E. Whitney, "Computer-Controlled Assembly," *Scientific American,* Vol. 238, No. 2, (pp. 62-74), February, 1978.

Swiftly moving from the laboratory to the shop floor for assembly, machine loading and other production assignments is the PUMA robot, possibly the most flexible industrial robot available. It features high repeatability, easy programming and quick changeover capabilities

PUMA Robots Find Cost Effective Jobs In Batch Manufacturing

The PUMA robot is the first truly programmable, computer controlled robot arm specifically designed for close tolerance assembly operations and other work on small parts. Commercially introduced in 1978 by Unimation Inc., Danbury, CT, it closely approximates a human arm in function and size. This allows the sophisticated unit to occupy about the same space as a human. But more

than that, the PUMA robot holds the promise of making robot technology economically applicable in light, batch manufacturing where dedicated automation cannot be applied and where more expensive, less flexible robots historically have failed rigorous return on investment analysis.

It is clear that the PUMA robot no longer is a laboratory curiosity. Deliveries began early this year.

Scores of manufacturers are working to install the robot arms. Unimation reports more than 50 orders have been placed.

The term "PUMA" has dual meaning. The acronym was coined by General Motors Corp. to stand for Programmable Universal Machine for Assembly—a new flexible assembly system which is made up of conveyors, robots and humans. The robot came to be known as PUMA because of its close association with GM's PUMA system. PUMA now is the tradename used by Unimation.

What Makes The PUMA Robot Different? Configuration and programming flexibility are two areas where the PUMA robot is different from other robots in its class. The PUMA robot features five revolute axes of motion, corresponding to a human's waist, shoulder, elbow, wrist and hand rotation.

Microprocessor-controlled electric servos position the robot arm. Overall control is by means of an LSI-11 microcomputer which allows the robot to be programmed using a spacial X-Y-Z and joint angle coordinate system—a powerful and unique feature of the robot. The robot can be programmed using either a "teach module" or an optional computer terminal. Both methods can be simultaneously employed where appropriate. Programming is effected using Unimations's VAL computer language—a high level language with access codes and commands in English.

The PUMA robot is extremely accurate—Unimation pegs repeatability at plus or minus .004 in. although tests at GM indicate much better repeatability. The load capacity of the pneumatically-operated gripper/end effector is rated at about five lbs, including the end effector. The unit has shown the capacity to handle loads close to seven lbs, however. Arm-tip velocity is quoted at 3.3 fps with maximum load.

A Long Time Coming. The PUMA robot was conceived nearly a decade ago at M.I.T. and Stanford University. But, it wasn't until Unimation and General Motors joined forces to develop a robot that the PUMA concept was shaped into a viable production tool. Richard Beecher, department head, Assembly Systems, Manufacturing

The PUMA system—a conceptual view of GM's stand alone batch assembly system consisting of conveyor, parts feeding devices, work holding fixtures, one or more PUMA robots and human workers. Mock-ups of such systems are now operating at GM's Technical Center in Warren, MI

Development, GM Technical Center, explains the history of PUMA: "A little over three years ago, we began investigating the feasibility of performing assembly operations using programmable devices. Our work was spurred by the need to increase mechanization of mid-volume, batch-type assembly jobs in which changeover times are long and costly and labor expense is high. The need for a flexible, relatively inexpensive assembly system capable of automating simple assembly tasks on light parts was highlighted by the results of an internal study which showed that 95% of auto parts weigh less than three lbs. The obvious solution was the application of industrial robots, but at the time, there were no commercially available robots which met our requirements."

As part of the project, Beecher says GM purchased two robots which came as close to company needs as anything on the market: a Japanese-made Kawasaki Unimate and a Swedish-built ASEA robot. The two robots were put to work

on a prototype assembly line at GM's Technical Center in Warren, MI. The line was designed to assemble small parts such as heater controls, brake cylinders, power seat transmissions and instrument clusters.

A New Assembly Concept. "We demonstrated the feasibility of using robots to assemble small parts," says Beecher. "But we also used the two robots as a basis of comparison in the development of performance specifications for a completely new robot which would be specially designed for assembly operations. We investigated such questions as: How fast should an assembly robot be? What kinds of motions should the robot have? What should be the work envelope of the new robot?"

The experiments culminated with the development of performance specifications. There was a need for improved repeatability—plus or minus at least .004 in. The robot needed to be light weight so it could easily be pulled off the line for repair. The robot needed to be compatible with adaptive con-

467

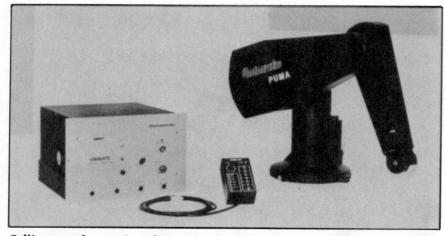

Selling at a base price of $35,000, the PUMA robot consists of a computer controller, teach pendant and five-axis manipulator. A computer terminal and a floppy disc system are optional

trols and have five degrees of freedom. And the robot had to be electric so it would be easy to move and operate quietly. In Beecher's words: "No assembly line worker wants to work eight hours next to a snarling hydraulic robot."

Floyd Holroyd, a senior project engineer at the Technical Center, reported: "Our studies produced a concept foreign to the thinking of most people in the industry at the time. We concluded that assembly robots should be capable of working side-by-side with humans on the assembly line—robots should be assigned tasks on the line much in the same way as humans are assigned specific tasks. At the time we first investigated the use of robots in assembly, most people in the industry were trying to develop highly sophisticated robots capable of assembling an entire product from start to finish. We reasoned that this approach was not workable because an expensive back-up robot would be required to keep things running in the event of a breakdown. Or equally impractical, a manual assembly line would be required as back-up. What we wanted was a small, yet highly flexible robot which easily could be moved and

reprogrammed to perform simple tasks at another location.

"The PUMA robot concept evolved out of this kind of think-

"Our studies produced a concept foreign to the thinking of most people in the industry at the time . . ."—Floyd Holroyd

---

ing. A robot built especially for assembly could be used on conventional assembly lines or be part of what we call a PUMA system—a stand alone assembly cell made up of one or more robots, a conveyor, work holding devices, parts feeding mechanisms and one or more human workers."

The Search for a Builder. Since no available robot fit GM's performance specifications, GM engineers approached nine industrial robot manufacturers with a proposal for joint development of a new robot. Only four builders responded. Of the four, Beecher says Unimation "came back with the most attractive proposal in terms of cost and delivery time."

Unimation was in an excellent position to supply the new robot. It recently had purchased Vicarm—a small, not-well-established West Coast-based robot builder whose product was a small, highly sophisticated robot arm not unlike the robot arm sought by GM. Vicarm was founded by a researcher who had helped develop experimental robot arms at Stanford.

On May 1, 1978, after 14 months of development work aimed at refining the Vicarm robot arm to suit GM specifications, the first PUMA robot was delivered to GM's Technical Center. A second prototype was kept at Unimation for further experimentation and development.

Intensive Testing. Since its delivery, the PUMA robot has "gone through intensive repeatability and durability tests." According to Beecher, the robot has performed at levels exceeding original specifications. For example, tests indicate it is capable of repeatability of plus or minus .002 in.—original specs called for repeatability of plus or minus .004 in.

Over the past year, the prototype was put to work in a variety of test production set-ups in the Machine

". . . The PUMA system, a highly flexible, economical system for the assembly of untold thousands of products . . ."—Richard Beecher

Perception Laboratories at the GM Technical Center. Planning now is underway by a number of GM divisions to assign PUMA robots to jobs on the shop floor. Five divisions will each receive at least one robot: Harrison Radiator, Buffalo, NY; Rochester Products, Rochester, NY: Delco Moraine, Dayton, OH; Oldsmobile, Lansing, MI and Delco Electronics, Kokomo, IN. Other divisions have purchased robots to experiment with; all totaled, GM will take delivery of 11 PUMA robots this year.

The first production application of a PUMA robot by GM probably will be at Delco Electronics, Kokomo, IN. It expects to use a PUMA robot to palletize radio speaker magnets. Startup should be by the end of summer. (At this time however, it appears GM will not boast the first production application of a PUMA robot. Beating it to the punch will be Chesebrough Pond's Inc. It plans to install a robot to unload cocoa butter bottles from a processing line. The robot's flexibility is the key in this installation because the line is frequently converted to produce different size and shape bottles.)

At Delco Products Div., Rochester, NY, a Puma robot will pick-up a 450 F motor armature, transfer it to a press where a commutator will be added, then place the assembly on a conveyor leading to a cure oven. This installation also is expected to be in place this year.

An application being studied involves the assembly of carburetors. A robot would be used to drop screws into a tooling plate at a high rate of speed. The plate would then automatically be shuttled under a multiple screw driver head which would torque the screws. With the current product mix, up to five different hole patterns would be used. A prototype system is now operating at the Technical Center.

Increasing Interest. Although the PUMA robot was developed for assembly operations, manufacturers

Original prototype PUMA robot at GM's Technical Center. Production mock-ups test feasibility of different applications

are discovering it has great potential for other applications. Companies like General Electric, Westinghouse, RCA and 3M have ordered PUMA robots for machine loading, pick and place, packaging

---

The value of the robot stems from increased flexibility in programming

---

and parts handling. The feasibility of using the PUMA robot for electrical harness assembly, circuit board manufacturing and spot welding also is being investigated.

Unimation is continuing to improve the PUMA robot. Plans call for introducing a model with six degrees of freedom late this year. The company also is working to develop a vision option, tactile and forcefeed back capabilities.

GM is committed to further development, too. Beecher says:

"Our most difficult assembly problem continues to be parts feeding and orienting. Initial PUMA robot applications will use dead reckoning and accurate fixturing to guarantee part pick-up and placement. With these applications successfully underway, attention will increasingly be directed toward refinement of the PUMA system and the addition of adaptive controls to the PUMA robot. Tactile and visual feedback, acoustics, lasers and a variety of proximity sensors are being evaluated.

"Potential applications for the PUMA system cover a broad spectrum of products. Each new application brings with it a unique combination of challenges. The PUMA system is not the answer to all assembly problems, any more than any other assembly system. It does have its place as a highly flexible, economical system for the assembly of untold thousands of products. The perfection and application of sensory feedback techniques will make possible even wider applications of the PUMA concept." *Thomas J. Drozda* □

Reprinted from Robotics Today, Summer 1980

SRI Reports on its Programmable Part Presenter

This mechanism takes randomly-oriented parts

and places them in a desired orientation

using computer vision and controlled tumbling.

No special tooling or fixtures are required

JOHN W. HILL
Senior Research Engineer
and
ANTONY J. SWORD
Research Engineer
Artificial Intelligence Center
SRI International

In programmable assembly systems being developed concurrently at SRI[1] and other research laboratories,[2,3] a robot picks up parts from a number of feeding stations and assembles them one by one into a product. The task is greatly simplified if each feeding station presents its part in a predetermined position and orientation for proper gripping and insertion. A cycle of one to four assemblies per minute is desirable.

Difficulties with programmable assembly arise in the required part feeders. Small parts can be fed from vibratory bowl feeders, but larger parts require very large and expensive feeders. Long changeover time for different parts is another disadvantage, requiring such proposed solutions as adjustable wipers or interchangeable track modules. Alternatives include hand-loaded magazines.[2,3] Birk and Kelley[4] are developing a robot with vision feedback to pick parts from bins. This system is programmable, but the problem of identifying and then grasping bin-contained parts is a major constraint at this time.

The major goal of the SRI programmable approach is to present parts without the need for special-purpose tooling. The method is as follows: separate out a single part on a known surface; use vision to sense stable state and orientation of the part; and reorient parts, applying previously developed techniques.

The system is based on controlled turning and tumbling. Tumbling an object can be visualized as pushing it over a step. A complete analysis of this technique for a given part is very complicated, involving such factors as the initial orientation of a part before tumbling, the points of contact as equilibrium is lost, free-fall dynamics, resiliency of a tumbling surface (bounce), and the dynamics of coming to rest on a flat surface.

A mathematical approach based on stable states and the probabilities of state transitions makes the problem tractable. Stable-state probabilities for some parts, based on random starting configurations, are given in a University of Massachusetts study.[5]

When a part in a known starting state is brought to a predetermined orientation, tumbling can become repeatable and less subject to chance. Such experiments are easily conducted by pushing parts by hand over a stack of magazines on a desk top. Though this problem requires a more

1. This overhead view shows elements of the experimental part presentation unit. Computer-controlled operation of the elevator, shuttle, turntable, and camera positions yoke castings to be properly grasped by the robot.

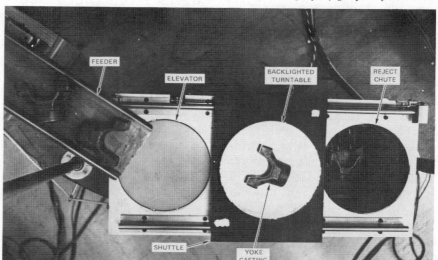

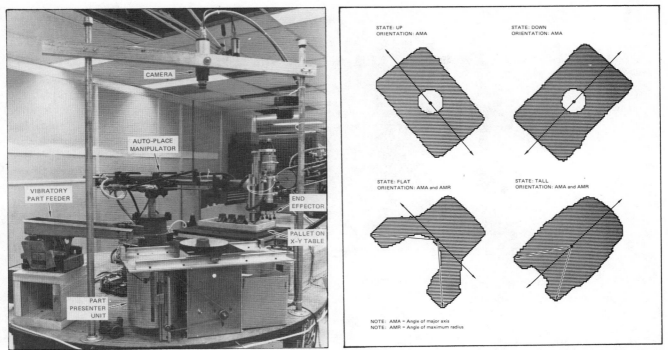

2. *At this part presenter station, the Auto-Place manipulator packs all the randomly-oriented parts neatly on the X-Y table.*

3. *Images of the four stable states of a yoke casting include only one suitable for pickup ("up"). Visual reorientation parameters are shown.*

general analysis, empirical results indicate that controlled-tumbling plans for many parts are simple to conceive and are reliable in execution.

The elements of an apparatus designed for experiments in programmable part presentation are shown in *Figure* 1. The system consists of a belt or vibratory feeder that dispenses parts individually; a movable shuttle, an elevator, and a turntable; a reject chute for undesired parts; and a solid-state camera for viewing the part on the turntable. Computer-controlled operations of the elevator, shuttle, turntable, and camera can bring a part to the desired stable state/orientation/position for grasping by a robot hand.

The three movable elements in the part presenter unit are each driven by a stepping motor. The elevator platform moves up and down, forming a bin of programmable depth of 0 to 6″ (0 to 150 mm). The translucent turntable rotates in 1.8° steps and is backlighted to present silhouetted images to a camera positioned above it. The shuttle, which transfers the part back and forth between the elevator and turntable, can also push the part to the reject chute. The shuttle and elevator each have a diameter of

7.87″ (200 mm); the shuttle has a total range of horizontal movement of 17″ (432 mm).

To experiment with part presentation, the part packing station shown in *Figure* 2 was developed. The system consists of an Eirez 20A vibratory feeder, the part presentation system, an SRI vision module with General Electric TN2200 camera, an Auto-Place Series 50 limited-sequence manipulator, and an *X-Y* table with a range of 10 x 10″ (250 x 250 mm). All the devices are interfaced to a PDP-11/40 minicomputer with 28K words of memory. The minicomputer's RT-11 operating system is currently in use, while user application programs are written in Robot Programming Language (RPL).[6]

To illustrate the flexibility of the programmable part presentation station, software was written in RPL to handle two different classes of parts—yoke castings and hydraulic couplings. These serve as examples to illustrate the general approach to part presentation.

The part presentation sequence of operation is as follows:

● *Part Acquisition.* The vibratory feeder is turned on. When a part falls onto the elevator, the elevator limit

switch is deflected and the vibratory feeder is turned off.

● *State Determination.* The shuttle pushes the part to the turntable and a picture is taken by the vision module. The image is processed to determine which of the previously-trained states the part is in, using the nearest-neighbor classification described by Agin.[7]

● *Stable-state Processing: Rejects.* If the image does not match any of the pretrained states, the shuttle pushes the part to the reject chute and returns under the feeder for a new part.

● *Stable-state Processing: Reorientation State.* If the image corresponds to one of the states to be reoriented, the part is turned to prepare it for tumbling, the elevator is lowered a predetermined amount corresponding to the existing state, and the shuttle pushes the part into the bin. The elevator is then raised and the shuttle pushes the part back to the turntable for another pass.

● *Stable-state Processing: Pickup State.* If the image corresponds to a pickup state, the part is turned to bring it to the pickup orientation. If the part needs centering, the shuttle pushes the part over the elevator and back for that purpose.

4. Each of the three parts of the hydraulic coupling has three stable states and one preferred pickup state.

whether centering is necessary.

The RPL program[8] was modified so that a mixture of the hydraulic couplings shown in *Figure 4* could be handled by the part presentation station. This set of parts differs from the yoke castings in that there are three different parts, each of which has three stable states and a single preferred pickup state. Visual images of the entire set of stable states are shown in *Figure 5*. As with the yoke castings, only three visual parameters plus state recognition are required to handle these parts.

In general, rigid parts with a finite

● *Palletizing.* Once the part is centered in the desired pickup state, the manipulator picks it up and the *X-Y* table moves to a new position. At this point, motions of the part presenter and manipulator overlap until the part is packed and the part presenter has acquired a new one. The foregoing operations are repeated until the manipulator has packed all the parts in an orderly array on the *X-Y* table.

It should be noted that this sequence is not peculiar to any one part. To handle a different part, only the code in the processing of the new state(s) needs to be modified. This modification is necessary for selection of appropriate visual features to use in reorienting the part, as well as to specify the depth of the elevator onto which the part must be dropped after turning.

Most parts can be tumbled to the desired state in one cycle. For the yoke casting, two cycles are sometimes required. If the part lands in the desired state immediately, no tumbling is required. Depending on their stable states, three or four parts per minute can be handled. Some of this speed is achieved by programming simultaneous operations of the manipulator, *X-Y* table, and part presentation unit.

A single yoke casting from an automotive universal joint can assume any of four stable states. *Figure 3* shows the images of these states as seen by the vision module. Of these four states, only the *up* state is suitable for pickup. The remaining three

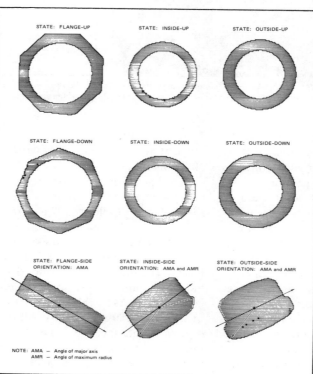

5. The entire set of nine stable states of the hydraulic coupling are shown with visual parameters for reorientation. A preferred pickup state for each set of parts is indicated.

states—*down, flat,* and *tall*—must be reoriented.

Also shown in *Figure 3* are the visual parameters used to rotate the turntable; only four are required for handling this part. The angle of the maximum radius (AMR) is used in conjunction with the angle of major axis (AMA) to turn the *flat* and *tall* states, so that the maximum radii point toward the elevator. The AMA is also used in reorienting the *up* and *down* states. Finally, if the part is in the pickup state, the distance between the centroid and the center of the camera's field of view is then used to determine

number of stable states, each identifiable by the vision module, can be presented by this system. Parts that do not fit in the shuttle cannot be handled, nor can round parts that roll or oscillate. The geometrical properties of parts that would preclude the feeding of parts with this technique have not yet been studied.

Determining stable state and orientation by the vision module is the principal limiting factor. Experience in the laboratory with the nearest-neighbor classification algorithm shows that the vision module can be trained to distinguish accurately among six to nine different shapes.

Often, however, it is very difficult to differentiate between two states. An example is a relatively flat symmetrical object with an identifying label on one face. In this case, a backlighted silhouette will not give sufficient information to distinguish between the two stable states. Alternative techniques for viewing part asymmetry include backlighting to show the silhouette of the part, diffuse or linear top lighting to show recesses and markings, a sheet of light projected from above, and side or oblique viewing utilizing any of these types of illumination.

Determining the orientation of parts is more difficult than ascertaining the stable state, since automatic methods for using local features are not yet available. The part presentation algorithm currently makes only three measurements of an image to determine its orientation—angle of major axis, angle of maximum radius, and center-of-gravity distance.

Expressions using these measurements have been developed for the parts described here. The problem of automatic determination of orientation by asymmetries based on local features is being investigated by Bolles.[9] The results of this work will be integrated into the vision module at a later date.

Feeding parts by means of the described techniques offers the following advantages:

• *Programmability.* Many different parts may be fed with the same hardware/software system; the only changeover required is the computer program.

• *Larger Parts.* Larger parts that are impractical to feed by conventional means can be fed with this system.

• *Implicit Inspection.* Parts are automatically inspected by the vision module, thus facilitating quality control.

This programmable approach to part presentation is new and offers definite advantages over conventional feeders and magazines. A detailed comparison is given in the accompanying table. The adaptation of sensors and active reorientation techniques to part feeding will receive growing acceptance in industry as more refined techniques become available. Application of these techniques to bowl feeders has been described by Nitzan.[8]

Further work is needed to accelerate the operations of programmable part presenters and to provide new active reorienting devices. Additional research is necessary to define the categories of parts that can be seen and presented with the aid of vision systems, as well as to develop visual operators that can determine orientation and symmetry automatically. ∎

This work was supported both by the National Science Foundation under Grant APR75-13074 and by 15 industrial affiliates.

REFERENCES

1. McGhie, D.F. and Hill, J.W. "Vision Controlled Subassembly Station." SME Technical Paper MS78-685. Dearborn, MI. 1978.

2. "Exploratory Research in Industrial Modular Assembly." Report R-1111, C. S. Draper Laboratory. Cambridge, MA. September 1977.

3. Abraham, R. G., Stewart, R. J. S., and Shum, I. Y. "State-of-the-Art in Adaptable-Programmable Assembly Systems." International Fluidic Services Ltd. Kempston, Bedford, England. 1977

4. Kelley, R. et al. "A Robot System Which Feeds Workpieces Directly from Bins into Machines." *Proc. 9th International Symposium on Industrial Robots,* SME. Dearborn, MI. March 1979.

5. Boothroyd, G., Poli, C. R., and Murch, L. E. *Handbook of Feeding and Orienting Techniques for Small Parts.* University of Massachusetts. Amhurst, MA. 1976

6. Park, W. T. and Burnett, D. J. "An Interactive Incremental Compiler for More Productive Programming of Computer-controlled Industrial Robots and Flexible Automation Systems." *Proc. 9th International Symposium on Industrial Robots,* SME. Dearborn, MI. March 1979.

7. Agin, G. J. "Vision System for Inspection and Manipulator Control." *Proc. Joint Auto. Control Conference.* San Francisco. June 1978.

8. Nitzan, D. et al. "Machine Intelligence Research Applied to Industrial Automation." Ninth Report, NSF Grants APR75-13074 and DAR78-27128, SRI Projects 4391 and 8487. SRI International. Menlo Park, CA. September 1979.

9. Bolles, R. C. "Symmetry Analysis of Two Dimensional Patterns for Computer Vision." *Proc. International Joint Conference on Artificial Intelligence.* Tokyo. August 1979.

—Adapted from "Programmable Part Presenter Based on Computer Vision and Controlled Tumbling." Presented at the 10th Industrial Symposium on Industrial Robots, March 5-7, 1980, Milan, Italy.

COMPARISON OF PROGRAMMABLE AND CONVENTIONAL FEEDERS

Feature	Programmable	Conventional
Method of Operation	Vision plus active reorientation	Mechanical reorientation of parts
Size	4 part lengths	25-50 part lengths
Part Feeding	Handles different parts	Designed for a single part
Setup Time	Fast—one minute for program change	Slow—about 30 minutes for mechanical setup and adjustment
Advantages	Simple part inspection and rejection	Inexpensive mechanism
Disadvantages	Vision module costly	Tooling, costly jamming, nonfeedable parts

Presented at the Robot II Conference, October/November 1977

Force Controlled Assembler

By John W. Hill
SRI International

A powered accomodation device is described that provides controlled forces between the end-effector of a Unimate 2000B manipulator and external objects being handled. Many assembly processes require that parts be pushed into place by the manipulator similar to the way human workers assemble them. Since industrial manipulators cannot control the forces applied, special purpose compliant mountings between the end-effector and working objects have to be designed. To overcome these limitations we have developed and demonstrated a general purpose force controlled manipulator which directly generates the forces to push parts together. Examples of bolt insertion and placing a retaning ring on a shaft are shown.

NEED FOR FORCE CONTROL AT THE WRIST

During the past year at SRI, we have developed a powered accommodation device for industrial use. The concept is to provide controlled relative motion between the end-effector and the manipulator wrist (see Figure 1). The desire for such a system stems from the inability of industrial manipulators to accurately position and push parts together. Positioning accuracy limitations of $\pm$ 1 mm ($\pm$0.040 inch) are typical in these manipultors and most assembly processes require better than 0.1 mm accuracy. Many assembly processes require that parts be pushed into place by a manipulator in a way similar to the way human workers assemble them. Since industrial manipulators cannot control the forces applied, special purpose compliant mountings between the manipulator wrist and the workpiece have been designed [1]-[3]. We have developed a general-purpose force controlled device which directly generates the forces to push parts together.

Industrial manipulation tasks require many types of control besides accurate positioning. It is often desired to push or pull parts with a certain force to see if they are correctly fastened. For example, after a part is set into a forging die, conforming to the correct orientation of

----

* This work was supported by the National Science Foundation (RANN) under Grant APR75-13074.

Figure 1. Role of Active Accommodation

the die by passive accommodation, we wish to maintain our grip on it as the hammer strikes, yielding to the motions of the blow, but not loosing our grip and measuring its new position relative to the immobile wrist. These and other control and sensing functions can be added to industrial manipulators by an active accommodation device. The desired characteristics of such a device for industrial use are listed below.

1. Position Modes

__Locked__: Transmit the motion of a manipulator directly.

__Positioning Offsets__: Generate hand ofsets measured in thousandths of an inch.

2. Force Modes

__Free__: Allow parts to fall into place by only reaction forces (passive accommodation).

__Force Generation__: Apply a controlled force in any direction to push parts together.

__Compliant__: Generate force proportional to distance (a programmable spring). The force may be nonlinear or may have time varying properties, such as neutral point or spring constant.

3. Sensing

__Position Sensing__: Measure the relative position in the force modes.

__Force Sensing__: Measure the force applied to the end-effector in the positioning and compliant modes.

EXPLORATORY WORK AT SRI

We have experimented with a single axis (the z axis) accommodation system mounted on the Unimate Model 2000A at SRI [4]. The system consists of a linear ball slide powered by a double acting pneumatic cylinder and position and pressure sensors. The unit, with an impact wrench mounted on it, is shown in Figure 2 and its schematic diagram is shown in Figure 3. A force capacity of 40 kg (88 lbs.) is provided over a stroke of 25mm (1 inch). The back driveability is only 200 g (7 oz.) or 0.5 percent of the full load. This single axis system has been successfully used to extract bolts from a feeder and insert them into threaded holes using the free mode with gravity pulling the impact wrench toward the hole and the locked mode with the unit driven upward against a mechanical stop.

THREE AXIS FORCE CONTROLLED ASSEMBLER

A three-axis unit, including x, y and z motions, has been developed and is shown in Figure 4. The construction of each of these axes is similar to the z-axis unit with exception of the pneumatic cylinders. Locking cylinders have been added to hold both the x and y axes at the centers of their range.

These pneumatic systems can provide a wide range of control and measuring functions for fitting and assembly simply by changing the control system configuration. For example, by controlling pressure with the valves and pressure transducer, the external force can be controlled. Making the generated force proportional to the displacement from zero (the center position), we obtain a spring equivalent. These changes can be brought about quickly during manipulation by simply switching in a new control system. A microprocessor based control system may be the most efficient way to operate such a multi-function system.

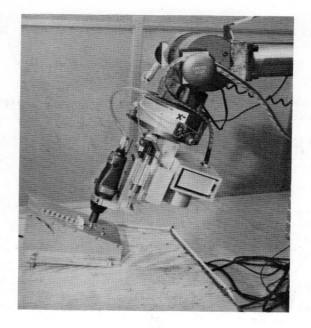

Figure 2.   End-Effector Picking up Bolt From Bolt Feeder

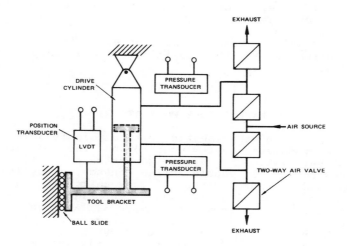

Figure 3.   Schematic Diagram of Single Axis Accomodation Uni

Three uses that have been demonstrated with the Force Controlled Assembler are shown in Figure 5. In the first example, a bolt is picked up reliably from a bolt feeder by pushing the socket of a pneumatic impact wrench into the end of a conically tapered guide. The two othogonal axes (x and y axes) are freed so that the force on the in-out axis (z axis) automatially drives the socket through the tapered guide and locates it precisely over the head of the bolt. The reaction forces backdrive the x and y axis as the forces on the z axis drive the socket into the hole.

The second example, shown in Figure 5(b), is inserting a bolt into the head of an air compressor. In this case the compressor is moving down a conveyor belt and is being tracked imperfectly by the Unimate manipulator. As the end of the bolt is spun over the hole in the compressor, ready for insertion, the z axis is made free to let the bolt descend. As the spinning end engages the hole within half a bolt diameter away, reaction forces are developed along the x and y axes and move the impact wrench toward the center of the hole. Freeing the x and y axes as the gravitation force is applied along the free z axis permits the impact wrench to follow the bolt into the hole, thus achieving centering

Figure 4. Three-Axis Force Controlled Assembler

(a) Bolt Pickup

(b) Bolt Insertion

(c) Retainer Ring Insertion

Figure 5. Uses of the Force Controlled Assembler

of the bolt. It takes 1.5 seconds to track, insert, and tighten a 1/4-inch diameter, 1-1/2-inch long bolt into the head of the compresser moving along the belt at 30 cm/sec.

The third example is placing a retaining ring in a slot on a cylindrical shaft, as shown in Figure 5(c). In this case the manipulator positions the three axis assembler in front of the shaft. In a series of five actions, the assembler first pushes the ring against the shaft below the slot ,centering the ring on the shaft by freeing the translational motion. Secondly, the ring is raised and slid along the shaft by a vertical force until it falls into the slot. Thirdly, the vertical axis is freed, eliminating the vertical lifting force which causes the ring to bind in the slot. Fourthly, the force against the shaft is increased to about 30 pounds, and firmly push the ring in the slot. Finally, this force against the shaft is reversed, retracting the ring holder and leaving the ring on the shaft. The entire ring insertion takes less than a second.

REFERENCES

1.   Mario Salmon, "Robot Technology at Olivetti - The SIGMA System," Technical Report, Ing. C. Ollivetti and C. S.p.A., 10010 Scramagno, Italy (April 1976).

2.   K. Takeyasu, T. Goto, T. Inoyama, "Precision Insertion Control Robot and Its Application", Journal of Engineering for Industry, Transaction of the ASME 76-DET-50 (June 1976).

3.   Paul C. Watson, "Multidimensional System Analysis of the Assembly Process as Performed by a Manipulator," CSDL Report R-996, Charles Stark Draper Labs, Cambridge, MA 02139 (August 1976).

4.   C. Rosen, et. al., "Machine Intelligence Applied to Industrial Automation," Sixth Report, NSF Grant GI-38100X1, SRI Project 4391, Stanford Research Institute, Menlo Park, CA (November 1976).

INDEX